Urban Environment, Travel Behavior, Health, and Resident Satisfaction

Anzhelika Antipova

Urban Environment, Travel Behavior, Health, and Resident Satisfaction

palgrave
macmillan

Anzhelika Antipova
The University of Memphis
Memphis, TN, USA

ISBN 978-3-030-08934-4 ISBN 978-3-319-74198-7 (eBook)
https://doi.org/10.1007/978-3-319-74198-7

Printed on acid-free paper

This Palgrave Macmillan imprint is published by the registered company Springer International Publishing AG part of Springer Nature.
The registered company address is: Gewerbestrasse 11, 6330 Cham, Switzerland

To my mom and family.

Acknowledgments

This book could not have been written without the support of numerous people. At the University of Memphis, Daniel Larsen, Jerry Bartholomew, and David Dye, who friendly encouraged me to apply for a sabbatical leave and provided moral and professional support. I am indebted to College of Arts and Sciences and Department of Earth Sciences at the University of Memphis for your generosity. I would not be able to focus on writing otherwise. Thank you for creating a stimulating atmosphere in which to work productively.

I would like to thank DAAD (Deutscher Akademischer Austausch Dienst e.V.) (German Academic Exchange Service) for granting me a research scholarship. It gave me an amazing opportunity to design and implement my research. This is where the idea was born. I want to express my gratitude to the entire Geography Department at the University of Duisburg-Essen (Essen, Germany) including a group of students who assisted with questionnaire administration and all faculty including Prof. Dr. Rudolf Juchelka, Friedrich Schulte-Derne, Ulrike Ohl, Ulrike Overbeck, Svenja Böttcher, and others. Thank you for your help with questionnaire design and translations into German, for hosting me, and for your sincere support—I truly needed your friendship.

Thank you, my anonymous reviewers who greatly contributed to this book.

I owe a great deal to my family. My mom's love has helped me at every stage of the project. I thank my husband, Yuri, for being there and keeping me focused. Lastly, I thank my daughters Xenia and Julia for being my friends and being very patient as I worked on the book.

Contents

List of Figures

List of Tables

CHAPTER 1

Introduction

1.1 Urban Growth

Prior to the 1820s, only a small percentage of the population lived in cities, which were restricted in physical area and mainly served as maritime trade centers for overseas shipping. In 1790, the first federal census showed that only 5.1% of the total population lived in cities (Miller 1973). In port cities, the occupations of merchants and skilled artisans and mechanics sustaining a shipping-based economy were important with most businesses and activities located close to the waterfront during the pre-1820s era. However, most people were primarily engaged in agriculture-related economic activities and lived outside the major cities (95%). Cities were growing very slowly at that time. New York City grew incrementally as the needs of the growing Atlantic economy and trade boom required construction of new docks, wharves, and piers and consequently a larger port. Starting in 1807, it finally experienced population growth so high that a layout was devised by a special commission authorized by the state of New York covering almost the entire area of Manhattan Island consisting of farms, villages, and open country terrain (Peterson 2003). In 1820, New York City's population reached 100,000 and became the first "great city." By 1840, two more cities acquired "great city" status—Baltimore and New Orleans (Fig. 1.1).

A. Antipova, *Urban Environment, Travel Behavior, Health, and Resident Satisfaction*, https://doi.org/10.1007/978-3-319-74198-7_1

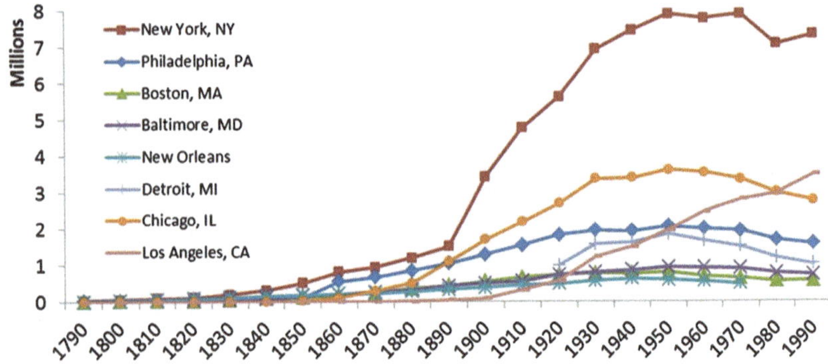

Source: Author, using data from U.S. Bureau of the Census, Population of the 100 Largest Urban Places: 1790-1990. At: https://www.census.gov/population/www/documentation/twps0027/twps0027.html

Fig. 1.1 Major city growth in the US in 1790–1990

Since the 1820s, urban growth notably exceeded population growth. As a national market economy developed, cities served as commercial, manufacturing, and financial centers. As inland markets opened up, port cities expanded their businesses giving rise to many new occupations including bankers, brokers, insurers, retailers, and so on. The urbanization of America has increased with the industrialization. From the 1840s onward, a new type of production emerged. Factories replaced old *cottage industries* (small shops producing cloth, garments, shoes, and other consumer goods). As the national market evolved and mechanization increased, factories achieved ever larger scale. Large-scale factory production required mass labor, which in turn required mass nearby housing. Mass production created a growing consumer goods market. Great urban centers emerged in the mid-nineteenth century.

In the United States, the rate of population increase during the nineteenth to early twentieth centuries was mostly determined by rapid natural increase (more births than deaths) and population migration (more immigration than emigration) increasing the labor force and the size and diversity of urban population. Immigrants, many of them of European descent, supplied the labor demand in manufacturing and construction. The Lower East Side of New York and the West Side of Chicago once provided the cheapest rent quarters for the newly arrived European immigrants, while later the worst buildings were removed by demolition. The populations of New York City, Chicago, and Philadelphia each surpassed the one million mark by 1890.

As cities grew, their physical form changed. All American cities, large and small, had a well-defined central core district. By the beginning of the twentieth century, the core business districts of the cities concentrated most commercial, financial, and retail uses including offices, banks, hotels, theaters, restaurants, department and specialty stores, and the like. Other functions of the downtown area included governmental and wholesale. All these downtown functions drew employees, clients, and visitors, with the city population relying upon the central core district for jobs, consumer goods, and services. The central core district kept its centrality advantage until the 1920s. Light industry and manufacturing was abundant near the central district and was also located along the freight railroads or major waterways or both. Industries included lumberyards, grain elevators, and bulk storage facilities (Peterson 2003).

Residencies also developed, and became diversified by class. Immigrants and common workers walked to the shops and factories and needed housing nearby. Crowded working-class neighborhoods consisting of cheap rental barracks, tenements, and wood-frame houses are seen in the photographs of the time blackened with coal smoke coming from adjacent industrial sites. Better apartments of the emerging lower middle class lined streetcar routes on both sides so that dwellers of these newer neighborhoods could ride the streetcar to work. Best lands near the central core and the edge of the city were occupied by single-family housing of upper-income class. The expensive inland enclaves and suburbs were often located along the streetcar lines or commuter railroads offering fast transportation.

1.2 Urbanization and Problems

Urbanization has brought with it many problems. With great city growth beginning since the 1820s, unregulated, unplanned growth soon produced unforeseen environmental stresses including dirty streets, poor air, recurrent fires, congested working and living conditions, and, at the worst, slums. Tenement houses promoted transmission of dangerous communicable diseases such as tuberculosis due to their crowded, congested conditions. Workers in large industrial cities in the mid-nineteenth century occupied the "railroad flats," with limited sunlight and fresh air, with no water supply or sanitary facilities provided.

From the 1840s onward, cities turned into big congested places characterized by concentration and density. In health this manifested in frequent epidemics such as cholera and high mortality and morbidity. High rates of infant mortality and low life expectancy impacted population in a devastating way. Levy (2013) blames congestion for the natural decrease experienced for much of the nineteenth century by most large American cities: there were more deaths than births, but in fact, many cities increased due to in-migration. Cities of the nineteenth century had no water treatment facilities or sewage disposal.

Although the early twentieth century saw abatement of the worst overcrowding in urban slum neighborhoods compared with the nineteenth century, a substantial amount of substandard dwellings were still found in many large American cities. For example, St. Louis, where many structures erected in the previous century had inadequate facilities (e.g., no indoor toilets were installed but merely outhouses, plus only cold water or no running water was provided) (Hoffman 2012). In the early twentieth century, the coming of the automobile made roads dangerous to pedestrians and exposed people to noise and gasoline fumes.

Implementation of urban housing programs (described later in the chapter), especially the creation of the public housing program in 1937 and the enactment of the *Housing and Urban Development Act* of 1968, resulted in renovation and repair of existing structures so that the share of residential buildings in need of substantial physical repair greatly declined, while living conditions greatly improved. To compare, in 1940, 45% of American homes had inadequate facilities (that is, no running water, a flush toilet, or a private bath); the percentage of homes lacking these indoor facilities was just under 3% in 1980.

The level of crowding has declined steeply over the years 1940 to 2000. According to the US Historical Census of Housing (2011), the overall share of crowded households (occupied housing units are considered crowded if there is more than one person per room) declined over time: the percentage of crowding decreased from 20% in 1940 to 4.5% in 1980; however, it rose to 5.7% in 2000. The level of severely crowded homes (homes are considered severely crowded if there is more than 1.5 persons per room) changed in a similar way: in 1940, 9% of all homes were severely crowded, while only 1.4% of occupied housing units in 1980 were considered as such. This percentage rose again in 2000 to 2.7% (2.9 million were considered severely crowded).

Geographically, in 1940, states with high crowding rates were largely concentrated in the South. New Mexico had a crowding rate more than twice the national rate: 48% versus 20% national rate, followed by Alabama

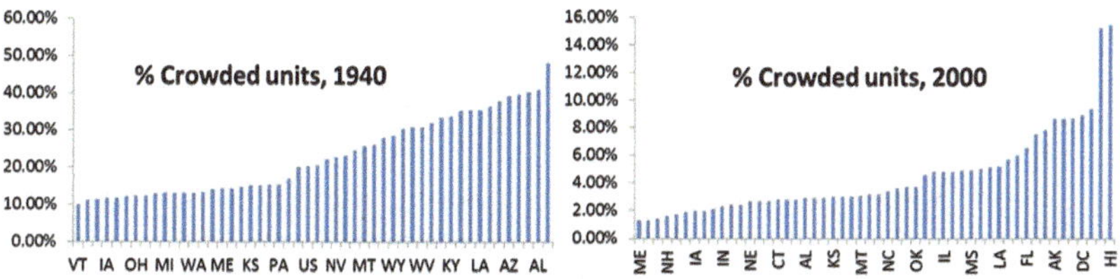

Source: Author, using Historical Census of Housing Tables: https://www.census.gov/hhes/www/housing/census/historic/crowding.html

Fig. 1.2 Housing crowding rates, 1940 and 2000

(41%), Mississippi (40%), Arizona (39%), Arkansas (37.8%), and Georgia (36.5%). The states in the Northeast and Midwest had lower than the national average crowding rate: Vermont had the lowest rate of about 10%, while New Hampshire, Delaware, Iowa, and Massachusetts each had a rate around 11%. The rates and geographical pattern have changed over time, for example, in 2000, the states with the highest crowding rates were Hawaii and California, each at about 15%, followed by Texas (9.4%), Washington, DC (8.9%), and Nevada, Arizona, Alaska, each with 8.6% compared with the national average crowding rate of 5.7% in the same year. The Northeastern states such as Maine (1.3%), West Virginia (1.3%), Vermont (1.4%), and New Hampshire (1.6%) each had low crowding rates (US Census 2011). Similarly, the number of people living in severely crowded conditions decreased from 9% in 1940 to less than 1.5% in 1980 (US Census 2011) (Fig. 1.2).

Recent rapid urbanization triggered many new environmental and socioeconomic problems. These include urban heat islands, energy consumption, air pollution, public health problems, deforestation, biodiversity, and the loss of high-quality agricultural lands (Clinton and Gong 2013; Zhang et al. 2013; DeFries et al. 2010). To lessen the impact of these and other problems cities are facing, understanding of urban growth is of critical importance to urban planners.

1.3 Urban Policy

While there is no simple definition, urban policy can be summarized as a

(1) Design and implementation of a set of complex activities in response to pressures of urban growth including the provision of essential public services and municipal infrastructure,

(2) Development of means to control and coordinate urban growth,
(3) Creation of social and environmental programs designed to improve the effects of decline and poverty.

Contemporary urban policy emerged during the nineteenth and early twentieth centuries as rapidly growing cities were industrializing. Within a single country, urban policy changes over time reflecting different goals and objectives (UN-Habitat 2014). Additionally, there is no agreement about what constitutes the most appropriate tools and techniques of urban policy.

The mid-twentieth century urban planners were seeking to develop a universal, rational, scientific research-based method to decision-making whereby different scenarios of the interactions between land use, transport, and housing within cities could be analyzed (a section on urban models is provided later in the chapter). Urban planners believed a scientific method should be employed for selecting the optimum solution for specified goals as well as predicting outcomes. On the one hand, research is determined by policy goals.

On the other hand, credible research can either support or disprove statements and assumptions about a policy's merit. With many countries having policies in place aimed at a reduction of private vehicle dependency, the design of compact walkable communities is considered an effective strategy. However, it is unclear what factors affect travel behavior (that is, commuting time and distance, modal choice, trip chaining, and the like), making it hard to formulate efficient urban policies targeted at changing travel behavior (Dieleman et al. 2002). For example, choice of a transport mode and distances travelled may be impacted both by urban form and design as well as personal characteristics including household income and size, and trip purpose (work, school, shopping, social, and recreation). Compared with low-income households, families with higher incomes have higher car ownership rates. Use of a private vehicle is also higher by higher-income families and those with children. What attributes of urban residential environments determine modal choice and travel distance? To explore these relationships in detail, generalizable and replicable research is needed. Rigorous research makes valuable contributions by helping formulate a useful policy tool.

Urban policy-making should be informed both by the direct and indirect effects of different urban decisions. For example, the decision-maker must have a good understanding of how investment in transportation (e.g., construction of a new bus station) will improve commute experience for the current travelers and those who may switch from car to public transit (direct effect), but also what changes in land-use patterns it may trigger, such as changes in the level and location of employment and increased apartment rents and land prices, both of which are indirect effects caused by improved (real and perceived) accessibility of a place. Therefore, modeling is useful in the research context (analysis of complex relationships) and in many urban policy-making contexts (analysis of urban policies) at the different levels, from local to federal. For example, local governments issue policies on zoning that permit specific economic activities in some urban parts but not in the others, while transportation planning policies contribute to a provision of a transportation system that is safe and convenient while impacting travel behavior of the existing commuters and addressing the needs of lower income families, those without cars, and so on.

1.3.1 Urban Policy of the Nineteenth Century

In the nineteenth century, cities experienced chaotic urban growth. Inefficient structural arrangement led to severe overcrowding of many urban activities (often incompatible with one another) in one location, which created unsanitary living environments. Municipal infrastructure was lacking: cities had no sewage disposal, so sewage polluted wells and other water sources while refuse was not removed.

For example, Chicago was growing rapidly between the late nineteenth and early twentieth centuries. The population of Chicago grew from 3820 in 1836 to 93,000 in 1857, quadrupling from1857 to 1873, then tripling from 1874 to 1899 (Hoyt 1939, p. 82) to becoming the third most populous city in the United States. However, the city was evolving without much state regulation and coordination.

Accompanying this population explosion was a sustained expansion and outward movement of high-grade fashionable neighborhoods farther from Chicago's downtown ("The Loop"). The speed with which high-grade neighborhoods move to new locations varies from place to place. Although many local factors determine the rate of neighborhood change and accordingly the rate of change in the internal structure of a city including residential characteristics, type of buildings, and so on, the rate of population growth is one of the most important determinants. The process of population growth and movement of the high-grade neighborhoods to new locations continued for more than a century in many other fast-growing American cities including New York, Detroit, Los Angeles, Seattle, and Washington, DC.

With time, local municipalities were created through legal framework. Urban policies were initially a physical reaction to problems of overcrowding, congestion, and public health and their consequences in fast-growing cities. Local governments were responsible for constructing buildings, building municipal infrastructure, and providing essential municipal services to improve citizens' welfare and public health. To cover costs, a taxation system was developed whereby property and other types of taxes are collected to generate revenues.

1.3.2 *Urban Agenda*

Since the early twentieth century, municipal governments became responsible for managing growth and regulating the character of urban development. The patterns of land uses in American cities have been changing. In addition to shifts in residential neighborhoods, there were changes in the pattern of commercial and industrial uses. With time, as the city's population increases, non-residential land uses in the central business district (CBD) tend to start expanding. Financial and business uses press outward and invade residential areas near the CBD and there occurs a subsequent replacement of residential area as the expanding retail commercial uses force people to move farther out. New technology (e.g., a steel-frame skyscraper, elevator) enables the vertical growth and leads to the intensification of land use within the CBD area with the height of the buildings changing the skyline of the downtown area.

As the middle- and upper-income class residencies relocate toward the periphery (simultaneously decreasing purchasing power of people who reside near the CBD), the stores and shopping centers also move away from the CBD in the direction of the high-class residential area. Retail commercial uses relocate to where two or more main highways converge and form outlying business centers, or along the principal highway. The outlying business centers are comparable in terms of services and facilities (e.g., chain stores) with those of the main business center and become increasingly more important regarding the total volume of retail sales. For example, in pre–Civil War Philadelphia, the downtown area was within easy reach for all the city's residents and most retail and banking was concentrated within the central business core. In the 1930s, as the city grew in area and population, outlying shopping centers developed outside the central district. Similarly, in Chicago during 1900–1915 businesses were concentrated in the Loop area, however, the development of new high-priced housing beyond the elevated lines led to the rise of outlying shopping centers after 1915 (Hoyt 1939, p. 109).

Besides the problems of physical growth, the urban agenda also faces social and environmental problems that are addressed by urban policy (Cochrane 2007). For example, a decline in manufacturing in late 1960s–1970s caused by an increase in international competition led to rising unemployment and environmental dereliction as many industrial plants and factories shut down, while abandoned buildings became run down, former industrial sites became contaminated, and underutilized infrastructure crumbled. Dissatisfaction with rising poverty and government's incapacity to make good on its commitment to full employment resulted in public unrest and street violence in many American cities. To improve poor housing and a series of linked social conditions such as low educational attainment, substandard housing, family breakdown, high crime and the like, the War on Poverty program was launched in the United States. This special initiative targeted low-income neighborhoods (UN-Habitat 2014).

1.4 Urban Problems of the Twentieth and Early Twenty-First Centuries

Urban problems of the twentieth and early twenty-first centuries are multifaceted and include the following issues: severe housing shortage of the Great Depression and post–World War II period, inner-city poverty concentration in many American cities, residential segregation, affordable housing, residential overcrowding, substandard overall housing conditions, fair access to housing and mortgage finance for minority and low-income families, and the like. These and other important issues have been the focus of urban housing policy. Some issues have been resolved. Others remain the primary debate issues including preservation of low-income housing, enforcement of fair housing laws, and the like.

1.5 Negative Impacts of Urban Policy

Urban policy can produce undesired effects. For example, after the US government's investment in a highway interstate system whereby a network of suburban freeways has been constructed starting in 1950s, the speed of deconcentration of American cities and decentralization to outlying centers increased greatly leading to unplanned growth and urban sprawl. To achieve a reversal of the direction of the "white flight" (that is, people who had moved to the suburbs to escape the problems of rising crime and social disorder of core city locations), the new urban policy seeks to attract families back to inner-city areas. The new urban policy of *gentrification* (which means the physical renewal, rehabilitation, and retrofitting of old buildings with the purpose of transforming city districts into viable commercial centers and reinvigorated residential areas) targets especially the middle-income classes, young professionals, and creative class workers (Turok 2009).

In the United States, urban sprawl is often characterized by its fragmented character, dispersed development, and its light regulation. The United States introduced the tradition of a particular style of housing, freestanding single-family houses (rather than attached houses such as terraces and semi-detached, which are more common in Europe) (UN-Habitat 2014), associated with rising car ownership, further promoting urban sprawl and low density neighborhoods. Many countries today have policies to shorten distances traveled by private car and reduce car dependency (Dieleman et al. 2002) (Fig. 1.3).

Fig. 1.3 The US conventional car-dependent, low-density suburban neighborhood where light trucks are the preferred personal vehicles (Author, 2017)

In response to suburban sprawl and separation of different land uses resulting in long commuting times, traffic congestion, and air pollution, as well as to promote better, more compact urban forms, higher density, and mixed-use developments, a "new urbanism" or "smart growth" approach got traction in Europe and the United States (UN-Habitat 2013). Other than private cars, modes of transportation are promoted including public transport, walking, and cycling. Higher density and mixed-use land uses can be achieved within transit-oriented development (TOD) where increased accessibility offered by mass transit tends to create higher-density housing around the stations (UN-Habitat 2014).

Other local urban policies promote job creation, childcare support, community services, poverty reduction, and the like. Lives of all urban residents are impacted, although to a different degree, by local urban policies: municipal zoning ordinances can affect property values, while local property taxation affects the quality of local schools.

1.6 Gautreaux Program

Municipal housing strategies are used to further the US Department of Housing and Urban Development's (HUD) fair housing commitment to eliminate housing discrimination and promote residential, racial, and economic integration. An example of one such local program is Chicago's 1966–1998 *Gautreaux Assisted Housing Program* developed to move households away from segregated and very low-income inner-city areas. A class action lawsuit, *Gautreaux v. Chicago Housing Authority*, filed in 1966, became a landmark event for housing policy. The Chicago Housing Authority (CHA) and HUD were charged with racial discrimination because public housing projects were concentrated in mostly Black neighborhoods and public housing tenants were segregated by race into neighborhoods of the same race. In order to deconcentrate the Black tenants of public housing, the US Supreme Court ruled the CHA to scatter new public housing and avoid their concentration in Black neighborhoods as well as provide opportunities for Blacks to live in predominately white Chicago neighborhoods (Basolo 2013). There is compelling evidence of the success of the Gautreaux program reflected in a range of improved socioeconomic outcomes including lower neighborhood poverty rates after residential relocation (DeLuca et al. 2010), increased job opportunities and higher employment wages, improved education achievements such as higher rates of high school graduation, as well as improved household satisfaction (Gabriel 1996).

1.7 Federal Urban Policies and City Impact

Governments in many countries take the initiative to regulate urbanization processes and manage city growth in a comprehensive way by formulating a national urban policy. Urban policy determines the form and character of urban growth, impacting in turn the well-being and life opportunities of their citizens. Urban policy addresses the multiple complex problems and challenges stemming from the growing concentration of population, businesses, infrastructure, and institutions within urban areas. Urban policy alleviates many urban ills including poverty, discrimination, inequality, traffic congestion, urban sprawl, and environmental degradation and pollution, while paving the way to well-functioning (economically, environmentally, and socially) cities with efficient spatial arrangement of activities and resources making them productive, livable, and sustainable.

National governments create the legal framework and financial instruments to implement urban policies effectively. To ensure long-term development, national governments actively collaborate with various actors and groups including local municipalities, the private sector (property developers, financial institutions), and other stakeholders. Cities need resources allocated to reduce the pressure on different urban systems including land and infrastructure networks due to growing population. Thus, national governments provide the investment for city infrastructure including water, energy, and sewage as well as road networks, and provide resources to address environmental health problems. National governments lack the local knowledge, including emerging problems, and therefore local decision-making is carried out by local governments.

Urban policies span multiple sectors including housing, transport, basic infrastructure, and other sectors.

Two examples of US urban policies will be discussed including housing and environmental policies as both affect urban areas in a significant way. Housing policy impacts household (population) location, which is interrelated with firm behavior, transportation network, and other subsystems of urban environment. Housing policy assists disadvantaged populations such as racial minorities and low-income families with affordable housing meeting their needs. Understanding household location decisions helps better understand city growth (therefore household location is an important component of many urban models discussed in Chap. 4). An example of environmental policies as they apply to brownfields is given due to increased risks of environmental damage from pollution exposure within towns, cities, and metropolitan areas.

1.8 The US National Housing Policy

Cities in the United States are impacted by the US national housing policy with the policy-making process linked to political ideology held by the dominant group presently holding power (and who can thus affect housing policy through laws, appointments, and the like) (Hays 2012). Following is a brief review of the US housing policy and the resultant changes within the cities in the United States.

1.8.1 Major Housing Programs and Impacts

Different housing programs have been used during the different periods of the twentieth century in the United States, each impacting American cities and the lives of the millions of Americans by increasing home ownership rates, improving housing conditions such as overcrowding, provision of multi-family units, and others. Hoffman (2012) in his chronological review of the US housing policy since the Great Depression distinguishes four major urban housing policy-making periods.

The major housing programs have been initiated by the US government and responded to several large housing-related crises. They include the following:

(1) The public housing program produced by Franklin D. Roosevelt's New Deal social program during the Great Depression in the US Housing Act of 1937 (also known as the Wagner-Steagall Housing Act) was a response to a huge loss of jobs and incomes accompanied by high foreclosure and eviction rates. It is still a major component of the US low-income housing policy.

The period was characterized by the severe problems of the housing sector, for example, the soaring number of mortgage foreclosures aggravated by high unemployment in the 1930s (by 1933, around 25% of workers were unemployed). To help solve the crisis, a new housing policy was designed that strengthened the private housing industry and the financial institutions (the commercial banks and insurance companies) that supported it. New governmental structures were created, for example, the 1932 Home Loan Bank System, which consisted of twelve Federal Home Loan Banks and a Federal Home Loan Bank Board (FHLBB) together with Federal Housing Administration (FHA) and banking and finance measures. Public housing had the appeal of a new program viable to clear and replace urban slums in the parts of American cities where the poor people lived with better homes.

(2) The postwar housing shortage of the mid-1940s, caused by the surging need for homes for returning armed forces members aggravated by an unprecedented increase in the amount of households who likewise needed homes, was addressed by the government by adding a housing provision to the G.I. Bill (the Servicemen‘s Readjustment Act of 1944) and provisions of the Housing Act of 1950, which expanded the G.I. loan program. The VA housing program became a very popular and long-term package for veterans.

The result of the VA housing program was mass inexpensive suburban single-family housing outside cities. The same act also created the college housing program where educational institutions got direct government housing loans to build dormitories. During the housing boom of the 1950s–1960s, cities experienced rapid suburban development with single-family houses comprising about two-thirds of the new housing. Not only did urban areas experience substantial growth, composition of residential neighborhoods also changed. Whites in ever-increasing numbers moved out of the inner urban areas to live in the growing suburbs, contributing to residential segregation.

(3) Two large low-income housing production programs based on subsidizing private industry, the Fair Housing Act of 1968 (which barred housing discrimination) and the Housing and Urban Development Act of 1968 (which created low-income housing), were enacted by Lyndon Johnson's administration to respond to the urban crisis of the late 1960s. Growing disorders and violence in Black neighborhoods across many cities in the United States during the civil rights movement beginning in New York and Los Angeles, in 1964 and 1965,

respectively, were rooted in racial inequality-based problems of America's cities including urgent issues of discrimination in residential housing, employment, education, and sustained poverty resulted in the national urban crisis of the late 1960s.

Since urban violence was believed to be caused by "environmental deprivations," including the bad housing of urban slums and ghettos, a solution to end the urban crisis and social unrest and solve massive urban problems was found by providing the lower-income classes with better homes. A commission named "the Committee to Rebuild America's Slums" was set up in 1967 to develop the policy to rebuild America's slums. During the Great Inflation period of 1968–1983, despite problems within the American economy such as relatively high unemployment and increases in energy prices, including the 1973 oil price spike, the rate of production of new homes as well as subsidized dwellings for low-income households was at an all-time high.

In 1970, new government-subsidized dwellings comprised 30% of all new housing. As a result, rates of low-income home ownership and multi-family renting grew rapidly. However, this huge supply of low-income housing by the large-production housing program did not correspond to an existing demand. Further, during the early 1970s, public housing ran into severe problems caused by poor management of low-income housing, declining incomes of their tenants (most of them Blacks), high vacancy rates, in turn, lowering rents and revenues, while inflated construction costs and overvalued properties led to an astronomical increase in maintenance expenses of housing projects and worsening conditions of low-income housing. These factors created an unattractive image of public housing (Hoffman 2012).

(4) Housing policy changed starting in the early 1970s and a different approach was employed by Richard Nixon's administration that introduced a new *voucher-based* strategy to help the poor and the *underclass* (that is, people who had been in poverty for generations) rather than continuing the strategy of massive low-income housing production of the previous period.

The new approach was based on the idea of housing allowances with the purpose to help low-income people move out of inner-city concentrations of poverty and find homes in better neighborhoods as well as help them to pay rent. A *housing allowance* is a direct cash subsidy to housing consumers based on a certain percentage of income; policy-makers establish an "appropriate" percentage level for a household's expenditure on housing; the amount of a voucher is pegged to this level. Thus, a cash housing grant makes better quality housing affordable to low-income households by allowing them to buy or rent better dwellings than they could otherwise (Hays 2012).

However, not all groups backed this new approach. There were criticisms of voucher-based assistance programs. The sociologist and urban planner Herbert Gans voiced concerns that displacing people from low-rent units into better housing has limited social and emotional effects. Although the potential benefits may include improved health and morale in occupants, better housing per se is not linked to better parenting, work habits, or emotional health, nor does it reduce poverty. According to Gans, the strategy of providing the eligible poor renters and the near-poor with vouchers will not work in many cities where inexpensive housing is scarce and will inflate local rents (Gans 1993, pp. 222–224).

The Housing and Community Development Act passed in 1974 created the Section 8 rental assistance program. This new program contained both elements of production (new construction and rehabilitation using rent supplements) and rental subsidies (with the rent supplement paid by the local housing authority). Studies evaluating the program have been conducted with results indicating that the housing voucher-related concerns of the opponents were not realized. However, it took twenty years before the program became fully adopted (Hoffman 2012).

The 1970s massive new residential construction was accompanied by two trends. The first trend was remodeling of existing buildings including installation of indoor facilities such as toilets, private baths, and running water (described earlier). The improvement of overall housing quality was enhanced by the elimination of substandard housing units: between 1974 and1985, about 6% of deficient units were removed from the housing stock contributing to the shortage of low-quality housing (Gabriel 1996, p. 675). Furthermore, the improvement in housing conditions undermined the reason for the mass production of new housing for low-income households and introduced a change in the federal policy. The second trend impacted cities in terms of size and form of settled areas as many middle- and working-class Americans, both Whites and Blacks, starting in 1970, left inner-city neighborhoods and moved to the outer parts of metropolitan areas, contributing to urban sprawl, while negatively impacting inner-city areas: depopulated, with multiple abandoned buildings becoming increasingly more impoverished (Hoffman 2012).

Despite the overall growth of the economy with several recessions in the early 1980s and the early 1990s, and increasing wealth, as well as land values and home prices, an increase in median income was only slight. At the same time, the number of people who were paid for shelter increased reflecting the great floods of immigrants. The last two trends added to the problems of low-income families.

1.8.1.1 Shortage of Affordable Housing

Gabriel (1996) writes about an increase in the shortage of affordable housing among very low-income households with pressing housing needs (that is, those who paid more than 50% of their income in rent or who lived in substandard homes) whose number steeply rose by 700,000 to 5.6 million between 1989 and1993 (Gabriel 1996, p. 675). Because the poor could not afford to rent or purchase standard quality units due to insufficient income, they were limited to substandard housing that did not meet their needs, so it was believed that a *housing allowance* in the form of direct cash would get poor families access to the private market where they could obtain standard housing of their choice (Hays 2012).

1.8.1.2 Housing Policy Change. Poverty Deconcentration

After interest in the big public housing production diminished, popular prior to 1970, federal policy aimed at *deconcentrating poverty* (that is, neighborhood racial and income mixing) in inner-city areas shifted to new approaches focusing on the promotion of mixed-income communities.

The Moving to Opportunity (MTO) program was a long-term policy experiment created in 1993 and implemented in 1994 to investigate the effect of voucher-holder moving from high- to low-poverty areas in five cities across the country. Section 8 vouchers could be used by low-income households to move out from areas of concentrated inner-city poverty to low-poverty suburbs. Although analyses of the MTO program indicate that residential quality measured by the poverty rates improved for the two treatment groups (an experimental treatment group consisting of voucher holders who moved from high- to low-poverty neighborhood, and a Section 8 (HCVP) treatment group) versus the control group, this effect diminished over time (Ludwig 2012).

However, the MTO program did not result in the anticipated widespread positive benefits; it had no effect on employment and mixed effects on school quality (Sanbonmatsu et al. 2011, 2012). In her work, Basolo (2013) studied employment opportunities for voucher holders and access to good-quality schools for their children in Orange County, California. Specifically, she examined (1) whether these outcomes are different for movers versus nonmovers in the voucher program, and (2) whether the movers had different socioeconomic attributes including neighborhood poverty levels, employment status, and school quality before and after residential relocation.

1.8.2 Public Awareness of Housing Discrimination

How much does the general public knows about the federal law pertaining to housing discrimination? Do people support the legislation and what is their behavior in response to practices of discrimination in the area of housing? We need to know the answers to these questions since the more homebuyers or renters know about the Fair Housing Act, the more the general public may benefit from it.

Homebuyers or renters should have enough knowledge to recognize cases of housing discrimination and if they have experienced discrimination, they should respond, for example, by filing a formal charge with HUD for investigation. This is an important enforcement provision of the Fair Housing Act that relies not only on the individual's knowledge and their ability to recognize discriminatory treatment but also on personal willingness to take action by the alleged victims of housing discrimination. Otherwise, the current public policy enforcement may be ineffective and discriminatory practices will continue.

An amendment to the act was passed in 1988 that added familial status and disability to the list of the established bases for housing discrimination. The amended Fair Housing Act took effect in 1989; it bans discrimination based on "race, color, national origin, religion, sex, familial status, and disability" pertaining to housing actions including rentals, home sales, mortgage lending, home improvement, and zoning activities (Abravanel and Cunningham 2002).

1.8.3 Housing Discrimination 2001 Survey

The federal Fair Housing Act that promotes equal opportunity in housing and prohibits discrimination in the rental or sale of housing was enacted in 1968, yet only in 2001 did the level of the US public knowledge about the act and its attitude toward aspects of this law become known. Additionally, a national HUD-sponsored survey that was conducted in late 2000 and early 2001 identified the degree to which people thought that they have suffered from housing discrimination when trying to purchase or rent an apartment or house because of different actions or decisions taken by owners of rental buildings, home sellers, real estate agents, and mortgage lenders.

For example, illegal actions are the situations when a real estate agent limits a client's housing search to a certain geographical area on account of neighborhood racial composition (steering). More specifically, a real estate agent may decide to limit the search for housing of a Black client to Black-only neighborhoods thinking that the homebuyer or renter desires to buy or rent in minority-only areas, thus steering them away from neighborhoods where other races are predominant. Other examples of illegal actions include differential treatment of families with children from that of other types of renters, such as when a family with children is restricted to a particular building by an apartment owner because younger children may bother other tenants due to their noise, or religion-based preference when an apartment building owner advertises a tenant preference (such as advertising "Christians preferred") and a landlord denies rental occupancy on the basis of an applicant's religion (discriminatory advertising), or ethnicity-based requirement of a larger down payment by a mortgage lender, and the like.

To measure fair housing, the survey respondents were provided with a set of ten hypothetical scenarios related to the sale or rental of housing with eight scenarios portraying some illegal actions by persons involved with housing transactions. The survey included questions on whether certain actions appear to be discriminatory or legal under Federal law. The results of the survey indicated that the overwhelming majority of the adult population was aware of fair housing law and that one in every seven adults thought they had experienced housing discrimination at some point (Abravanel 2006).

1.8.3.1 Investigating Housing Discrimination

The federal government is responsible for investigating cases of alleged housing discrimination and taking legal action against a discriminatory practice on the basis of race, religion, or nationality. An administrative hearing or trial resulting from filing formal charges with HUD as authorized by the law occurs at no cost to the complainant. HUD is the federal agency that promotes and increases public awareness of fair housing law as well as enforces the Fair Housing Act where housing discrimination is believed to occur. To increase the degree of public awareness of and approval for fair housing law, the department collaborates with state and local agencies and nonprofit groups to organize fair housing outreach and education programs over many years.

Fair housing proponents suggest using housing consumer education as a tool when dealing with discrimination. Educational programs include information on what housing discrimination involves, who the law protects and under what circumstances, and what actions can be taken if discriminatory practices occur. To illustrate, in 2003–2004 public service advertisements such as "Accents" and "Do You Still Like Me?" aired on television and radio and appeared in print media in both English and Spanish. These PSAs provided information in a creative way on how to recognize and report housing discrimination. To eliminate housing discrimination, the HUD reports enforcement effort cases that resulted in charges of housing discrimination including the local media (Abravanel 2006).

1.8.3.2 "Do We Know More Now?" Survey

To learn whether running media campaigns on the problems of housing discrimination lead to improvements in the extent of the general public's use of Fair Housing Act provisions as well as awareness and support of the law, a second HUD-sponsored survey called "Do We Know More Now?" was conducted nationally in 2005.

Regarding the 2005 survey methodology, a random digit dial telephone survey was administered in the forty-eight contiguous states and the District of Columbia. Besides the general population survey, four categories of the nation's populations have been targeted including Blacks, Hispanics, families with children under 18 years of age, and households with persons with disabilities. To generate representative samples and increase incidence of targeted households, telephone numbers were randomly chosen from zip codes containing at least a 50% population who are Hispanic or Black, while persons in families with young children and in households with disabled individuals were randomly selected from zip codes with at least 50% of households with children under 18 years of age and from the base sample, respectively. This sampling method resulted in over-sampling of each targeted subgroup. A sample size of at least 384 respondents was thus achieved with a 95% confidence level and a + or – 5% margin of error. Each sample was then proportionally stratified using the US Census-based information for each category.

1.8.3.3 Housing Discrimination Survey Findings

The results of the survey indicate that public support increased with respect to opposing the race-based restrictions of home sales and actions by real estate agents restricting client home searches based on racial concentration.

Among many characteristics, *race* was identified more frequently than any other attribute as the basis of the perceived housing discrimination: as many as one-fifth of all Blacks claim having experienced housing discrimination at some point due to their race/ethnicity; the rate is considerably higher when compared with other groups. The results of perceptual data suggest only 6% of Hispanics stated they have experienced housing discrimination because of their race/ethnicity. It is noteworthy that despite disability-based discrimination being one of the most frequent complaints received by HUD, a lower than expected level of 2005 survey respondents (just under 1%) stated disability as a basis for the perceived discrimination. However, the results of the survey of the perceived housing discrimination require additional attention and should be treated with caution; it was suggested that actual housing discrimination experienced by Hispanics and persons with disabilities versus the perceptual data on these groups may be considerably higher.

The survey revealed that only a small percentage of the public believe that filing a fair housing complaint would have been worth it and anticipated good results. The survey also found that perceived discrimination rarely produces a response and most people choose to do nothing when confronted with what appeared to be discrimination. Many people who believed they have experienced housing discrimination were disinclined to complain in response due to many reasons including not knowing where or how to complain, or believing filing a complaint would be costly, or expecting retaliation against them.

1.8.3.4 Housing Discrimination: Public Awareness as a Public Policy Tool

There is no agreement among fair housing proponents as to the importance of public awareness with some questioning this enforcement approach. The fact that many people choose inaction with discriminatory practices unchallenged explains why the relative value of consumer education as a tool for addressing discrimination is downgraded by some fair housing proponents. Because housing consumers have only limited knowledge about discriminatory acts, and not all discriminatory actions are observable to the general public, fair housing proponents suggest a more proactive approach when offenders are sought out, prosecuted, and punished.

Despite the overall public backing for the federal government's role in fair housing enforcement, at the same time, 40% of the adult population either does not think that examining claims of housing discrimination constitutes a legitimate federal function or has no opinion on the matter. About 25% of the population do not approve of federal responsibility for inspecting claims of housing discrimination or taking action on behalf of injured parties. Higher degree of public support is needed to both secure funding and implement fair housing enforcement efforts; increasing that level of support is essential. Therefore, in order for future efforts to educate the general public about housing discrimination to be successful, it is important to understand what factors inform this unfavorable attitude (Abravanel 2006).

Other programs designed to address social problems in high-poverty areas included Housing Opportunities for People Everywhere (HOPE VI) (public housing rehabilitation program) and the Housing Choice Voucher Program (HCVP).

1.8.4 Air Pollution Program and Brownfields

The Air Pollution Control Program Support by the Environmental Protection Agency (EPA) provides funds to local municipalities to evaluate transport impacts, assess local air quality, develop attainment plans, and enforce control strategies including regulation of pollution-emitting stationary and mobile sources and trip-reduction measures. Many real properties formerly used by industries and often found close to the downtown areas in many American cities were abandoned, however, whose redevelopment or reuse is often complicated due to the contamination with hazardous substances or pollutants, urban communities with the brownfield sites may benefit from the federal Brownfields Assessment and Cleanup Cooperative Agreements program. This federal program evaluates brownfield sites and implements brownfield-related planning including cleanup and redevelopment efforts. Local brownfield communities are given new life through economic redevelopment: reuse of cleaned-up areas increases local tax bases, creates jobs, utilizes existing infrastructure, and at the same time improves the environment.

A report by the EPA titled "Anatomy of a Brownfields Redevelopment" (2006) illustrates the brownfields cleanup and redevelopment process from the perspective of real estate development. One success story is the city of Emeryville located between Oakland and Berkeley, California. A formerly industrial city, most of its industry left during the 1970s. By the mid-1990s, more than one quarter of the land within the city was either underused or vacant, while more than 90% of this unused land had been contaminated with heavy metals, hydrocarbons, and other soil and ground water contaminants. Emeryville was awarded a $200,000 Brownfields Assessment Grant in 1996 to perform assessment, cleanup, and redevelop ten area brownfields. In addition, the EPA grant funds were used toward a database development for potential public and private investors and developers who can find out about the city's brownfields targeted for reuse and other essential facts. One such former brownfield site in Emeryville was sold to a private corporation that built mixed-income housing units on the property.

Another success story is about a former gas station property in the city of Prineville, Oregon, which revealed severe soil and ground water petroleum contamination during environmental assessment after the gas station operations discontinued in 1997. The property was identified as a potential site to accommodate municipal services including administration, planning and community development, police, and other public services. Prineville was awarded a $200,000 EPA Brownfields Petroleum Cleanup grant in 2003 to assist with the cleanup efforts. New Prineville's City Hall and Community Plaza opened in 2005 (Anatomy of a Brownfields Redevelopment 2006).

The book is structured as follows. Chapter 2 discusses distinct urban differences between the United States and Europe resulting from varying settlement time, industrialization, infrastructure development, land ownership, planning, and regulation. In vigorous debates on "good city form," a sprawling urban form in the United States is often contrasted with European compact form. The contemporary US metropolitan form is characterized by low development densities, dispersed population and employment, and a larger role for the private car. Compared with European cities, US cities have lower population density and lower compactness, but higher complexity and lack of centrality as well as larger open space compared to the total urban area. European metropolitan areas have higher densities and more centralized land-use patterns, with lower levels of car use. Many countries have centralized political power and exercise top-down planning. Other urban characteristics in Europe include development in rail corridors, while in the United States a stronger emphasis is on suburban development, which reinforces vehicle dependence and in turn leads to the inefficient use of scarce urban space. Chapter 2 starts with a definition of urban form and measures to quantify it. Then follows a discussion of factors that influence urban form, and a comparison of urban form between Europe and the United States.

Chapter 3 examines the impact of local urban amenities on neighborhood perceptions and satisfactions by examining "the importance" of various local amenities including green spaces and parks, children's playgrounds and educational facilities, and parking and commercial facilities, among others, within close proximity to the residential neighborhoods. There is a lack of studies capturing the subjective dimension of accessibility of different elements of urban environment and travel behavior, as well as travel-related negative outcomes. Chapter 3 contributes by integrating an analysis of satisfaction- and perception-related indicators at an intraurban level. Specifically, the case study offered in Chap. 3 investigates the various aspects of urban environment from the perspective of resident perceptions and satisfactions.

It is necessary to study people's perceptions and evaluations to avoid the mismatch between objective (i.e., spatial characteristics including urban form, transport infrastructure, socioeconomics) and subjective (non-urban form characteristics expressed by residents as attitudes and perceptions towards urban aspects) indicators of the aspects of urban environment (Klinger et al. 2013). Further, recent research on public transit quality and green space availability in Vienna, Austria found an overall high degree of agreement between the subjective responses and objective measurements (Haslauer et al. 2015). The individual-level primary data were collected during the fall of 2010. The author and a small group of students of the Geography Department at the University of Duisburg-Essen (Germany) conducted a survey and collected information on specific variables related to travelers and their stated preferences. The detailed questionnaire was designed and passers-by were approached, explained the survey's purpose, and requested to answer questions. The questionnaire had stated preference and satisfaction questions related where appropriate to the following categories: (1) individual-level; (2) local access to various amenities at the neighborhood- and city-level; (3) daily travel and activity pattern; and (4) negative travel outcomes and travel-related health impacts.

Neighborhood amenities perceived by residents as desirable, and therefore important, include public services such as schools, neighborhood life cycle, physical maintenance, commercial facilities such as restaurants, grocery stores, and nearby parks, culture, socioeconomic characteristics of its residents, ethnic make-up, and crime levels. Urban planners and developers are challenged with designing "development with zero growth" appealing urban environments to attract new residents and retain existing ones. Understanding these features is not possible without careful examination of residents' valuation of these urban features.

The premise of Chap. 3 is that residents' perceptions and satisfactions regarding their neighborhoods are shaped by the spatial arrangements of the various urban amenities and are conditioned by socioeconomic and demographic characteristics. For example, a resident may deem living close to good public schools important, but might reside far away from this amenity, therefore expressing dissatisfaction with their actual proximity to good schools. To illustrate further, regarding access to the CBD, results of the 2010 questionnaire demonstrated that most people value access to the city center, with 46% perceiving it as important and 27% perceiving it as very important. There were fewer respondents for whom it was either not important or did not matter (15% and 12%, respectively). When asked about the actual satisfaction with their access to the CBD, about 62% and 27% stated being satisfied and very satisfied, respectively. About 8% of all survey respondents were dissatisfied with their actual CBD access.

Residents' perceptions of their neighborhoods, measured here by the degree of perceived importance, may be influenced by the availability of various urban features including ample green areas, nearby shopping and services, access to work, recreation, education and sports facilities that enhance the quality of life. In order to confirm this hypothesis, an empirical study was done in Essen located in the Ruhr area in Germany.

We used the following categories for the stated preferences, satisfaction, and perceived health impact. Preferences were coded by four categories including "Not important," "Less important," "Important," and "Very important." On a scale of 1 to 4 the survey participant evaluated what elements of the built environment were of importance. Analogously, we also asked travelers their actual satisfaction with urban amenities as well as with their neighborhood. On a scale of 1 to 4, satisfaction was coded as "Very unsatisfied," "Unsatisfied," "Satisfied," and "Very satisfied."

Chapter 4 talks about understanding locations of various components of the city including the central and outlying business districts, residential areas, the factories, the blighted areas, and so on, at any given period of time within the city. Understanding spatial location of various economic activities including household location decisions, employment, infrastructure, and so on, helps better understand city growth. Location of the center, industrial areas, commercial districts, residential zones, parks, and other components of urban structure can be generalized. Therefore, household and industrial locations are important components of many urban models discussed in Chap. 4. Methods for analysis and models of city structure and growth have been developed. Chapter 4 reviews urban models of the internal structure models developed by urban sociologists including the concentric circle, sector, and multiple nuclei models.

Ernest Burgess developed his concentric zone model with five successive rings (zones) based on the observation of Chicago. The rings are differentiated not only spatially but also by the types of areas. However, the concentric circle theory has certain limitations. Many American cities reveal big variations in the patterns of land uses. Nevertheless, despite its limitations, the concentric model of the internal city structure is a useful starting point for studying a pattern of land uses within a city.

Inner structure of the city (a configuration of the various city components) does not remain fixed as various land uses can decline or increase in importance and shift their location away from points of origin. Techniques for studying dynamics of city growth include comparing a series of maps of settled area visualizing the form of city structure at successive time periods. Using this method for Washington, land economist Homer Hoyt (1939, p. 112) noted that the best residential section with the highest rents in the mid-1930s moved beyond the limits of the settled area of the late 1880s; a new area was vacant at that time. Hoyt developed another model of urban land use explaining inner structure and growth of urban areas. Among different land uses, several types of residential areas have been distinguished. Using essential housing characteristics at the city block level, Hoyt analyzed the distribution of high, low, and intermediate residential rental areas within the city. According to the sector model, land uses seem to conform to a pattern of sectors where areas of high rent form clusters with areas of lower rent surrounding the high-rent poles; the pattern repeats itself with a downward progression of rents in all directions finally reaching the worst slums. So city growth was explained not by ring but by axial development.

As the city grows new elements are introduced (e.g., new industries emerge, in turn attracting more population), again modifying the structure. Industrial uses formerly located close to the heart of the city near the workingmen's homes in most cities prefer to move to the periphery of cities where they locate in a specialized industrial district with the advantages of less taxes, cheaper land allowing to devote big tracts of land to a single use, single-storied building construction enabling uninterrupted industrial process (and therefore savings on factory operations), and access to limited-access freeways. Thus, the relatively simple structure of a city having a single main business center is replaced with a more complex pattern. To reflect the changing city structure, other models have been developed that state that urban areas may grow around not a single but several centers. The multiple nuclei model was described by Harris and Ullmann in 1945.

Employment concentrations are important local and regional economic development resources. Chapter 4 further reviews spatial structure models developed by urban economists modeling the location of firms and residencies (Lowry 1964; Harris and Wilson 1978). Population and employment suburbanization, and improvements in transportation and communication, have resulted in de-densification and horizontal expansion of urban areas (Rode et al. 2014) leading to a polycentric urban form in many cities (Gordon et al. 1998; Ingram 1998). Urban form and transport systems are interrelated (Smith 1984; Holtzclaw 2000; Rode et al. 2014), with high-density areas experiencing higher transport intensity including vehicle-miles-travelled (VMT) even when other factors are taken into account (Ewing, Bartholomew et al. 2008). Changes in urban structure such as the spatial distribution of activities also result in changes in travel patterns, especially those related to the journey to work. Accordingly, Chap. 4 concludes with a case study of the impact of urban structure on commuting times and distances considering Baton Rouge, Louisiana as an example. The case study demonstrates the impact of metropolitan structure on commuter travel and investigates the commuting patterns under the monocentric and polycentric assumptions. Two measures of the spatial separation between home and work locations have been utilized including commute time and length.

Chapter 5 provides information on the relationship between urban environment and travel behavior. Extensive studies on the link between transport/travel and urban form using various measures, often incomparable among different studies, did not yet reach consensus. Several studies used objective indicators including residential and housing density, road network design, and land use diversity (3Ds) (Cervero and Kockelman

1997), as well as average speed and VMT among others. Yet other objective factors are socioeconomic, such as rising incomes and higher demand for private space resulting in more car owners and longer distances traveled, which causes sprawling highway-oriented cities (Brueckner 2000). Other studies analyze preferences and attitudes towards mobility. People's perceptions of local transport systems need to be analyzed to better understand urban mobility as a factor shaping urban form. Klinger et al. (2013) assessed the subjective dimension of urban mobility cultures across forty-four cities in Germany and integrated satisfaction and perception-related indicators at a city level.

Chapter 6 examines the relationship between urban environment and travelers' health. Urban sprawl is an urbanization-related concern that threatens human health and sustainable development. Urban sprawl is conducive to the use of motorized vehicles but not walking and thus is associated with public health issues including obesity. Environmental interventions can potentially influence health outcomes and cardiovascular risk (Hirsch et al. 2014). Developing countries undergoing fast urbanization advocate using a combination of walking and public transport and avoiding using more private cars, especially in the context of China's current population and limited land use (Tong et al. 2016). China is examined in the chapter as it represents the most populous country in the world undergoing an extraordinary urbanization over a short time, expanding urban areas and consuming agricultural land accompanied by substantial environmental degradation.

Rapidly developing countries enduring tremendous urban growth, especially in a short time, experience severe environmental degradation. Air pollution is a major environmental concern with estimated huge health impacts: it causes deaths and various diseases in many large cities in both developing and developed world countries. Van Donkelaar et al. (2010), using satellite-based data, report the fine particulate matter, or PM2.5 concentrations exceeding 100 $\mu g/m^3$, for the major industrial regions in eastern China. Further, PM2.5 concentrations are higher in many urban regions of China than in other countries documented by satellite imagery (van Donkelaar et al. 2010). Urban environmental quality, including air and water pollution, contributes to disease occurrence while a steady increase in vehicles and associated traffic-related accidents represent another major public health threat (Gong et al. 2012).

In the US context, major roadways have been often routed through low-income communities so that exposures to several stressors are spatially correlated. A study on an adult US Puerto Rican population residing in the Greater Boston area demonstrated the link between residential exposure to low traffic volumes and the elevated risk of cardiovascular disease. Especially susceptible to the effects of traffic exposure were individuals with elevated body mass index (Rioux et al. 2010). Body mass index (BMI) is defined as weight in kilograms divided by the square of height in meters. Rigorous research on the health impacts of urban growth and environmental degradation as well as the interactions among urban form, health, and environment can provide better evidence for public policy promoting greater equity in health and social outcomes.

The book concludes with Chap. 7, which describes perceptions of the general public towards urban policy, and more specifically, the federal government's involvement in social policy on housing, as well as general attitudes towards Transit-Oriented Developments (TODs) such as what motivates businesses and households to locate within such urban forms. It also discusses the changing demographic profile of urban areas in terms of the older members of the population accounting for a substantial share and thus the need to meet their mobility needs. One way is by encouraging new development in places with access to public transportation (TODs) since their residents reveal less automobile dependence, drive less, and make shorter (but more) trips.

Livable communities mean environmentally healthy places. Chapter 7 accordingly explores the role of the urban built environment in pollution exposure and what intervention measures (such as afforestation) may decrease airborne particle concentrations and volatile organic compounds (VOCs). The chapter proceeds with the trends in urban air pollution and its health effects, including how riding a school bus adds to air pollution exposure. Since commuting largely occurs in the urban environment, the links between travel, health, and work experience are examined. Chapter 7 discusses the health contributions of active travel and measures to prevent obesity. Finally, it provides information on the top walk- and bike-commuting cities by their size and geographic location.

Bibliography

Abravanel, M.D. 2006. *Do We Know More Now? Trends in Public Knowledge, Support and Use of Fair Housing Law*. Prepared for the U.S. Department of Housing and Urban Development. HUD USER. Washington, DC: The Urban Institute. At: https://www.huduser.gov/Publications/pdf/FairHousingSurveyReport.pdf. Accessed 7 July 2017.

Abravanel, M., and M. Cunningham. 2002. *How Much Do We Know? Public Awareness of the Nation's Fair Housing Laws*. Washington, DC: The Urban Institute. At: https://www.huduser.gov/publications/pdf/hmwk.pdf. Accessed 7 July 2017.

Anatomy of a Brownfields Redevelopment. 2006. *Environmental Protection Agency (EPA)*. EPA-560-F-06-245. October. At: https://www.epa.gov/sites/production/files/2015-09/documents/anat_bf_redev_101106.pdf. Accessed 7 July 2017.

Baker, W. 1989. A Review of Models of Landscape Change. *Landscape Ecology* 2: 111–133.

Basolo, V. 2013. Examining Mobility Outcomes in the Housing Choice Voucher Program: Neighborhood Poverty, Employment, and Public School Quality. *Cityscape: A Journal of Policy Development and Research* 15 (2): 135–154.

Batty, M. 1997. Cellular Automata and Urban Form: A Primer. *Journal of the American Planning Association* 63: 266–274.

Batty, M., and P.A. Longley. 1994. *Fractal Cities: A Geometry of Form and Function*. San Diego: Academic Press.

Bettencourt, L.M.A., J. Lobo, D. Helbing, C. Kühnert, and G.B. West. 2007. Growth, Innovation,. Scaling, and the Pace of Life in Cities. *Proceedings of the National Academy of Sciences of the United States of America* 104: 7301.

Brueckner, J.K. 2000. Urban Sprawl: Diagnosis and Remedies. *International Regional Science Review* 23 (2): 160–171.

Cervero, R., and Kockelman, K. 1997. Travel Demand and the 3DS: Density, Diversity, and Design. *Transportation Research Part D: Transport and Environment* 2 (3): 199–219.

Chowell, G., J.M. Hyman, S. Eubank, and C. Castillo-Chavez. 2003. Scaling Laws for the Movement of People Between Locations in a Large City. *Physical Review. E, Statistical, Nonlinear, and Soft Matter Physics* 68: 066102.

Clarke, C. 1951. Urban Population Densities. *Journal of the Royal Statistical Society, Series A* 114: 490.

Clinton, N., and P. Gong. 2013. MODIS Detected Surface Urban Heat Islands and Sinks: Global Locations and Controls. *Remote Sensing of Environment* 134: 294–304.

Cochrane, A. 2007. *Understanding Urban Policy: A Critical Approach*. Oxford: Blackwell.

DeFries, R.S., T. Rudel, M. Uriarte, et al. 2010. Deforestation Driven by Urban Population Growth and Agricultural Trade in the Twenty-First Century. *Nature Geoscience* 3: 178–181.

DeLuca, Stefanie, Greg J. Duncan, Micere Keels, and Ruby M. Mendenhall. 2010. Gautreaux Mothers and Their Children: An Update. *Housing Policy Debate* 20 (1): 7–25.

Dieleman, F.M., M. Dijst, and G. Burghouwt. 2002. Urban Form and Travel Behaviour: Micro-Level Household Attributes and Residential Context. *Urban Studies* 39 (3): 507–527.

Ewing, R., K. Bartholomew, S. Winkelman, J. Walters, and G. Anderson. 2008. Urban Development and Climate Change. *Journal of Urbanism: International Research on Placemaking and Urban Sustainability* 1 (3): 201–216.

Gabriel, S.A. 1996. Urban Housing Policy in the 1990s. *Housing Policy Debate* 7 (4): 673–693.

Gans, H.J. 1993. *People, Plans and Policies: Essays on Poverty, Racism and Other National Urban Problems.* New York/Chichester: Columbia University Press.

Gong, P., S. Liang, E.J. Carlton, Q. Jiang, J. Wu, L. Wang, and J.V. Remais. 2012. Urbanisation and Health in China. *The Lancet* 379 (9818): 843–852.

Gordon, P., H.W. Richardson, and G. Yu. 1998. Metropolitan and Non-Metropolitan Employment Trends in the U.S.: Recent Evidence and Implications. *Urban Studies* 35 (7): 1037–1057.

Hays, R.A. 2012. *The Federal Government and Urban Housing.* 3rd ed. Albany: State University of New York Press.

Harris, B., and A.G. Wilson. 1978. Equilibrium Values and Dynamics of Attractiveness Terms in Production-Constrained Spatial Interaction Models. *Environment and Planning A: Economy and Space* 10 (4): 371–388.

Haslauer, E., E.C. Delmelle, A. Keul, T. Blaschke, and T. Prinz. 2015. Comparing Subjective and Objective Quality of Life Criteria: A Case Study of Green Space and Public Transport in Vienna, Austria. *Social Indicators Research: An International and Interdisciplinary Journal for Quality-of-Life Measurement* 124 (3): 911–927.

Heppenstall, A.J., A.T. Crooks, L.M. See, et al. 2012. *Agent-Based Models of Geographical Systems.* New York: Springer.

Hirsch, J.A., A.V. Diez Roux, K.A. Moore, K.R. Evenson, and D.A. Rodriguez. 2014. Change in Walking and Body Mass Index Following Residential Relocation: The Multi-Ethnic Study of Atherosclerosis. *American Journal of Public Health* 104 (3): e49–e56.

Holtzclaw, J.W. 2000. *Smart Growth—As Seen from the Air: Convenient Neighborhood, Skip the Car.* Air and Waste Management Association's 93rd Annual Meeting and Exhibition.

Hoyt, H. 1939. *The Structure and Growth of Residential Neighborhoods in American Cities.* Washington, DC: Federal Housing Administration.

Hunt, J.D., D.S. Kriger, and E. Miller. 2005. Current Operational Urban Land-Use-Transport Modelling Frameworks: A Review. *Transport Reviews* 25: 329–376.

Ingram, G.K. 1998. Patterns of Metropolitan Development: What Have We Learned? *Urban Studies* 35 (7): 1019–1035.

Klinger, T., J. R. Kenworthy, and M. Lanzendorf. 2013. Dimensions of Urban Mobility Cultures—A Comparison of German Cities. *Journal of Transport Geography* 31: 18–29.

Levy, J.M. 2013. *Contemporary Urban Planning*. 10th ed. Upper Saddle River: Pearson Education.

Lowry, I.S. 1964. *A Model of Metropolis*. RM-4035-RC. Santa Monica: The Rand Corporation.

Ludwig, J. 2012. "Guest Editor's Introduction," Symposium on Moving to Opportunity. *Cityscape* 14 (2): 1–28.

Makse, H.A., S. Havlin, and H.E. Stanley. 1995. Modelling Urban Growth Patterns. *Nature* 377: 608.

Matthews, R.B., N.G. Gilbert, A. Roach, et al. 2007. Agent-Based Land-Use Models: A Review of Applications. *Landscape Ecology* 22: 1447–1459.

Miller, Z.L. 1973. *The Urbanization of Modern America. A Brief History*. New York: Harcourt Brace Jovanovich, Inc.

Ohls, J.C., and P. Hutchinson. 1975. Models in Urban Development. In *A Guide to Models in Governmental Planning and Operations*, ed. S. Gass and R.L. Sisson. Maryland: Sauger Books.

Parker, D.C., S.M. Manson, M.A. Janssen, et al. 2003. Multi-Agent Systems for the Simulation of Land-Use and Land-Cover Change: A Review. *Annals of the Association of American Geographers* 93: 314–337.

Peterson, J.A. 2003. *The Birth of City Planning in the United States, 1840–1917*. Baltimore: Johns Hopkins University Press.

Rioux, C.L., K.L. Tucker, M. Mwamburi, D.M. Gute, S.A. Cohen, and D. Brugge. 2010. Residential Traffic Exposure, Pulse Pressure and C-reactive Protein: Consistency and Contrast Among Exposure Characterization Methods. *Environmental Health Perspectives* 118 (6): 803–811.

Rode, P., G. Floater, N. Thomopoulos, J. Docherty, P. Schwinger, A. Mahendra, and W. Fang. 2014. *Accessibility in Cities: Transport and Urban Form*. NCE Cities Paper 03. LSE Cities. London: London School of Economics and Political Science.

Sanbonmatsu, Lisa, Jens Ludwig, Larry F. Katz, Lisa A. Gennetian, et al. 2011. *Moving to Opportunity for Fair Housing Demonstration Program: Final Impacts Evaluation*. Washington, DC: U.S. Department of Housing and Urban Development, Office of Policy Development and Research.

Sanbonmatsu, L., J. Marvakov, N.A. Potter, F. Yang, E. Adam, W.J. Congdon, G.J. Duncan, L.A. Gennetian, et al. 2012. The Long-Term Effects of Moving to Opportunity on Adult Health and Economic Self-Sufficiency. *Cityscape* 14 (2): 109–136.

Sante, I., A.M. Garcıa, D. Miranda, et al. 2010. Cellular Automata Models for the Simulation of Real-World Urban Processes: A Review and Analysis. *Landscape and Urban Planning* 96: 108–122.

Smith, W.S. 1984. Mass Transport for High-Rise High-Density Living. *Journal of Transportation Engineering* 110 (6): 521–535.

Tong, X., Y. Wang, and E.H.W. Chan. 2016. International Research Trends and Methods for Walkability and Their Enlightenment in China. *Procedia Environmental Sciences* 36: 130–137.

Turok, I. 2009. The Distinctive City: Pitfalls in the Pursuit of Differential Advantage. *Environment and Planning A* 41 (1): 13–30.

UN-Habitat. 2013. *Urban Planning for City Leaders*. Nairobi.

———. 2014. *The Evolution of National Urban Policies*. The Global Overview. Nairobi. At: http://www.citiesalliance.org/sites/citiesalliance.org/files/National%20Urban%20Policies.pdf. Accessed 4 July 2017.

U.S. Census Bureau. 2011. *Housing and Household Economic Statistics Division*. Census of Housing, Historical Census of Housing Tables: Crowding. At: https://www.census.gov/hhes/www/housing/census/historic/crowding.html. Viewed 1 June 2017.

van Donkelaar, A., R. Martin, M. Brauer, R. Kahn, R. Levy, C. Verduzco, and P. Villeneuve. 2010. Global Estimates of Ambient Fine Particulate Matter Concentrations from Satellite-Based Aerosol Optical Depth: Development and Application. *Environmental Health Perspectives* 118 (6): 847–855.

von Hoffman, A. 2012. History Lessons for Today's Housing Policy: The Politics of Low-Income Housing. *Housing Policy Debate* 22 (3): 321–376.

Wilson, A.G. 1970. *Entropy in Urban and Regional Modelling*. London: Pion Press.

Zhang, G.J., M. Cai, and A. Hu. 2013. Energy Consumption and the Unexplained Winter Warming over Northern Asia and North America. *Nature Climate Change* 3: 466–470.

CHAPTER 2

Urban Environment: The Differences between the City in Europe and the United States

2.1 Urban Form Definitions

Urban form is the way human activities, buildings, and infrastructure are organized across the space. Urban form is mainly affected by urban land use and transportation (Anderson et al. 1996). Urban form describes the spatial structure of land use and transport system within an urban area (Ou et al. 2013). Urban form influences urban functioning such as commuting, the effectiveness of resource utilization, and the amount of produced pollution (Bereitschaft and Debbage 2013). For example, compared with residents of low-density areas, residents of denser urban areas travel less on a daily basis due to shorter distances between origins and destinations in high-density areas, better transit access, and because greater congestion and parking fees act as disincentives to driving (Marshall 2008). Urban form is also defined in terms of geometry, such as the size, shape, and layout of the built-up area (Bechle et al. 2011). Urban forms are sometimes viewed as "vague and ambiguous" and defined as facts on the ground directly contributing to the built environment (Conzen 2001, p. 3).

2.1.1 *Urban Form Measures*

Different measures are used to quantify urban form. Urban form metrics commonly used include *density*, *street accessibility*, *land use mix*, *centrality* (also known as *degree of centering*), and local-level urban design. Due to

A. Antipova, *Urban Environment, Travel Behavior, Health, and Resident Satisfaction*, https://doi.org/10.1007/978-3-319-74198-7_2

conceptual simplicity and the ease of measurement, residential/workplace population and housing density (the number of people or households distributed over an areal unit) is one of the most common measures. Land-use mix complements density, measuring how many uses can be found in a given area (Frumkin et al. 2004); it represents land-use diversity. Land-use mix can be vertical (standard in European cities and common in older areas of US cities but largely outlawed since the advent of zoning) and horizontal (various land uses are adjacently located). When applied to neighborhood, land-use mix and diversity can be measured with Simpson's Diversity Index (Lowry and Lowry 2014). Centrality can be represented by average distances to commercial areas, parks, K-12 schools, and transit bus stops, while accessibility can be measured by street connectivity and street pattern (calculated as the number of streets-to-intersections and cul-de-sacs-to-streets ratios), as well as median perimeter of residential blocks and length of cul-de-sacs (Banai and Antipova 2016). In their landmark study, Ewing et al. (2002) used four urban form factors including residential density, land-use mix, street accessibility, and degree of centering to develop a composite index to measure urban sprawl of the big US cities. The least sprawling areas are located in the Northeast, California (San Francisco), and Hawaii (Honolulu), while areas in the South and Southeast and a few others in California are the most sprawled. In the updated and refined report, *Measuring Sprawl* 2014, four primary factors including residential and employment density; neighborhood mix of homes, jobs, and services; strength of activity centers and downtowns; and accessibility of the street network have been used to analyze development patterns and assign a *Sprawl Index* score to each of the 221 US metropolitan areas. As index scores rise, so do several quality-of-life factors, including greater economic mobility experienced by individuals in compact, connected metro areas reflected in lower combined costs of housing and transportation; greater choice of transportation; lives that are longer, safer, and healthier, with fewer frequencies of obesity and fatal car crashes compared to residents of metro areas with sprawl (Measuring Sprawl 2014).

Sprawl is a type of urban form with both land use and transportation fundamental to sprawl (Frumkin et al. 2004). Ewing (1997) defined sprawl as development that is leapfrog or scattered with vacant land between developed uses; has commercial strip development where the consumer has to pass commercial uses along busy arterial roads in order to do shopping rather than to accomplish shopping with a single stop; or large areas have low density or single-use development with segregated land uses; sprawl is also indicated by poor residential accessibility (measures of accessibility include mean travel time and mean travel length). Auto dependence is caused by four features: low density and poor land-use mix resulting in low proximity as well as low connectivity. Additionally, walkable "town center" neighborhoods (indicating the vibrancy of a center and thus sense of place) and public open spaces are relatively scarce.

2.1.2 Urban Form and Travel

Urban form is commonly used by *travel behavior* studies, which investigate how different aspects of travel are related to various features of urban form such as compactness versus sprawl. Urban form is believed to have the potential to impact travel behavior. Urban sprawl index, developed by Ewing et al. (2002), mentioned earlier, besides measuring sprawl is a useful metric for predicting travel behavior, specifically, four measures including person hours of travel, person miles of travel, hours of delay, and number of annual vehicle fatalities (Ewing et al. 2003).

Measures of urban form and location explain the three selected aspects of commuting with urban form variables better explaining commuting modes than commuting time and commuting distance (Grunfelder and Nielsen 2012). Studies on the links between compact urban form and travel behavior (e.g., driving) are asking whether less driving occurs in communities designed according to the principles of compact development, with findings commonly reported in the form of statistical estimates of the average reduction in driving as a result of increases in the population density of a community, mixed land uses such as commercial land uses mixed with residential, and so on. Driving is measured in different ways. The most common measure is the vehicle miles traveled (VMT) expressed as the total distance a person/household drives over a certain time period. Other measures include number of driving trips made, time spent driving, and the number of automobiles owned. *Compact development* is commonly measured using five features referred to as *D-variables* due to the each variable's name starting with D—density, diversity, design, destination accessibility, and distance to transit (the latter is low in compact developments) (Stevens 2017). Compact environments are those that are high on the D-variables (except for distance to transit, which in compact developments is low, noted above) while sprawling built environments are those that are low on the D-variables (Ewing and Cervero 2017).

Compared with lower density places, higher density is linked with shorter trips, a greater number of home-based trips, greater mode choices (an increase in transportation options available to a traveler), and decreased auto ownership rates. The proximity of destinations determines how people travel: higher land-use mix in theory means a greater variety and more destinations (activity sites) in a community so that one's needs could be met within this area. Households in communities with fewer land-use mixes generate longer car trips and fewer trips using alternative modes (Frumkin et al. 2004).

Planners need to know *how much less* people of compact communities drive to help them determine whether it is worth promoting compact development as a way to get people to drive less. Stevens (2017) found that although people do tend to drive less when D-variables indicate more compact development, only modest goals for reduced driving can be achieved by encouraging compact development. With compact development features having a small influence on driving, planners and municipal decision-makers should not rely on a single strategy such as compact development to reduce driving (e.g., decrease VMT) (Stevens 2017). To illustrate, if the distance from downtown decreased by half (e.g., if a house moved from 12 miles to 6 miles from downtown), then a household is estimated to drive 32% fewer miles. However, the statement caused other researchers to voice a concern over Stevens' conclusions being "over-reached" and inconsistent with his own results, or with those by earlier studies (Ewing and Cervero 2017), and missing an important point (Handy 2017). As other researchers point out, even small decreases in VMT is an important benefit (Handy 2017). The decrease in VMT is often accompanied by "co-benefits" generated by such development including increased physical activity, better health (mental and physical), economic benefits (e.g., better job performance, less health care costs), sustainable livable environment as well as intangible ones such as stronger neighborhood or social cohesion, human capital, better quality of life, and others, which when taken together may be substantial (Sallis et al. 2015). Additionally, even though not much reduction in driving can be achieved just via promotion of a compact urban form alone, without it driving cannot be reduced, that is, a comprehensive strategy is needed that includes a creation of the conditions for driving less (i.e., compact form), among other elements (Handy 2017).

While there seems to be no large statistical association between driving and urban form, other factors including the price of gasoline, economic growth, and changes in technology (e.g., online shopping, social media) may determine VMT in a much more significant way (Knaap et al. 2017). Clearly, the growing VMT dropped during late 2007 to late 2014 as observed in Fig. 2.1 showing the period 1970–2017. Peaking in November 2007 at 3,038,886 million, total VMT fell until December 2014, after which it recovered due to growing economy and changes in energy prices, demographics, and consumer preferences. However, the studies on the relationship between travel and the built environment while focusing on D-variables ignore the problem related to a massive amount of land set aside for vehicles in urban areas such as off-street parking; with driving occurring on valuable public land, this land should not be given away for free (Manville 2017). Further, while the use of fuel pricing rather than land-use regulation is often advocated to curb driving and its social impacts, urban planners cannot control the former, but the built environmental variables can be modified as the *combined effect* of the Ds on travel could be large (Ewing and Cervero 2010). Finally, the benefits of per mile tailpipe reductions since the 1960s are greatly undermined by the annual increases in national VMT, however, land-use strategies promoting more compact urban forms have potential to limit the annual growth, resulting in additional benefits to regional and local air quality (Stone 2008).

Similarly to the evidence of the influence of urban form on travel activity, housing in the US locations with walkability and proximity to various destinations including employment, retail, and service amenities is resilient in terms of retaining or appreciating in value even during market downturn, compared to distant, auto-dependent locations (Yen 2012). In the United States, urban form is generally automobile dependent, impacting various economic activities including retail and shopping. In a recent study, Banai and Antipova (2016) explored whether the beneficial relationship between retailing and housing works both ways by examining the effect of urban form on retail-center viability at urban (Hickory Hill) and suburban (Bartlett) locations in Memphis, TN, using Walmart supermarket as an example. Sales data was used as an indicator of business viability. Since the literature of new urbanism and smart growth often cites land-use mix, density, and proximity of residential to "core commercial areas" at a convenient, walkable distance as desirable neighborhood features, the determinants of urban form in the study included density, land-use mix, land value, and proximity to housing, businesses, and roads at the finest US Census block scale. Land-use mix was measured with three land-use indices, including the *employment-to-residential workers ratio* (computed

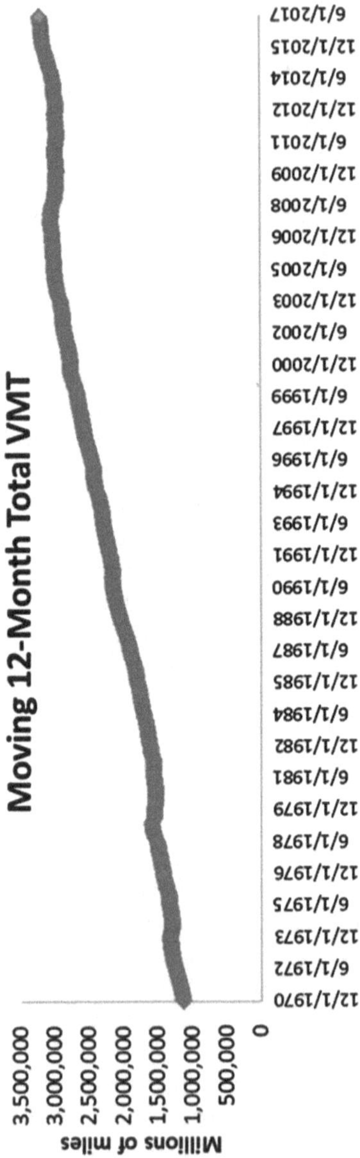

Source: Author, using U.S. FHA, Moving 12-Month Total Vehicle Miles Traveled, retrieved from FRED, Federal Reserve Bank of St. Louis; https://fred.stlouisfed.org/series/M12MTVUSM227NFWA

Fig. 2.1 Moving 12-month total vehicle miles traveled, 1970–2017

by dividing total employment in an area by a number of resident workers) indicative of the neighborhood's "self-containment," the Shannon Index (the entropy index) assessing the diversity of land use, and a total number of land uses per a given areal unit as a direct measure of land-use diversity.

Banai and Antipova (2016) compared the two neighborhood shopping centers that had similar urban forms regarding pedestrian orientation with proximity to residential land use, nodal location of businesses, retail mix (office, retail shop, and restaurant), sidewalks with pedestrian connection to the supermarket, and ample parking space; however, these two centers were distinguished by variable land-use mix and density, reflective of the urban versus suburban locational distinction. Urban form was explored at three gradations of distance: from 0.25, 0.5, to 0.75 miles of the road intersection where the shopping center is located. More specifically, the hypothesis was tested whether the retail center sustains viability due to proximity to residential land use. The relative importance of land-use type, density, and mix that varies by distance from the retail center was determined. The Wilcoxon test showed the statistically significant differences in sales volume between the two communities, Hickory Hill (urban) and Bartlett (suburban). Businesses in the urban settings had consistently higher sales across three buffers confirming the linkage between retailing viability and urban form (Banai and Antipova 2016) Fig. 2.2.

In addition to traditional urban form measures, ***landscape (spatial) metrics***, developed in the late 1980s, are useful for the analysis of urban environments and urban land-use patterns; for instance, to quantify the shape and pattern of vegetation in urban areas. Thus, urban form can be measured using composite measures including ***urban continuity*** and ***shape complexity***. These landscape (spatial) measures are used to describe the temporal evolution of urban land patterns (Luck and Wu 2002), model future urban growth (Herold et al. 2005), characterize and model urban sprawl (Wei et al. 2006; Bhatta et al. 2010) as well as analyze the spatial distribution of urban elements and characterize the urban forms (She et al. 2017). ***Urban continuity*** represents the extent to which the urban landscape is fragmented and can be interpreted as a measure of "leapfrog" development (i.e., when new growth areas lack physical connectivity between them): less fragmented urban areas with more contiguous and less "leapfrog" development have higher urban continuity. ***Urban shape complexity*** measures urban area-to-perimeter ratio and indicates the

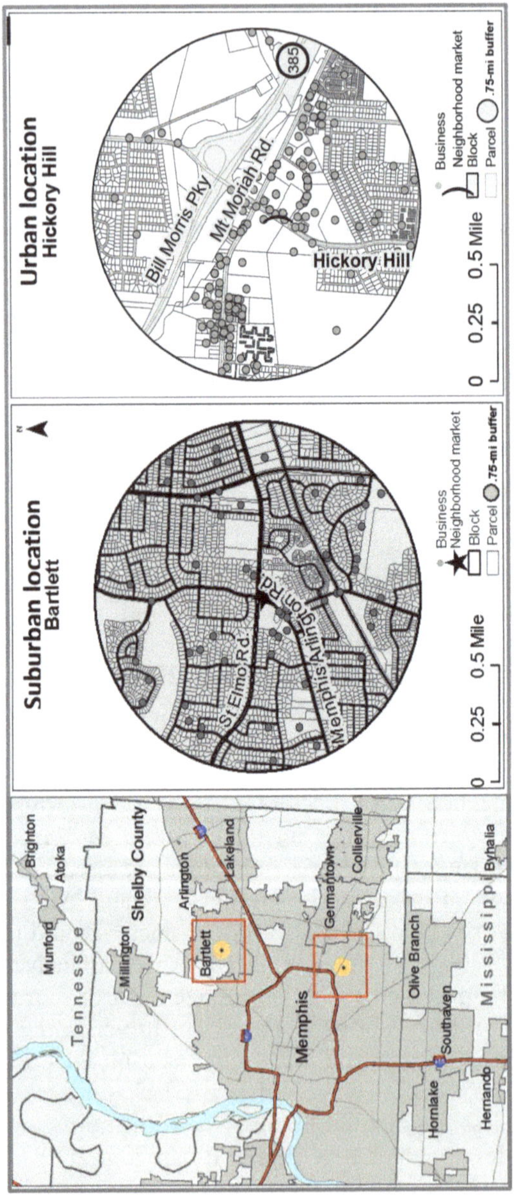

Fig. 2.2 Urban form and retail viability relationship in the Memphis metropolitan area (Source: Author)

"jaggedness" of the urban boundary: suburban and exurban areas often have highly irregular boundaries (Bereitschaft and Debagge 2013).

Urban form is often used by environmental studies calling to lower the ecological footprint of cities. These studies explore the impact of urban form on air quality. Air quality outcomes are modeled based on projected urban growth and various land-use scenarios. For example, in an investigation of features of urban design that can improve or worsen the urban environment and public health, besides the dimensions typically used to classify urban form such as population density and *urban extent* (defined by built-up area), other satellite-derived urban form metrics are used including the *contiguity* and *compactness* (Bechle et al. 2011). *Contiguity* measures urban patchiness and describes a degree of "leapfrog" development, analogous to continuity noted above; it is calculated as the ratio of the main contiguous built-up area to the total built-up area of the city. *Compactness* measures the circularity of the city (i.e., the built-up area), calculated as the ratio of built-up area to total buildable area within a circle around the main built-up area of the city. Geometrically compact and contiguous cities, which are sufficiently dense, are expected to diminish travel and vehicle emissions and improve air quality (Bechle et al. 2011). For example, a more contiguous urban landscape may offset a population increase. Urban circularity ("compactness") and contiguity, thus, associated with energy-, or pollution-efficiency, may be "better" urban forms.

Cities that are more compact and dense (also referred to as *smart growth*), through greater physical proximity to opportunities, enable the physical concentration of people, services, and economic activities, and thus facilitate *agglomeration economies* (Rode et al. 2014). In contrast, cities with sprawl due to increased automobile dependency and the respective greater travel are expected to have higher transportation-related $CO2$ emissions; similarly, less contiguous built-up areas are linked to higher $NO2$ concentrations, thus making sprawl-like urban areas examples of less desirable urban form. Urban design strategies searching to decrease air pollution including $NO2$ should develop policies supporting contiguous urban growth rather than isolated development (Bechle et al. 2011).

Bereitschaft and Debbage (2013) used a combination of traditional urban form measures and landscape-based metrics of urban form. Traditional urban form measures consisted of overall sprawl score (*Smart Growth America (SGA) Index*) developed by Ewing et al. (2002) and various measures of urban form. The SGA composite urban sprawl index developed by Ewing et al. (2002) combines four measures of urban form including residential density, land-use mix, degree of centering, and street accessibility. The SGA is a composite index obtained from the four measures of urban form using principal component analysis. Other traditional urban form measures included residential density, *land-use mix* (that is, land-use diversity measured as a jobs-to-population ratio), *degree of centering* (essentially, it is a density gradient indicating the proximity of housing and jobs to the central business district), and *street accessibility* (or, *street network connectivity* represented as average block size). Additionally, by using the principal component analysis (PCA), two composite measures of urban form that are not captured by the urban sprawl score, *the continuity* and *shape complexity* of urban development, have been obtained. The PCA, a variable reduction technique, was used to reduce nine original landscape metrics (contagion, contiguity index, percentage of like adjacencies, largest patch index, the clumpiness index, area-weighted mean shape index, landscape shape index, area-weighted mean patch fractal dimension, and edge density) to just two *uncorrelated* urban form components such as *urban continuity* and *shape complexity*. The two components explained almost 80% of the total variance among the nine original urban form measures. The former measure quantifies aggregation or fragmentation of urban development, while the latter is based on a perimeter-to-area ratio and reflects the compactness of the urban area and the intermixing of urban and nonurban land cover (Bereitschaft and Debbage 2013).

She et al. (2017) define urban form as the land-use configuration of the urban landscape across the space. As noted earlier, population density is commonly employed by studies as a measure of urban form, however, when used alone it may turn out insufficient to comprehensively characterize urban form.

Urban forms can be represented with *spatial* (or *landscape*) *metrics*, noted earlier, which represent *independent* measures of the urban socioeconomic landscape and a wide range of urban spatial characteristics (Ou et al. 2013). Accordingly, to characterize urban form, researchers start first by analyzing the spatial distribution of urban land use and transportation network, using the land use and transportation data to develop a set of spatial metrics. The spatial distribution of urban land use can be obtained through land-use classification. Then, a set of spatial metrics are developed to quantify urban form as well as urban land use and transportation network data (for example, see Ou et al. 2013). Several common spatial metrics quantify urban forms in different dimensions: size, shape, density, regularity, fragmentation, and traffic coupling factor. Specifically, spatial metrics include Total (urban) Class Area, Number of Patches, Edge Density, Mean Perimeter-Area Ratio, Percentage of Like Adjacencies, Patch Cohesion Index, Largest Patch Index, Urban Road Density, and Traffic Coupling Factor.

2.1.3 Global Urban Form Differences

Huang et al. (2007) compared regional differences in urban form across 77 metropolitan areas in Asia, the United States, Europe, Latin America, and Australia by utilizing satellite images. Their comparative analysis of urban form was supplemented with socioeconomic indicators for national wealth such as *gross domestic product (GDP) per capita* (defined as all goods and services produced in a country in a given year, divided by its population) using purchasing power parity (PPP US$), transportation ownership measured by national vehicles per 1000 population, and telecommunications/national lines ownership measured by national main telephone lines per 1000 people.

2.1.4 Remote Sensing Data Website

A good source of remote sensing images is the Global Land Cover Facility (www.landcover.org/), a website that offers comprehensive, free earth science data including satellite imagery and products derived from satellite imagery around the world to study land cover and land cover change from the local to global scales. The images can be used to extract the urban areas defined as the built-up or urbanized as observed in the images. Spatial (landscape) metrics can be used to describe urban form's complexity, compactness, centrality, density, and *porosity* (the ratio of open space compared to the total urban area). *Complexity* index measures the irregularity of the patch shape, while *centrality* typically describes the degree of closeness of the urban development to the central business district (CBD). In the study, *the centrality index* was calculated by taking the mean distance between dispersed patches and the city center defined as the centroid of the largest patch. Further, centrality characterizes the overall city shape, if it is elongated or circular: the more elongated overall shape is assigned greater centrality index (Huang et al. 2007). The *compactness index* measures both the individual patch shape and the entire urban landscape fragmentation (the more regular shapes of a patch have a greater compactness index value). Open space is an important amenity for

Fig. 2.3 Open space is an important amenity: Madison Square lawn in New York (Author 2013)

residents and for city sustainability; *porosity* measures the ratio of open space versus the total urban land. A generally recognized dimension of compactness or sprawl is measured by *population density* calculated by dividing the population of the urban agglomeration by the extracted urban area Fig. 2.3.

2.1.5 Landscape Spatial Analysis Website

These five urban form dimensions (that is, complexity, compactness, centrality, density, and porosity) are captured by seven spatial metrics such as *the area weighted mean shape index* (represents the shape irregularity of the patches, higher value indicate more irregularity of the shapes), *the area weighted mean patch fractal dimension* (describes the raggedness of the urban boundary), centrality, compactness index, compactness index of the largest patch, ratio of open space, and density. The first two metrics can be calculated using the *Patch Analyst* extension to the ArcGIS software system that assists with the spatial analysis of landscape patches and the modeling of patch attributes (www.cnfer.on.ca/SEP/), the website also provides other landscape analysis tools.

Regarding regional differences in urban form, metropolitan areas of the developing world have been found more compact and dense compared to urban agglomeration in either Europe or North America. Other striking differences across regions have been identified. In contrast to developing regions, the cities of the developed world display more complex character, less compactness, greater porosity, and the least dense urban forms. Developed regions exhibit greater sprawl than the developing areas of either Asia or Latin America, while the latter have greater compactness, density, and regularity (Huang et al. 2007). Spatial metrics between the two regions of the developed world can be compared using the regional averages. Within developed world, the United States, Australia, and New Zealand have been grouped together as they followed a similar path of urban development, while European and Japanese cities comprised another group. As a region, the European/Japanese region has urban forms that are more centralized, more compact, denser, and less irregular in shape than the US region (Huang et al. 2007).

On all spatial metrics, except compactness index and density values, the United States, Australia, and New Zealand have greater scores compared with the cities in Europe. For example, overall higher *complexity* values indicate that the cities in the region such as the US cities are more irregular in their shape with more ragged urban boundaries, and with the greater average distance between the dispersed parts and the urban center. Compared with the cities of the European region, smaller values of the *compactness index* for the cities of the US region indicate the greater fragmentation of the overall urban landscape, or the leapfrog development. The cities of the US region overall have greater porosity with more open space than their counterparts in Europe. As reflected by a higher *centrality* (reflected by the greater centrality index), the overall shape of the cities in the US region is more elongated rather than circular suggesting that they are less compact and more spread out, also indicated by the greater average distance of the dispersed parts to the city center. Finally, compared with the cities in the European region, the US cities are much less dense as indicated by their much lower residential density.

Regarding differences in urban form among cities, first, to make meaningful comparisons among them, cities need to be grouped based on a variable of interest. An entire sample consisted of 77 metropolitan areas in the world; cluster analysis has been applied based on spatial indicators. As a result, all cities have been classified into four groups that share similar urban forms. In this chapter, we will only consider here those groups that fall in either the European region or the North American region. The first group combines a few European cities such as Lyon, Berlin, Milan, Madrid, and Kiev with most Asian cities and Latin American cities. This set of cities is characterized by moderate centrality, density, complexity, centrality, and a moderately low level of open space.

Another group consists of the US cities such as Baltimore, Boston, Chicago, Dallas, Denver, Little Rock, Milwaukee, New Orleans, Oklahoma City, Phoenix, Seattle, and Washington, and many European cities such as Paris, Hamburg, Rome, Moscow, Barcelona, Glasgow, London, and Manchester. Despite the differences between the regions, the city clusters suggest that many European cities and US cities can be grouped together into a single group. European cities, particularly in the west and north of the European continent, have more similarities with the cities in the United States than existing transatlantic arguments would suggest. The group includes the most characteristic cities of the developed world. The cities in this group have significantly lower density than the other group, while open space averages are much higher. It is interesting that many European cities fall into this single group (Huang et al. 2007).

Regarding automobile dependence, 32 major cities in North America, Europe, Australia, and Asia were grouped by another study into five categories, with most US and Australian cities being within categories with a high or very high automobile dependence and a very minor role for public transport, walking, and cycling. Most European cities are in categories with moderate or low automobile dependence and an important role for public transport. Further, Munich and Paris, very wealthy cities in Europe, have a very low automobile dependence with public transport, walking, and cycling more important than cars (Newman and Kenworthy 1989). Many prosperous European cities such as Vienna, Copenhagen, and Stockholm have several times less use of cars; fuel consumption levels in those cities account for one-fifth to one-tenth the level of the US cities. Many cities with high levels of automobile ownership including Zurich, Copenhagen, and Freiburg in Europe, and Portland in the United States, have curbed automobile dependence by controlling automobile use. On the other hand, automobile dependence is not controlled in Los Angeles, where in fact individual mobility is fostered, inevitably resulting in the increased use of private automobiles (Newman 1996).

Cities are places where contemporary social processes including polarization between "haves" and "have-nots" are concentrated; growing cultural diversity leads to a changing understanding of community. Globally, urban landscapes are undergoing changes at an unprecedented pace reflected in both the urban form and spatial range of new landscapes (Conzen et al. 2012), with contemporary urban landscapes becoming increasingly fragmented, and new urban forms including "mutant" forms such as highly car-dependent *edge cities* (Newman 1996) and "privatopias," while a crisis of sustainability led to the emergence of a consciousness about the need of a sustainable development (Dear 2005). Though each city is a unique place, cities are increasingly integrated into the international urban system via globalization, although not all cities are impacted to the same degree. Distinct urban differences result from varying settlement time, industrialization, infrastructure development, land ownership, planning, and regulation.

In vigorous debates on "good city form," a sprawling urban form in the United States is often contrasted with European compact form (Ewing 1997; Nivola 1999). The contemporary US metropolitan form is characterized by dispersed population and employment, and a larger role for the private car (Giuliano and Narayan 2003). Compared with European cities, US cities have lower population density. Presently, the average metropolitan population density in the United States is more than three times lower and the share of the metropolitan population residing in the inner areas is 1.6 times lower than in Europe (van de Coevering and Schwanen 2006). US cities exhibit lower *compactness* (more irregular shape of individual patches of urban land and the overall urban landscape fragmentation) but higher *complexity* (lands have more irregular shapes and the overall urban perimeter is more rugged), lack of *centrality* (the average distance of the dispersed lands to the city center is higher resulting in the overall shape of a US city being more elongated than circular) as well as larger open space compared to the total urban area.

European metropolitan areas have higher densities and more centralized land-use patterns, with lower levels of car use. In addition, many European countries employ stronger land-use controls contributing to the preservation of the compact urban form, such as the British *Town Centre First* policy. Other urban characteristics in Europe include development in rail corridors while in the United States a stronger emphasis is on suburban development, which reinforces vehicle dependence (Salomon et al. 1993), which, in turn, leads to the inefficient use of scarce urban space (Rode et al. 2014).

2.1.6 *Factors Influencing Urban Form: Land-Use Patterns*

It is instructive to compare the contemporary urban structure with that of an earlier period. Regarding patterns of city structure, writing as early as the 1930s, Hoyt (1939, p. 9) noted that the amount of vacant land or yard space within the city block increases with the distance from the center of the city to the periphery center. The ground space is intensively utilized at the city center with business structures almost solidly occupying the blocks, and the amount of vacant spaces gradually increases farther out from the center of the city to the periphery of the settled area where some blocks are partially developed and others are vacant beyond which large tracts of rural land lie (Hoyt 1939).

Buildings and houses are common elements in the cultural landscape and are a useful tool of geographic analysis. Houses can be used for a better comprehension and analysis of the structure of the urban landscape because residential occupancy takes up far more land than all other urban land uses together (Rickert 1967). House types can be used in identifying the growth rings from each construction cycle (Adams 1970). The changing occupancy patterns in turn help to better understand the processes of urbanization and suburbanization. The identification of potential physical obsolescence factors can be used to predict future blight patterns. For example, the dominant architectural style may differ steeply from styles of the previous and following periods and, thus, may add to obsolescence problems in a neighborhood. For example, costs are high (between 10% and 20% of the market value) for repainting the exterior of a Victorian-era house, such as the Queen Anne house. So neighborhoods with this obsolescence factor (e.g., a certain architectural style) may be affected, for example, with a large fraction of housing stock falling into disrepair and becoming blighted (Adams 1970). These obsolescence factors might be used to delineate areas of potential land-use change so that the area is rezoned (e.g., land uses with obsolete one-family houses can be changed to multi-family or commercial uses in order to avoid value depreciation, etc.) (Rickert 1967).

Given that residential land represents the largest land-use category in contemporary American cities, it is helpful to break it down according to the predominant *type of structure* such as single-family, two-family, and multiple apartment areas to comprehend better the present land-use pattern in comparing urban form between American and European cities. That individualism triumphs over collectivism most often stands out in urban settings with Americans most decidedly preferring detached over multi-family housing (Conzen 2001). Urban area in the United States is occupied by many more single-family residences than multi-family residences. The physical form of a city reflects its underlying sociocultural values (Conzen et al. 2012). Homer Hoyt (1939, p. 24) notes: "Single-family structures predominate in American cities... and 63 percent of the American urban families lived in such structures...and two-family structures and apartments are too few in number." Compared with Hoyt's observations made nearly eight decades ago, in the United States, the level of single-family detached homes has changed little over the 60-year period, ranging aroundthe 60%, with single-family attached homes (row houses and townhouses) comprising 5.6%, and apartment housing with two to four units in the building accounting for less than one-tenth of the inventory in 2000 (U.S. Census Bureau 2011). In American city building, dwelling units in single-family structures tend to use more land compared with the same size unit (a single) in a multiple structure; thus, if single-family units rather than doubles or multiples are the preferred construction style during a certain residential development period, the result is the larger mean lot size for all residential structures constructed during that time period. Accordingly, the growth ring for that period is also larger in size (Adams 1970). Additionally, a general feature in American cities is the high prevalence of wooden structures with buildings having relatively short life (Conzen 2001) Fig. 2.4.

Fig. 2.4 Many more Europeans live on far less space than in the USA. Residential zone in Essen, Germany (Author 2010)

2.1.6.1 Other Factors Influencing Urban Form

Urban form, or the built-up area, is influenced by topographical features such as land elevation (although most US cities developed on largely flat sites), water bodies, and transportation (discussed in a greater detail later). The contour of water bodies influences the shape of settlement such as Lake Michigan, which affects the form of Chicago. The topography forces a certain form of development: a narrow river valley with hills on both sides resulted in string-like urban form of Charleston, WV Further, in the hilly terrain the land is used more sparingly, for example, built-up areas are more restricted in the mountainous regions (Nemes 2011), so relief negatively affects urban sprawl. The presence of steep-sloped terrain (defined as slope greater than 15% or steepness with an angle greater than 8.5°) effectively limits residential development (Saiz 2010) Fig. 2.5.

Other factors include political, economic, and social development (Berry 1973; Hawley 1986; Hall 1997). Generally, socioeconomic indicators such as purchasing power, vehicle ownership, and the like correlate positively and significantly with irregular shape of land and abundant open space, while negative association exists between these indicators for national wealth and land density and compactness (Huang et al. 2007). For example, higher purchasing power (i.e., more wealth) is associated

Fig. 2.5 The contour of Lake Michigan affects the shape of Chicago (Author 2016)

with higher land complexity and larger open space as well as with lower density and less compactness. Wealthy countries have more highways and higher private motor vehicle ownership. These conditions promote living in the remote suburban areas far from work places that are less compact, less dense urban form, and a fragmented urban fringe and large proportions of open space.

2.2 Transportation and Urban Growth

This section discusses how changes in transportation technology cause changes in urban form most conspicuously affecting its shape and size. Changes in internal transportation result in dramatic changes in city configuration. American cities were built during a series of intraurban transport eras with certain eras promoting concentric circle urban spread (i.e., when the transport systems provided a *movement surface*) while other eras distorted the concentric circle growth and fostered star-shaped development with the transport system relying on a *movement route network* (Adams 1970). Three main types of transportation could be distinguished in the evolution of internal transportation in the United States, including horse-car lines of the pre-1880s supplemented with steam railroads and cable cars (after 1882) providing fast transportation in some cities; electric streetcars (also referred to as trolleys or trams) of the 1880s to World War I supplemented with elevated railroads and subways in a few large cities; and the introduction of the automobile in 1920, which dramatically affected the character of urban areas and lives of urban residents. Different types of transportation affected the shape of the residential growth rings as well as densities and spatial arrangement in each of the rings during that era. Respectively, Adams (1970) in his seminal work "Residential Structure of Midwestern Cities" developed an urban residential structure model and identified residential growth rings, which he described referring to four significantly different urban transport eras Fig. 2.6.

The four transport eras distinguished include the following:

1. Prior to the 1880s: The walking/horsecar era;
2. 1880s–World War I: The electric streetcar era;
3. 1920s–1941: the recreational auto era; and
4. Post–World War II–present time: The freeway era.

Fig. 2.6 Invented in 1873 by Andrew Smith Hallidie, historical cable car in San Francisco (Author 2016)

The transport systems of the first (foot travel) and third (recreational automobile) transport eras provided a movement surface resulting largely in a compact, concentric circle pattern of urban spread. The transport systems of the second (streetcar lines) and fourth (freeways) transport eras provide sharply defined movement networks that promote transport surfaces and a star-shaped development pattern with most growth occurring along arterials of fast movement (Adams 1970).

Each type of the prevalent internal transportation favors a particular form of urban growth in certain periods. City growth dynamics (or the form of city growth) can be studied by comparing a series of maps of settled area (that is, showing the boundaries of urban development) visualizing the form of city structure at successive time periods (Hoyt 1939, p. 96). The walking/horse car transport era resulted in early urban growth. As cities were small, all getting around was by way of walking focused on a high-density center of nonresidential activities while surrounding the core was a high-density residential zone where densely arranged three- to five-story tenements made of stone, brick, or wood housed low- and middle-income groups, and upper-income class during the post–Civil War period occupied popular two-story Gothic-style houses with a tall, pointed look, and French-style houses with mansard roofs decorated with the "cupola that was the cachet of wealth and respectability; the prerogative of the merchant prince" (Rickert 1967, p. 226). Once cities started growing in the 1830s and 1840s, this transport system was supplanted by the primitive horse car-based transportation. The omnibus lines were installed in a few cities but soon were replaced with horse-car transportation, including horse stagecoaches and horse-drawn streetcars, with horse-car transport becoming the main mode of internal travel within the city from the 1850s to 1880. The horse-car-based transport systems were used for several decades in many cities; however, the systems were important on a small scale (Adams 1970). Horse-car transportation forced concentration of all economic activities as close to the central business core as topographical features allowed. Therefore, central growth was accentuated during the period of horse-car transportation.

Central growth implied high-density pattern within the former built-up area with housing of that era represented by multiple family structures with up to 1000 people per acre for low-income class and lower density for the middle-income class who lived in three- and four-story row houses on lots of 15, 18, and 20 feet, whereas the rich could afford to move away from the congested urban core to the countryside (Adams 1970). For example, six-story buildings comprised the urban core of 1880s Chicago, and four- to five-story walkup apartment buildings were constructed in Harlem, Brooklyn, and South Chicago, visible for mile after mile.

Central growth is enabled when all means of transportation have roughly equal speed from the city center in all directions to the city edge so that urban growth proceeds in compact concentric circles around the original starting point. As a result, many American cities had a compact circular form including pre-1840 era cities of New Orleans, Boston, Philadelphia, Baltimore, and late nineteenth-century cities of San Francisco, Kansas City, MO, Salt Lake City, Dallas, and Charleston, WV. Exceptions were New York City and Chicago where axial growth occurred along cable-car lines and suburban steam railroads. New York City and Chicago had to supplement low carrying capacity of downtown streets through elevated rail systems, introduced in 1868 and 1892, respectively. While not important everywhere else, in Chicago elevated railroads are still an important form of intraurban transportation. Cable-car lines inaugurated in the 1880s almost doubled the horse rate of speed. Suburban railroads facilitated scattered development with isolated nuclei of houses growing along the railroad stations at intervals along the line beyond the periphery of the city (Hoyt 1939) Fig. 2.7.

The introduction of the electric streetcar in the 1880s–1890s was nothing short of revolutionary. Similar to the mechanical cable cars, streetcars run on steel rails, but they are propelled by onboard electric motors and use a trolley pole to get power from an overhead wire. Although the urban core (which still was the most accessible for all city residents) continued being the dominant location for commercial and industrial activities, faster transportation from the city center to the periphery afforded by electric streetcar affected all American cities with the resultant axial residential growth. The electric streetcar changed the compact circular shape of settled area to spread out star-like settlement reaching far into the surrounding country with new development tied to the fixed line of streetcar rails rather than in the intervening areas between these radial lines during the period of 1890–1910 (Hoyt 1939).

The streetcar networks opened many new areas for residential development (Rickert 1967). For example, the radius of Boston was doubled with the coming of streetcar, as a new transportation means transformed the old walking city into a modern metropolis (Warner 1968). However, before the full impact of the streetcar-based transport technology could be felt, city residents still mostly walked in the 1880s (Adams 1970).

Streetcars "directed" residential development with streetcar neighborhoods built along and in the vicinity of the streetcar lines relieving high densities of the old residential districts, while the land that was not served by local public transportation formed intact unsettled interstices between the rays of development. Large-frame, two-story and taller houses constructed in streetcar neighborhoods on relatively narrow lots resembled farm houses built in the streetcar era; houses had large kitchens and dining areas, but no garages or sheds. The height of the Victorian gingerbread house, the typical residential style of the period, and the exterior chimney (indicative of an open fireplace used for decor rather than for heat) were symbols of the owner's prestige and status in the community. By the end of the era, story height decreased to one and one-half- to two-story buildings. The ornamental Queen Anne-style row houses with brick, stone, or terracotta walls and decorated with turrets and balconies, were fashionable in the Gingerbread period of the 1880s and 1890s; these turn-

Fig. 2.7 Elevated rail lines in Chicago are an important form of intraurban transportation (Author 2016)

of-the-century houses are a feature of the contemporary urban landscape in Georgetown, Washington DC. Real estate brochures of the period advertised the back-to-the-country suburban movement: "Buy large comfortable houses in park-like communities with all the best residential amenities …." (Rickert 1967, p. 226). With women still not part of the paid labor force, and larger-sized families, houses built during the trolley network era had many rooms. Regarding land prices, these were higher near the streetcar lines (Adams 1970). Meanwhile, subway lines, which provided the most rapid transportation, were installed in a few large cities including Washington, DC, further extending urban reach.

The introduction of streetcars and metro and regional rail systems started the process of the de-densification and horizontal expansion. However, because public transport needs urban density, horizontal expansion enabled by transit systems during this period continued supporting urban environments with compact and dense urban forms (Rode et al. 2014). Deviations from concentric pattern are introduced due to streetcar lines. When the city's transport system is in a form of a network, the urban areal growth no longer follows a concentric circle pattern, but residential growth advances faster along routes leading to the center where movement costs are lower, thus, a star-shaped pattern of urban spread is promoted by movement network (Adams 1970). In contrast, in interstitial areas where movement-cost gradients are steeper (i.e., higher costs), residential growth is restricted.

After 1917, with the arrival of mass motorization and the introduction of privately owned vehicles on a widespread basis, automobiles increasingly became universal means of internal transportation, while streetcar traffic declined in importance. To illustrate the growing importance of automobile, the total population in the United States grew from roughly 100 million in 1915 to 123 million in 1930, while the total number of vehicles increased from about 5 million to about 25 million during the same period, starting the period of moderately paced suburbanization in the 1920s (Levy 2013). Figure 2.8 presents the annual number of cars (in the millions!) bought after the post-war Depression of 1919–1920 was over, between 1922 and 1929, the year of the stock market collapse. The increasing role of the car and the introduction of hard-paved surface roads allowed to fill in the formerly undeveloped areas of the star-shaped cities with a massive residential development boom (Rickert 1967). Growth rings were created by filling in the formerly undeveloped areas between the "spokes of growth" that were concentric, while residential densities of

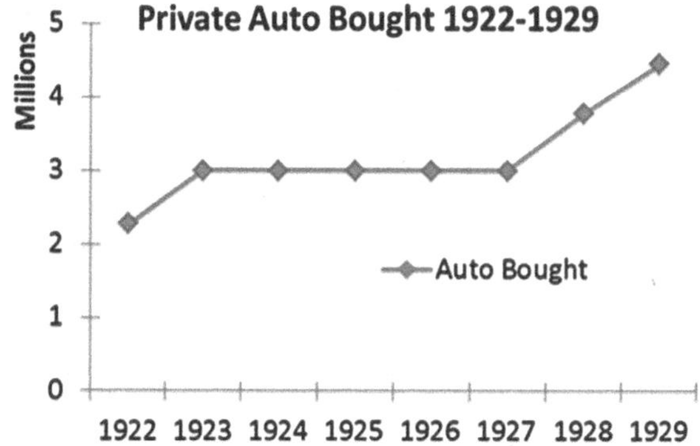

Source: Author, using auto purchase data by year in Scott (1995, p.186).

Fig. 2.8 Number of cars bought by year, 1922–1929

the recreational automobile era became lower than during the previous streetcar era due to greater intraurban mobility (Adams 1970). The residential construction that peaked in 1925 and sharply declined in 1933 differed substantially from the previous eras. Median income per capita rapidly increased after the World War I, impacting the consumption habits of urban Americans so that more income could be spent on consumer nondurables and services. Purchasing power increased while land prices declined as the amount of land made available by automobile for new urban development did not match existing demand for use of these available sites; the supply of many scattered vacant lots allowed buying spacious lots in the 1930s. Due to new building regulations, lot layout was standardized with nearly identical, mass-produced houses in residential subdivisions. Housing characteristics included a smaller size of houses constructed. Women became a part of the paid labor force, and with smaller families there were fewer rooms in the houses; kitchens and dining areas were reduced in size, and garages and paved alleys were added indicating the increasing importance of the family automobile (Adams 1970).

Automobiles owned by middle- and upper-income classes was not important for everyday living; cars were largely used for recreational rather than commuting purposes in the 1920s (thus, the name for the era). During the era of the recreational automobile, transit lines extended with buses affecting the spatial arrangement of residential development in the same fashion as the private car did, since in contrast to inflexible streetcar systems, flexible routes permit non-restricted bus movement on a transport surface of urban streets servicing residential developments where demand existed (Adams 1970). Since the late 1930s, automobiles increasingly began being used by commuters as high-speed limited-access highways were introduced shortly before World War II. The process of urban de-densification and horizontal expansion reached a different, much greater scale whereby access by physical proximity was substituted with access by automotive movement enabled by substantial reduction in mobility costs relative to incomes (Cervero 1998). Since 1940, only a small fraction of housing was constructed during the first six years of the decade due to the need for building materials required for the strategic war effort; however, as soon as the World War II ended, residential construction reached scales never seen before and accompanied much strengthened intraurban population movement during the post-war period (Rickert 1967). Housing of the era consisted of prefabricated structures that were reassembled on local sites and delivered to municipalities.

Because a speed of a car on open highways several times exceeded a speed of electric streetcars, extensive areas on the periphery opened up for urban growth. Thus, the automobile encouraged suburban development with extremely low density: single-family housing occupied lots up to an acre in size. This type of residential development was only made possible by movement requirement provided by high-speed automobiles and limited-access highways (Adams 1970).

The automobile was the most effective means for expanding the city over a big area. Cars can reach anywhere where hard-paved roads are available, thus a new type of development favored proximity to automobile highways, so that many formerly vacant interstices of the streetcar era became available for development and were filled in during the automobile age. Detroit, Los Angeles, and Miami grew especially rapidly with the coming of cars (Hoyt 1939).

Automobile transportation is conducive to a wide scatter of settlement. The city cannot grow equally in all directions since the transport system does not allow equal ease of movement in every direction. Instead, high-speed limited-access arteries, along which residential development occurred in the form of growth rings, were limited in number. Growth rings are narrow in width in those areas difficult to access. During the freeway automobile ear, a network of major highways directed the spatial spread of the city, with the growth rings deviating from concentric growth patterns fostering once again (as in transport era of streetcar lines) star-shaped deviations from concentricity (Adams 1970).

Urban sprawl is associated with the high-speed automobile and freeway era; any US city that grew substantially after World War II reveals irregularly shaped low-density residential suburban development with housing supply consisting of single-family units constructed on the edge of the cities and complemented by auto-oriented shopping centers. For example, in Memphis, TN, after World War II, and also in the 1970s and thereafter, population mobility patterns resembled those of many US urban areas, with middle- and upper-income class residents, mainly White, moving away from the central areas towards the urban-rural fringe forming suburbs including the cities of Germantown, Bartlett, and Millington (Antipova et al. 2017).

Transportation networks increasingly influence the evolution of urban form and guide urban development. In many European and US cities, urban development follows arterial roads, giving rise to "strip" and "ribbon" patterns (Huang et al. 2007). This process changes traditional monocentric urban form.

2.3 Costs Associated with Road Vehicles

There are huge health, social, and environmental costs directly or indirectly generated by motor vehicles borne mostly by those people not using them, such as pedestrians or cyclists who are most often killed or injured, children being particularly vulnerable. People die or are seriously injured and permanently handicapped every year from traffic accidents (Newman 1996). Low-income groups are also vulnerable because they walk or bicycle most often. Other costs associated with road vehicles relate to air pollution and noise. Road and highway construction are among the most disruptive activities within cities often causing forced evictions. Climate is impacted with cars being major and ever-increasing sources of greenhouse gases. In automobile-oriented cities, the poor, the elderly, children, and youth are particularly disadvantaged as they either cannot drive or cannot afford to buy and maintain motor vehicles.

As early as the 1920s, the automobile also contributed remarkably to financial and environmental tensions choking the downtown streets with merchants and with the city administration initiating and expanding major traffic projects that needed street widening, extensions, traffic signals, and the like (Scott 1995, p. 187). Other problems that emerged in the 1920s related to the accident-caused deaths and injuries; by 1925, the annual deaths and injuries numbered about 24,000 and 600,000, respectively (Scott 1995, p. 187).

In contrast to any other prior means of transport, travel mode by car requires substantially more space to operate (Rode et al. 2014). Rode and Gipp (2001) on the example of Berlin, Germany, calculated space consumption per person in urban environment that private cars need at speeds of 30 km/h and 50 km/h, assuming average car occupancy of 1.4 person per vehicle. Cars need 75 square miles of space per person in the first case and 169 square miles per person in the second case, compared to 4 square miles for buses (assuming typical bus occupancy levels at 40% full with a speed at 30 km/h).

Further, cars are not used for most of the time; for example, a car in the United States is idle 96% of the time (Heck and Rogers 2014), thus additional parking space, very substantial in aggregate, is required. For example, about one third of the total land is off-street parking in Los Angeles County (Chester et al. 2015). Parking costs are embedded in housing costs: rent or mortgage includes parking for more than 90% of American households (Manville 2017).

2.4 European Cities

Comparative geographies need to be embedded in historical context because a city is a part of a historical process (Max Weber 1921, City). Historical perspective helps to understand the evolution of the urban system. Though each city is a unique place with every city having developed its own unique urban spatial structure (Rode et al. 2014), a number of parallels can be identified in European urban forms.

In contrast to the United States, urban development in Europe has a long history, with much of historical geography evidenced in the contemporary urban landscapes (Conzen et al. 2012). Many European cities are dated back to the Roman Empire, which by the second century AD extended over Southern Europe. Cities served several functions such as trade, religion, and defense with defensive walls surrounding many cities and towns. Many boulevards and other streets follow the lines of former walls, while many walls are still part of today's urban landscape in Europe (Nelson 1961; Conzen et al. 2012). For instance, Como, Italy, dating back to the early second century BC and strategically located between major European regions, has the twelfth-century city wall (*Porta Torre*). The city is located to the north of Milan, the largest metropolitan area in Italy. Defensive walls influenced long-term urban form and growth as well as the layout and internal arrangement of many contemporary European cities. City walls initially sharply divided built-up land from the surrounding rural area, and later gave a distinct non-residential character to the expanding urban fringe area consisting of industrial, recreational, and civic uses. Streets and blocks were closely adjusted to the wall; even where walls disappeared, some streets took on the alignment with the earlier wall marking its former location (Nelson 1961). City walls, castles, and other fortifications, which were the necessary urban form elements in medieval Europe, are contemporary tourist attractions. Traditional city forms have been defined by gridded streets: the early cities were laid out using the gridiron pattern, which impacted later developments. In European pre-industrial cities, the center played a much more important social role than the periphery. Other features figuring prominently at European city centers include the marketplace, churches, and cathedrals. In Europe, market squares often survive and are used as an important open space continuing the tradition of open-air markets and other commercial purposes, while areas surrounding them developed into modern business districts with vibrant city centers in many cities (McCarthy and Johnson 2012) Figs. 2.9, 2.10.

Fig. 2.9 Historical buildings are important urban form elements in European cities. A palace in Potsdam, Germany (Author 2016)

Fig. 2.10 Cathedral in Berlin, Germany (Author 2016)

Centuries-long urban settlement and a consequent development based on these legacies resulted in *compact urban forms* in Europe (Vance 1990). Urban growth was most importantly fueled by rural-to-urban migration during the industrial period accompanied by uncontrolled growth to which Europe responded during the industrialization by introducing modern city planning (discussed below). Post–World War II is characterized by economic and demographic growth, which resulted in urban expansion. The effects of planning and land-use regulation during the period of urban growth included denser and bigger urban cores and more massive and more regular urban periphery (De Roo and Miller 2000; Yokohari et al. 2000).

European cities reflect different social, political, and economic systems. In Europe, many more people live on far less space than in the United States. Because of the pressure on limited land, urban development that contributes quality versus quantity is encouraged (Siy 2004). To illustrate, since the mid-1990s in the United Kingdom, which is made up of Great Britain (England, Scotland, Wales) and Northern Ireland, to ensure sustainable urban development, the state and local governments have been adopting a proactive approach and encouraging retail concentration in existing town centers (Cheshire et al. 2011), referred to as the *Town Centre First* policy with the government aiming to create quality places and prospering communities. The policy's objective is to focus development towards town centers rather than encouraging the growth of out-of-town supermarkets while promoting shops and markets diversity in town centers Fig. 2.11.

Intense agricultural activities still take place on land outside urban areas (Scholz et al. 2012). If these areas are located close to existing urban areas, due to low cost and pressure to be transformed into residential or commercial areas, they are prone to construction and soil sealing (i.e., imperviousness) (European Environment Agency 2016). Thus, development supporting the *green economy* and associated technologies as well as sustainable environment is preferred; thus, *brownfield* (i.e., derelict and underused land) infill re-development is preferred over new *greenfield* development (that is, undeveloped semi-rural land, currently used for agriculture or in its natural state).

Urban growth control measures include urban renewal and the revitalization of declining urban areas that are crucial to securing quality of life. This can be achieved by converting existing dwellings and non-residential properties as well as releasing land held for other uses. For example, in

Fig. 2.11 Residential zone in Berlin (Author 2016)

Stockholm, Sweden's capital city, former industrial and harbor sites are converted into mixed-use areas that combine commercial and residential development (Siy 2004). Growth consolidation is referred to as an "urban renaissance." Other strategies include increasing development densities in existing urban areas and making cities more livable. In Stockholm, to optimize the living environment and to help determine the location for new parks and green areas, ecological, cultural, and social factors are being mapped.

2.4.1 Automobile Use Is Limited

To ensure a compact land-use pattern that is favorable to public transport, bicycling, and walking, European cities regulate development and limit automobile use. Even in high per capita income cities, it is possible to accommodate the needs of the growing population without having to resort to automobile-based decisions. For example, in the 1970s, Zürich,

Switzerland improved its old tram system by expanding and upgrading the services (e.g., via reduced waits, total right of way at traffic lights) resulting in a remarkable increase in the use of public transit (Newman 1996). Additionally, public lands were put to better use as amenities such as pedestrian malls and outdoor cafés were permitted to use road space and parking lots, thus successfully solving the problem of the automobile and improving the quality of life of urban residents. Similarly, Copenhagen, Denmark responded to the growing use of private automobiles as in 1960s urban space increasingly incorporated separate isolated homes while more and more people were driving. To solve the automobile problem, parking space was reduced annually and more streets became pedestrianized and beautified through landscaping, sculptures, and seating (e.g., at sidewalk cafés). Street life was introduced including markets, festivals, and the like so that the city experienced reductions in traffic and revitalized street life with tripling of social and recreational activity in Copenhagen's major streets (Newman 1996). Freiburg, Germany, with high car ownership in the 1990s, stopped the growth of car use by steeply limiting auto use in the city and pedestrianizing the city center, calming traffic (i.e., enforcing a speed limit of 30 km per hour in the city's residential areas), increasing parking fees, and providing good alternatives to auto use, such as bicycling and public transit consisting of an extended and upgraded light rail system and buses that are used as feeders to the rail. Land-use regulations included restriction of the overall amount of land available to development and strict zoning of land for non-developed lands (agriculture, forests, etc.) Fig. 2.12.

2.4.2 Measures to Halt Sprawl: Area Regeneration

Area regeneration is an important measure to halt sprawl. In the United Kingdom, reusing already urbanized lands is regarded as a key tool to save the countryside with 60% of new housing development taking place in already built-up urban areas rather than greenfields and rural areas. The British initiatives include Enterprise Zones (EZ), geographical areas established in 2012 in England. In contrast to the earlier generation of EZs of the 1980s, the current EZs are focused on areas of "untapped potential" where sites ready for development or vacant sites near large corporations are designated as EZs with an objective regenerating an area through greater economic output and job creation. Unique, local area-based EZs support key industries of the country, such as manufacturing, technology, and renewable energy. Barriers to private sector growth are removed and

Fig. 2.12 Speed limit is 30 km per hour in a residential area in Berlin (Author 2016)

business investment is incentivized (LCCI 2012). The EZ is made attractive to *private capital* (that is, builders and private firms) by providing various tax breaks and government support. Among other numerous benefits that EZs offer businesses is automatic planning permission for certain developments, for example, creation of new industrial buildings or change in the use of existing buildings. In addition, EZs offer benefits to local communities and the country's economy by boosting local economic performance and attracting foreign investment into the United Kingdom, as well as new jobs and businesses across England. As of March 2016, there were 732 companies with private sector investment totaling £2661 millions located in 30 designated EZs. Several examples of the EZs include Lancashire, Leeds, Liverpool, Manchester, Sheffield in the North, Birmingham and Nottingham in the Midlands, Newquay and Bristol in the Southwest, Discovery Park, Science Vale Oxford, and Royal Docks London in London and Southeast, and several EZs in largely rural areas (HM Government 2016).

For example, located in Kent, Southeast area, and 90 minutes by high-speed rail to London, *Discovery Park EZ* focuses on science, technology, and enterprise. Specifically, the sectors include energy, industrial biotechnology, pharmaceutical research and development, and healthcare. It has world-class laboratory facilities; the drug Viagra was discovered and developed there. The pharmaceutical giant Mylan, which produces drugs in about 145 countries, has laboratories and office space in *Discovery Park EZ*. Business and professional services consist of various business services support firms ranging from recruitment to accounting to legal consulting (HM Government 2016).

In Britain, EZs were first adopted by Margaret Thatcher's government in the 1980s to address high unemployment and dereliction problems in post-industrial areas impacted by economic decline after traditional industries such as steelmaking and shipping had been closed (LCCI 2012). Therefore, the earlier EZs were appointed mostly out of town (e.g., Rotherham near Sheffield and Dudley in the Black Country) or in rural areas (e.g., Allerdale in Workington), however, some EZs were in urban areas (e.g., London's Isle of Dogs). Prior to 1996, 38 EZs in total were designated across Great Britain and Northern Ireland. A package of incentives included tax breaks, investment allowances, and flexible planning restrictions. The EZs were administered by urban development corporations (UDC), public bodies responsible for revival of an appointed area through effective use of land and buildings, provision of development support to existing and new industry and commerce, creation of an attractive environment, and provision of housing and social facilities to attract people to live and work in the area (LCCI 2012).

The Isle of Dogs in the East End of London, mentioned previously, a part of the London Docklands, had been a dilapidated area with mass unemployment and social deprivation after the docks closed in 1980. It had been designated the *Isle of Dogs Enterprise Zone* in the early 1980s to stimulate area redevelopment. The outcome was the prestigious Canary Warf development, modeled on Battery Park City in Manhattan (LCCI 2012). Canary Warf became one of the largest office complexes in Europe. The developments were attracted to the area due to relaxed, simplified planning (although in the city of London at the time planning had been restrictive), and financial incentives, such as the enhanced capital allowance on industrial and commercial buildings and the like, accompanied by transport infrastructure improvement, such as the new Docklands Light Railway, London City Airport, and others, which improved access into the

EZ and integrated the area with the rest of London. A large amount of floor space was created offsetting capital investment against tax, thus incentivizing a large-scale development (LCCI 2012). After the new Canary Wharf business complex was completed, many financial and business services were attracted to the Isle of Dogs. This approach is regarded as a substantial feature of sustainable development (Siy 2004).

2.4.3 *Measures to Halt Sprawl: Greenbelts*

Yet another device used in the United Kingdom to halt sprawl, check urban growth, and preserve community character is known as the "Greenbelt," which dates back to 1955 when it was first used in the form of a government circular presented to lower-level governments. Since then, local authorities from London to Aberdeen have been designating greenbelts around their towns to curb urban sprawl and prevent cities and towns from merging into one another resulting in a creation of greenbelts throughout the country. Greenbelts, which account for about 12% of England, effectively encourage the recycling and reusing of abandoned urban land (that is, urban regeneration).

Not all infill development schemes are success stories though. For example, in Como, situated in the north of Italy, the center of the county's silk industry since the late nineteenth century, the industrial site of the former Ticosa silk factory was closed in 1982 and the buildings demolished in 2007. The ex-Ticosa site, after two decades of remaining undeveloped, received redevelopment proposals to use existing urban space as a high-density mixed-use district consisting of variable-height commercial and residential buildings (Conzen et al. 2012). According to the multifunctional development proposal, designed by a Dutch company, residential dwellings were to be integrated with the features of a modern European inner city including retail, public and private offices, public services, gardens, bars and restaurants, and the like, adding international feeling to the cityscape. However, the plans so far have not been realized; the ex-industrial Ticosa site is still the land of rubble. Francesco Angelini (2017), 35 years after the complex's closure, describes it a deserted and degraded area, reduced to being merely a parking lot. Similarly, in the United States in the 1960s, revitalization was successfully implemented in one of the Savannah, GA communities, while a similar initiative in the mid-1970s failed to restore the city's Victorian District with vacant and deteriorating structures in the area in the early 1990s due to an insufficient market demand. Other key factors affecting success of an urban planning project include location, design, financing, entrepreneurship, and time (Garvin 1996, pp. 10–11).

2.4.4 *Urban Sprawl and Contemporary Europe*

In Europe, many more people live in self-contained small- to medium-sized cities within larger city-regions. However, some cities are expanding outward forming large conurbations. Patrick Geddes, a Scottish urban planner, coined the term "conurbation" in 1915 in his book, *Cities in Evolution*. The challenge is to prevent these city-regions from merging into agglomerations. There are about 50 *conurbations* (i.e., formations of extensive urbanization areas consisting of several cities, towns, and other urban areas merged into a single developed area) in Europe. The largest ones are London-Newcastle in England, Rhine-Ruhr and Frankfurt-Stuttgart in Germany, and Randstad Holland in the Netherlands. Therefore, managing urban growth and sustainability, including environmental protection and urban transportation, are key urban strategies.

By the end of the twentieth century, regulations on zoning and property ownership rules and trends in infrastructure and middle-class consumption patterns similar to those in the United States resulted in parallel urban growth (Dargay and Gately 1997; Giuliano 1999; Giuliano and Narayan 2003). For example, contemporary large-scale, regular suburbs are similarly located in the fringe area comprising a large proportion of the urban area in most European cities (Organisation for Economic Cooperation and Development (OECD) 2000; Hoffmann-Martinot and Sellers 2005), while urban sprawl has been recognized as among the most essential land-use changes in Europe reflected in the decreased quality of life in some European parts evidenced by segregation, changes in lifestyle, and neglected urban centers.

Urban sprawl consists of three components: the number of people living in a given area, per capita land uptake (the size of built-up area per capita), and the distribution of their housing over space (European Environment Agency 2016, p. 105). Urban sprawl impacts large parts of Europe. Urban sprawl can be measured in urban permeation units per square meter of landscape: low values were found for large parts of Scandinavia, and high to very high values were identified in Western and Central Europe (European Environment Agency 2016, p. 34). The researchers found two major high-sprawl hotspots including northeastern

France, Belgium, the Netherlands, and a large area of Western Germany comprising the first cluster while the other was mostly found in the United Kingdom.

2.4.5 *Built-Up Data Website*

Built-up area data for Europe can be retrieved from the HRL IMD dataset provided by the European Copernicus Programme offering data for monitoring the terrestrial environment. The website (http://land.copernicus.eu/) offers full, open, and free access to Pan-European High Resolution Layers (HRL) produced from 20 m resolution satellite imagery that provides information on specific land cover characteristics. The European Environment Agency (EEA) is an EU agency providing land cover/land use mapping such as CORINE Land Cover (CLC) datasets (www.eea.europa.eu/data-and-maps).

2.5 North American Cities

From a global perspective (considering the period after the arrival of English colonists), most urban areas are new creations that date from the eighteenth and nineteenth centuries, the period of merchant capitalism. Most US cities were never walled, although a small number of early cities and towns did have protective walls including Boston, Charleston, Savanna, New York, Albany, Schenectady, Detroit, New Orleans, St. Louis, Salt Lake City, and Ogden; most walls were built from wooden pickets. The old walls were largely removed or destroyed (Nelson 1961). Due to a different historical development of urban areas within the United States and their orientation to commercial purposes rather than town defense, there are neither "visible" pre-urban nuclei or urban fortifications, the characteristic features of European cities (or at least, comparable to their European counterparts); monarchical (i.e., royal palaces) or religious complexes on the scale comparable to those in Europe are lacking in American cities (Nelson 1961; Conzen 2001). White settlers initiated urban expansion suppressing indigenous groups. Urban growth was a result of cheap land and building materials (Jackson 1985; Gipps et al. 1997). The industrialization experienced by the US cities since 1843 was accompanied by exurban development as a middle class invested accumulated resources in exurban property (Huang et al. 2007). Regarding the physical organization of urban areas, for example, the degree of fragmentation, the contemporary US cities manifest the irregular, dispersed patterns associated with urban sprawl with a reliance on personal automobiles.

Many contemporary US cities tend to have a polycentric structure rather than monocentric with employment scattered across an entire metro area rather than being compactly located in a single central area. To illustrate, using *OnTheMap*, a web-based mapping and reporting application located at https://onthemap.ces.census.gov/, 64 employment concentrations, ranging from the small job clusters to medium job sub-centers to large job centers have been identified in Memphis, TN, capturing more than 83% of total employment in Shelby County (Antipova et al. 2017). The Memphis Aerotropolis (defined as the biggest employment area located around the Memphis International Airport with industries depending on a close proximity to the airport and using freight facilities and warehousing services) with good access to physical infrastructure, including freeways and major roads, railroad, storage, and distribution facilities as well as proximity to the airport and the CBD, is the largest employment center in the area accounting for 38% of total jobs. The central downtown location offers the highest accessibility and many financial and banking institutions alongside professional, scientific, and technical services within the downtown area (Antipova et al. 2017).

2.5.1 Change in Urban Shape in the Nineteenth- and Twentieth-Century United States

To understand urbanization, it is instructive to compare urban population growth to that of total population. During the nineteenth century, the US population grew 15 times from 5 million in 1800 to 76 million in 1900, while urbanized population increased 100 times from about 300,000 to 30 million during the same period (Levy 2013). The forces behind urban growth of the nineteenth century include population growth, agricultural productivity, factory production, and low-cost transportation.

In the United States overall, the rate of population increase during the late nineteenth to early twentieth centuries was mostly determined by ***natural increase*** (i.e., more births than deaths) and ***population migration*** (with more immigration than emigration). Arriving immigrants mainly of European descent have been sustained by expanding urban labor markets that often required low-skilled labor. To illustrate, the population of Detroit increased 450% between 1900 and 1930 due to the rise of the automobile industry. However, the same employment-related mechanism that led to a huge population increase during the first thirty years of the twentieth century, many years later resulted in a dramatic population decline: the loss of automobile manufacturing employment in Detroit resulted in the city's losing its population between 2000 and 2010 (Levy 2013).

Figure 2.13 portrays both the absolute number of *immigrants* (also referred to as the *foreign born*, defined as people residing in the United States who were not US citizens at birth, including naturalized citizens, legal nonimmigrants on student or work visas, refugees, and illegal persons) residing in the United States and their percent of the total US population between 1850 and 2015. Figure 2.14 illustrates immigrants legally admitted for permanent residence from 1820 to 2015. During the 10-year period, 1900–1910, the country absorbed 8,202,388 immigrants or 50% the total population increase during the first decade of the twentieth century (population grew by 15,905,927 over 1900–1910), representing the third largest number of immigrants legally admitted to the United States

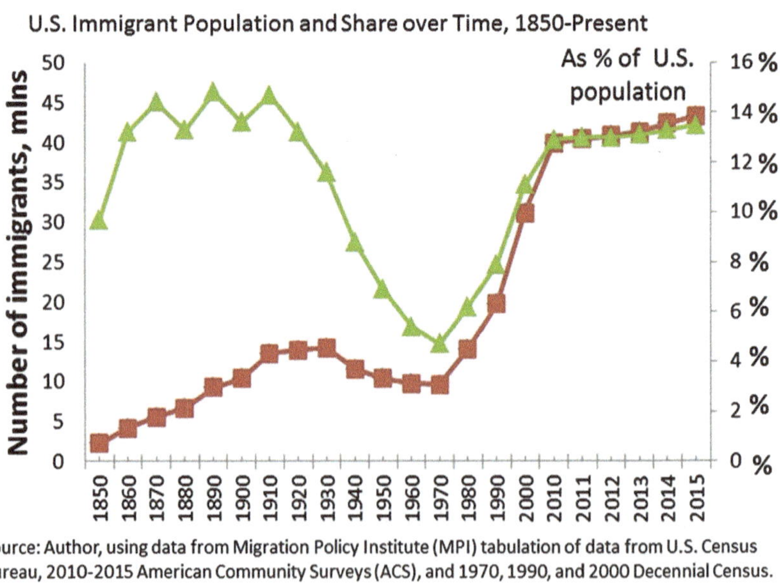

Source: Author, using data from Migration Policy Institute (MPI) tabulation of data from U.S. Census Bureau, 2010-2015 American Community Surveys (ACS), and 1970, 1990, and 2000 Decennial Census.

Fig. 2.13 US Immigrant population and share over time, 1850–present

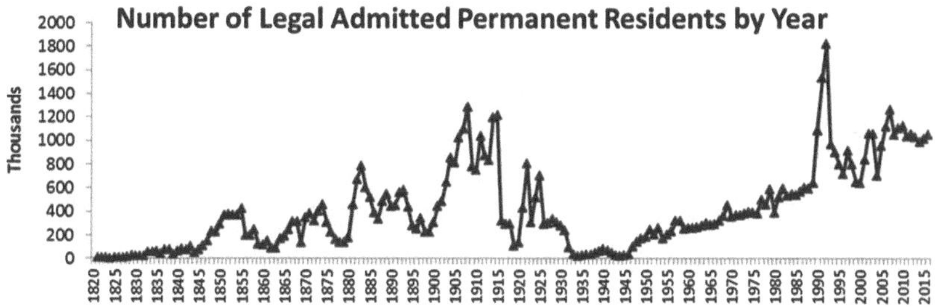

Source: Author, using Migration Policy Institute tabulations of U.S. Department of Homeland Security, Office of Immigration Statistics, Yearbook of Immigration Statistics (various years). At: http://www.dhs.gov/files/statistics/publications/yearbook.shtm.

Fig. 2.14 Number of legal admitted permanent residents by year

(the largest was 10,299,430 foreigners admitted between 2000 and 2010, followed by 9,775,398 immigrants who entered between 1990 and 2000).

Additionally, many American cities grew due to the great internal rural-to-urban migration during and after World War I and between the 1950s and early 1970s. Since 1910, the rural-to-urban movement decreased population in a third of the 3000 national counties (Scott 1995). Farm employment peaked before the United States declared war on Germany in April 1917, with wartime conditions accelerating rural migration to the cities. This mass wave of internal migration toward the cities mostly consisted of Blacks and other minorities pursuing economic opportunities. The rural-to-urban migrants were driven out of agricultural areas and small towns in the South due to the mechanization of farm labor and the tremendous increases in farm worker output together resulting in the decline in agricultural employment affecting poor Black farmers especially hard (Levy 2013). The displaced tenant farmers moved to the central cities en masse during the time when manufacturing and other low-skill and manual-labor industries were declining adding to central cities' stresses and problems of urban poverty, unemployment, and racial discrimination.

2.5.1.1 The 1800s: Dense Urban Cores

The nineteenth-century cities in the United States were remarkable in terms of concentration and density. While the colonial city was distinguished by the open pattern with buildings separated by spaces, the spaces disappeared and taller structures were built. Levy (2013) describes 1900s New York City with 2.2 million residents living on an area of just 22 square miles on Manhattan Island with an average density of 100,000 people per square mile; its densest wards on the Lower East Side had densities several times that high (Levy 2013, p. 11). The main reason behind high concentration was cost differences between land and water transportation necessitating the growth of port cities and concentration of economic activity in places with water access with housing concentrated nearby.

2.5.1.2 Era of Skyscrapers

Since the 1820s, as railroad technology began developing, this concentrating effect reflected in the concentration of industrial activities, including manufacturing and wholesale. Desire to save on transporting freight and cargo handling necessitated their location close to rail lines (Levy 2013). Similarly, land developers utilized the minimum land for a certain use, thus bringing into being compact multi-story buildings for manufacturing and commercial purposes as well as tenement dwellings ("railroad flat"); the latter commonly containing rooms without sanitary facilities, water supply, or windows for light and air. The cheap tenements were widely acknowledged to produce the "vice, crime, and disease" (Howe 1911, p. 15). Other reasons contributing to the popularity of lofty buildings and higher urban density included steam engine technology for power generation at the time, which could not send power over a long distance, and the introduction of the steam and electric elevator and steel-frame construction promoting expansion upward.

Skyscrapers sprouted up everywhere, including in New York, Chicago, Detroit, San Francisco, and Miami, contributing to many urban ills, including housing problems (noted earlier) and crime. Charles Mulford Robinson (1910), an American urban planner and writer of a series of municipal improvement articles and author of *The Improvement of Towns and Cities* published in 1901, proposed restricting the building heights in business districts because they generated unbearable traffic conditions along with a suggestion of prohibiting manufacturing in residential areas. Soaring office buildings were cited as breeders of clotted streets, congestion, and traffic problems in the report by the Commission on Congestion of Population in 1911, which made recommendations to impose immediately height restrictions in the New York metropolitan area; although, it took three-and-a-half years to introduce height limitations (Scott 1995). Even in Los Angeles, where structures were limited by a zoning ordinance to a maximum height of 150 feet due to the earthquake hazard, a 27-story city hall with its pyramid-shaped roof resembling Turkey's Mausoleum at Halicarnassus, one of the seven wonders of the ancient world, was constructed in 1928 (Scott 1995).

2.5.1.3 Population and Manufacturing Decentralization

The introduction of the electric streetcar in the 1880s began city decentralization, effectively doubling the radius of the old walking city. Other technologies introduced in the twentieth century favoring decentralization of population include mass-produced automobiles since 1915, which made suburban residence for city workers feasible, and trucks, which allowed retail, wholesale, and light manufacturing to follow their customers and the labor force. In fact, trucks were as important for wide-scale decentralization of these economic activities as the automobile was for suburbanization of population (Levy 2013). Due to the greatly increased efficiency of trucking with the provision of a road network, manufacturing shifted from being rail-borne to truck-borne. Trucking, improved communications, and new better roads provided conditions enabling reduction of expansion rates of warehousing and its substitution with direct factory shipments. In the period after World War II, labor- and capital-intensive producer and business services, including finance and insurance, advertising, telecommunications, corporate headquarters, and the like, activities serving large regions and requiring frequent face-to-face contacts with each other, rapidly filled newly built skyscrapers in the downtown area. However, other economic activities, including factories and manufacturing facilities, increasingly decentralized changing from multi-story to one-story buildings requiring vast tracts of land to reflect the shifts in their configuration and for employee parking (Adams 1970). The electric transmission lines provided factories with electrical energy by drawing power from the remote sources one hundred to two hundred miles away allowing for a new pattern of manufacturing distribution away from the few congested old industrial zones, including ports and railroad yards, to multiple new suburban locations sometimes served by superhighway transportation and both rail and highway at other times (Scott 1995).

Following industrial decentralization were shopping centers and strip commercial developments, as well as local governments, and elementary and secondary schools, which followed their customers and patrons away from the central business district (Adams 1970). The United States was officially declared an urban nation in 1920 with 51.4% of the population living in incorporated towns and cities with at least 2500 residents (Scott 1995).

Since World War II, urban areas have been growing while becoming increasingly more oriented around the mobility patterns of the car where "conventional urban development" lacking in diverse neighborhood units is the basic settlement unit. This reflects a move away from "traditional neighborhood development" compactly built around the mobility patterns of the pedestrian, containing various house types, and incorporating commercial and social uses into a residential community including schools, day care, grocery stores, post office, banks, library, parks, and so on, amenities that improve the quality of life of the residents (known as "traditional urbanism") (Duany and Talen 2002a). Unsustainable urban practices proceed for various reasons such as low-density housing preferences, racism, white flight, federal subsidies, and construction practices as well as planning implementing land use separation and spatial fragmentation of urban land uses (Duany and Talen 2002a).

The US suburban areas, especially in the urban fringe area, display the network of circuitous streets and cul-de-sacs. The curvilinear street pattern following the natural topography's contours and creating more public space goes back to the 1870 plan of Riverside, a planned commuter suburb west of Chicago, planned by Frederick Law Olmsted and Calvert Vaux. Other design techniques commonly used today include avoidance of right-angle intersections, fitting the street network to the topography, preserving green areas, and separating local traffic from through traffic (Levy 2013, p. 34).

2.5.2 Urban Growth Management

In the United States, growth management programs have been instituted in hundreds of towns, cities, and counties aimed at regulating how much, when, where, and the way development will occur. The motivations differ, with the most common being environmental considerations, preservation of existing community features, ensuring adequate facilities for future needs, and the like. To discourage urban development, land-use policies are developed at the local or regional levels of government including "urban growth boundaries" and "form-based codes" (see Sect. 2.7.6) recently administered by state and municipal governments to control growth (Stone 2008). These two planning strategies have been used outside of the United States for a long time to protect natural areas and prevent urban zones from spreading uncontrollably. Once growth boundaries have been established, urban population and employment densities are permitted only within these boundaries while beyond it urban development is discouraged through planning policies such as zoning ordinances, which may establish large minimum lot sizes (e.g., 40 acres) and use restrictions beyond the designated zone. Alternatively, public infrastructure and services including sewer lines and fire protection are permitted within the boundaries while beyond it their use is limited. Presently, cities in Oregon, Washington, and Tennessee are required to establish growth boundaries that appear effective in preventing peripheral growth and fostering density (Song and Knaap 2004). Form-based codes put an emphasis on street network connectivity associated with less vehicle travel and emissions, and therefore, have the potential of yielding benefits to regional air quality (Stone 2008). Additionally, urban growth control policies, including growth boundaries, have been found to be beneficial for racial segregation reduction, specifically, a decrease in segregation has been found between White and Black households (Nelson et al. 2004).

2.5.2.1 Government Home Programs and Suburban Development

Spatial expansion of metropolitan areas and suburban development were largely sponsored by the US government. The US housing industry and home ownership rates were greatly impacted by several government home programs. To compare the home ownership statistics, as of July 2017, the homeownership rate was 63.7% (U.S. Census Bureau 2017) while only about 46% of all housing units were owner-occupied in the United States in 1910 (Hordon 1973). The Federal Housing Authority (FHA), founded to boost the construction industry and home ownership, insured or guaranteed home mortgage loans at low interest rates to the lender by strategically lowering down payment requirements and increasing the average length of mortgages up to 25 years. As a result, lending institutions were encouraged to increase lending, and prospective homebuyers enjoyed both reduced down payments and average monthly mortgage payments. The Veteran's Administration's housing program guaranteed mortgage loans and insurance to veterans of World War II and the Korean War who could borrow up to 100% of the house price without down payment while taking up to 30 years for mortgage payment. Additionally, home construction and purchase were stimulated by the Federal National Mortgage Association (FNMA), also referred to as Fannie Mae, that could both borrow and relend funds at lower-than-private-market interest rates to individual borrowers and private home builders, thus, raising the supply of mortgage money (Hordon 1973).

So, with little or no down payment requirement, individual ownership for most citizens became economically feasible. Thus, the federal government fostered suburban development by increasing the supply of funds in the private-home mortgage market and decreasing mortgage interest rates. As a result, due to increased life of mortgages and therefore lowered average monthly mortgage payments, millions of American families could afford to buy and maintain their homes.

2.5.2.2 Technological Factors of Suburbanization

Another set of factors influencing the spatial expansion of urban areas and decentralization of population and economic activities is technological innovations. Improved telephone communication made the need for face-to-face contact less important for certain economic activities, such as manufacturing, contributing further to their decentralization. Outlying residential areas became more attractive as the central place lost its "monopoly of entertainment" with the development of motion pictures and commercial radio (Levy 2013). However, in the 1930s and 1940s, due to the effect of the Great Depression and World War II, the process of suburbanization slowed down. For example, Hordon (1973, p. 36) reports that the number of private homes built in 1934 was just one tenth the number of homes built in 1929. Besides low residential construction, civilian auto production was suspended, and gasoline was rationed during war years. A great expansion of limited-access highways, with construction starting in the 1920s and surging after the National Defense Highway Act of 1956, tremendously enhanced the suburbanizing effect of both population and economic activity. Rapid land development, sustained housing construction, and increasing ownership of personal transportation were made possible due to attractive terms of mortgage finance, high employment, and quickly increasing incomes.

Further improvements in electronic communications, including e-mail, fax, social media, long-stance direct dialing, shared carrier links, and the like, allow for a long-distance interaction and facilitate decentralization of economic activities (Levy 2013). By decreasing the need for face-to-face contact, many office activities, such as data processing, computer programming, income tax preparation, legal work, call center operations, and diagnostic medical imaging, relocated from the central city to other locations, including lower-wage nations overseas.

2.6 Comparison of Planning in the United States and European Countries

Planning is a laborious, ambitious, continuous process where goals are set and policies to implement these goals are developed. Although there are numerous definitions of the term, many agree that it is a process that seeks to respond to both existing and future problems. At the very simplest, it includes forethought, consideration of limitations and several alternatives, and proposed solutions. In the United States, the term on its own relates to land-use planning, including land use, the link between land uses, the interaction between land uses and environment, and land-use development (Cullingworth and Caves 2014). While there are diverse planning practices, planners are involved in "meso-level institutional design" in many areas, such as economic development, energy, land-use planning and control, housing, transportation, environment, and social and human services, including education and health. Planning also underscores public participation and interaction due to producing more effective and better outcomes (Alexander 2017).

2.6.1 Western European Planning

Land use is regulated differently in Western Europe and the United States. Western European cities have many desirable characteristics, including mixed land use, compact forms, more vibrant and healthier communities with lower rates of obesity, diabetes, and hypertension (Siy 2004), higher levels of walking and cycling, and smaller ecological footprint, than the United States. Noted previously, these quantities resulted from various complex factors, including markets and culture as well as the European experience in city planning.

During the late 1800s to the early 1900s, European planning focused on urban functions, but since then it started regulating land-use development by using different tools. Since the 1980s, planning has been impacted by the process of European unification. Shared by many Western European countries is the notion that to achieve sustainable spatial development today's spatial planning should be accomplished within a larger coordinating framework (both horizontal and vertical) and in greater cooperation among other countries and communities rather than in isolation. Yet another planning objective relates to the need to retain a multinational diversity to preserve the character and a cultural mosaic of communities.

2.6.1.1 Overview of Western European Planning: Centralized Planning

Western European spatial planning is similar to smart growth in that it addresses common growth and development concerns. Land-use planning and growth management are aimed at achieving "a balanced and sustainable development." This includes stricter environmental controls in Europe to secure quality of life for the long term. Most countries in Western Europe have centralized political power and a strongly centralized planning system including strict land-use control. In contrast to the United States, national governments, with few exceptions, play a stronger role relative to local governments making planning a national matter. Local plans are required to conform to the national plan. For example, the Netherlands, where the political power is centralized, is a recognized leader in controlling land use. The Dutch planning uses a top-down approach: the national government develops the laws and a national spatial plan. The Dutch government is involved in the housing market by directly regulating rental and owner-occupied parts of the housing market. Government intervention has its roots in the post–World War II reconstruction period when the huge housing shortage caused the Dutch government to set rents and construct planned, wide-level subsidized housing. Although this housing policy mitigated the housing shortage by the late 1950s, it remained in the 1960s and 1970s leading to a one-sided housing stock: a fast development of the public renting sector and near-disappearance of the private renting sector (Van der Vlist et al. 2001). During the 1980s and 1990s, when affordable subsidized housing was no longer needed, the policy was replaced by a local housing policy turning into a market-oriented system, which exists presently with cooperation among local authorities, housing corporations, and market parties. Additionally, the intervention by the central government included rent controls and rent assistance programs (Van der Schaar and Hereijgers 1991), resulting in flatter rent gradients compared to other countries (Van Ommeren et al. 1999).

In Europe, the national plans regulate development and include the methods that the public sector can use to influence the future distribution of activities in space and ensure a rational organization of land use so that the needs of economic development and protection of environment and cultural heritage could be met simultaneously. In the Netherlands's capital and the nation's largest city of Amsterdam, for example, land and property market are government-regulated allowing preservation of the historic core consisting of medium-height buildings, pre-automobile-designed street system, and a system of concentric canals used for scenic and recreational purposes.

Even in Western European countries where local governments are more powerful than the national governments, such as Sweden, planning practice differs from that in the United States. Municipal and regional Swedish governments greatly influence the pattern of development. First, these authorities exercise tight control of the timing of land development through *land-banking* (substantial public ownership of developable land within their borders). Second, municipal and regional governments regulate construction of the entire housing stock. Similar to other Western European countries, new development focuses on new towns built in Stockholm and other metropolitan areas that are surrounded by greenbelts. The towns where housing largely consists of multi-family structures are often placed near transit stops allowing easy access to downtown employment (Levy et al. 2013).

2.6.1.2 Regional Balance Is Encouraged

European planning supports reduction of the dominance of the major metropolitan areas and a more even distribution of economic activities between the regions than would otherwise be created by market forces, and a regulation of the land- and property-use conversion. In Britain, for example, since the late 1930s, the growth of the fast-developing London metropolitan area and other major cities has been discouraged using regulations and a system of greenbelts around major urban areas where all new development was prohibited, while promoting the economic and population growth in outlying areas by building "new towns" (Levy et al. 2013). Similarly, in France, to address regional imbalances, efforts are made to decrease the domination of the Paris region, which disproportionately concentrates population, manufacturing employment, and cultural resources, while developing other regional centers by diverting public funds into these "growth poles," investing in higher educational establishments, regional high-speed passenger rail system, and limited-access highway system.

2.6.1.3 Property-Led Development

Yet, in one way, Western European planning is becoming similar to that in the United States. Governments, since the 1980s, have increasingly sought private capital investment in property in a number of Western European countries. Under "property-led" development, urban development projects depend on private investment and occur within a context of some political power decentralization and fiscal pressures when revenues from the national to the local governments decrease and the latter are forced to compete for commercial investment due to the lack of the necessary funds. Such property-led development unavoidably sees some decision-making power shifting from public to private hands. Following are several examples of planning systems in Europe.

2.6.2 British Planning System

Some planning systems are unique, including the current British system. Central government develops national planning policy. To ensure accomplishment of the policy locally, local authorities are engaged through the detailed directives called Planning Policy Guidance Notes (PPG), which deal with many development priorities, including greenbelt establishment, recycling urban lands, assessing development potential with urban capacity studies, out-of-town shopping centers, renewable energy, planning and pollution control, and enforcement.

Initially, the British planning system used the zoning instruments similar to those in Germany and later in America. For example, the early suburbs and early land-use separation were created by the desire to escape industrial-related problems. In the 1700s, as upper-class people moved to the periphery of large industrial cities, single-use housing was created in England (Fishman 1987). To prevent lower-income residents from moving into the area, private deed restrictions soon enforced this land-use separation (McKenzie 1994).

English towns developed planning schemes as zoning tools that designated areas for different uses (Cherry 1996). The early twentieth century marked the start of a *centralized approach* to controlling growth and development. Later, the Town and Country Planning Act of 1947 made the planning system part of the English common law with an emphasis on land management. National legislation requires local governments develop individual local development plans that allow for flexibility. Central government is able to object to these local plans in case of non-compliance with national and regional policy. Thereby, rights of land development, including physical construction and change of use, were no longer private rights of an individual (even as long as planning schemes rules have been complied with), but instead the development decisions are subject to planning control and are made at the discretion of the planning authorities on a case-by-case basis, therefore called the *discretionary system* (Booth 2003). Permissions to develop land take into consideration the policies of generalized urban plans and Planning Policy Statements, reflecting the strong influence of the central government (Newman and Thornley 1996).

However, the United Kingdom is facing severe spatial pressures due to the anticipated increase of 4.4 million new households by 2021. The UK planning environment is being complicated by delegating more power to four member countries comprising the United Kingdom, including England, Wales, Scotland, and Northern Ireland, who develop separate planning policies. Following is the Wales Spatial Plan where sustainable development is a key feature. The definitions of sustainable development and new housing development policies are provided. Another key element of the plan is environmental protection, including measures to reduce eco-footprint.

2.6.2.1 Wales Spatial Plan: The Emphasis on Sustainability

The Wales Spatial Plan, adopted by the Cabinet of the Welsh Assembly Government in 2004 and updated in 2008, promotes the common standards of social justice, equality, and sustainable development. At the regional level, six Wales Spatial Plan Areas have been identified, including Central Wales, North East Wales, North West Wales, Pembrokeshire, South East Wales, and Swansea Bay. Each Area Group sets out their priority outcomes, actions, and the partnerships required to deliver them. Local authorities are involved with the work of the Area Groups. At the local level, planning authorities prepare a local development plan (LDP), which must have regard to the Wales Spatial Plan, however, unlike equivalent plans in England, should not conform to it.

The government of Wales' legislation, including the 1998 Act and the subsequent 2006 Act, give the Welsh Assembly Government a statutory duty to promote sustainable development ("The Sustainable Farming and Environment: Action towards 2020" report, 2007). *Sustainable development* is defined as improving well-being and quality of life. Sustainable development is delivered through integration of social, economic, and environmental objectives and more efficient use of natural resources. The growth is seen as increasing Wales' competitiveness and prioritizing assistance in less well-off areas and decreasing environmental impacts. Sustainability is integrated into the many overlapping themes of the Plan, including communities, economy, and accessibility. Sustainable communities are attractive places to live and work where deprivation including skills deficits, economic inactivity, and child poverty are eradicated, while community safety, public health, and environment are improved. Sustainable communities provide good quality housing to accommodate growing population.

New housing growth should be located in places near public transportation networks, including cycling and walking, be mindful of environmental constraints such as flood risk, and be energy efficient. Similarly, employment-related development should be linked to public transport and be near housing and infrastructure developments. Priority should be given to *brownfield* (that is, abandoned industrial or commercial or underused land that is often contaminated) rather than *greenfield* site development. Development of out-of-town retail should take into account various risks involving undermining town center *regeneration* (the upgrading of an area) and viability as well as unsustainable traffic patterns (Welsh Assembly Government 2013).

The Wales Spatial Plan describes the elements that deliver sustainable development including services, land use, and investment. For example, the *third sector* (that is, voluntary and community organizations) plays an important role in delivering the objectives needed for sustainable community building, with groups and local partnerships helping develop initiatives that tackle inequalities between communities, such as poverty concentrations, child poverty, employability, better health and well-being, and access provision to services and facilities in hard-to-reach places. The Plan calls for locations of providers of healthcare, education, social care and well-being, and other public services within or close to key settlements that are easily accessible by public transport. The promotion of good access is especially important to ensure equality of opportunity among the more vulnerable populations, including young and old people, those on low incomes, disabled people, and those living in rural areas.

Emphasis on Environmental Protection and Healthy Environment

The Wales Spatial Plan incorporates environmental valuation and calls for the protection and regeneration of natural and built environment, the key community assets. Since most people in Wales tend to live close to the sea or rivers, because of climate changes such as rainfall patterns and sea levels, there are flood-risk issues and consequences for land use, water resources, biodiversity, and wildlife. The energy and other resources that are required by humans to live, work, and move around need to be reduced.

A healthy and enjoyable environment can be delivered by various means including:

1) provision of safe and clean open spaces and nature sites where people can enjoy wildlife and biodiversity, particularly in urban areas;
2) improvement of air quality by employing traffic management approach or other measures;
3) sustainable management of waste, water and soils;
4) waste re-use and recycling;
5) encouragement of more walking and cycling.

As part of the environment protection and enhancement, the Plan requires each of the six Area Groups in Wales to develop practical plans going beyond the reduction of "eco-footprint" and carbon emissions across the range of its activities, and include plans to achieve a low-carbon region status. Given that the predicted changes in annual average daily temperatures may exceed 1 °C from as early as the 2020s in some parts of Wales, these areas are subject to higher emissions. The Wales Spatial Plan supports co-location of jobs, housing, and services to reduce the need to travel, as well as gives preference to green travel modes such as car sharing, public transport, walking, and cycling.

Sensitive land management is needed to encourage conditions for biodiversity, and ensure air and water quality as well as soil structure and carbon storage. For the agriculture to become a carbon neutral sector, the ways to achieve it include self-sufficiency in energy supplies and reductions in transport costs. For example, organic milk when processed locally using environment-friendly farming and food production methods and with higher regard for animal welfare, due to local production, requires shorter food miles between producer and consumer. Using locally produced feeds versus bought-in ones results in less transportation and reduced costs. Also, producing hardy nursery stock locally in Wales rather than importing plants from places such as Holland to meet the demand for the pot grown stock within both Wales and in the United Kingdom may also decrease the carbon footprint by reducing the road miles to transport plants while sustaining local farmers and allowing diversification ("The Sustainable Farming and Environment: Action towards 2020" report 2007). Biodigestion of animal wastes can help reduce emissions and farm costs. Local farms and community may utilize local biomass from existing woodland such as wood chips for heat and energy production and for animal bedding.

To conserve water, winter rainfall can be stored. Ways to offset the likely frequent and extensive flooding from increased precipitation due to changing climate may include creation of coastal and riverine wetlands, small-scale new tree and shrub planting, and natural landscapes where vegetation can slow down the rainwater run-off.

The public must be prepared to pay for these environmental goods. New ways are identified in which farmers are rewarded with "environmental land management payments" for the provision of the considerable economic and community benefits valued by the public, such as high-quality water, public access to countryside and recreation, wildlife management, and carbon storage from farming and forestry.

Emphasis on Sustainability of Rural Communities

There is recognition that long-term planning is needed even if short-term decisions may provide a better return through increasingly intensive commercial activity but little environmental benefits. In Wales, which is mostly rural, planning policies must be sensitive to the needs of the local rural communities. Similar to many rural areas across the developed world, there is an ongoing trend of "ageism," where the average age of people who make important farm-related decisions is 58 years, accompanied by in-migration of elderly population in the search of quality of life, while sons and daughters of local farmers are leaving the countryside and families tend to become smaller. In this light, opportunities need to be created to retain and attract young people by improving local education and developing key skills needed in the off-farm knowledge-based economy. In order for rural communities to be sustainable, rural planning policies related to the supply of affordable housing, diversification of rural economy such as organic production, horticultural cropping, and forestry development, and jobs in the public sector outside mainstream agriculture that support rural services such as tourism and leisure business, food processing, energy production, and waste management, need to be reviewed (Welsh Assembly Government 2013).

Provision of Wider Society Benefits

Traditionally, farmers and foresters have been delivering a range of public goods and services as a by-product of primary food and timber production including clean water, air, carbon capture, the management of biodiversity, and landscape and wildlife conservation, among others, collectively referred to as "environmental goods and services." Past human actions as well as an application of intensive farming practices (land drainage, use of pesticides and fertilizers, etc.) contributed to environmental degradation and greenhouse gas emissions, leading to climate change and the resulting impacts on livestock, plants, landscape, and ecology. Therefore, another central concern addressed in Wales' planning system is environment protection because changes in land use may impact the provision of these wider society benefits. Innovative measures may facilitate the creation of environmentally friendly and sustainable conditions and may improve farming's impact on the environment. Coordinated research carried out by the appropriate research institutions, specialist consultancy, and sustained collaboration and genuine partnership at every level from farmers to the government are vital if agricultural industry is to remain competitive and deliver environmental goods and services ("The Sustainable Farming and Environment: Action towards 2020" report 2007).

2.6.2.2 Danish National Planning System

Similar to the United Kingdom., development policy directives are periodically issued by Danish national government to local governments. The Danish national planning system employs spatial planning practices to promote sustainable development by integrating national, regional, and local plans that focus on sprawl containment and land-use regulation. The country is divided into three zones, including urban, recreational, and rural. The Planning Act regulates the development in the urban and recreational zones. Development in the rural zone, which accounts for about 90% of total Denmark's land area, is highly restricted to forestry- and agriculture-related development. About 85% of the people are accommodated within urban areas, which represents about 5% of the country (Siy 2004).

2.6.2.3 German Planning System

Planning is decentralized. Although federal (national) laws provide legal framework for and consistency of spatial planning and building code for the states and municipalities, municipalities have their regional planning systems. Urban and regional planning systems are autonomous and use different planning instruments dependent on economic development of a region. For example, planning for growing cities focuses on urban growth management while protecting natural resources. Above all, German planning practice is characterized by the objective of sustainable development favoring protection of the natural environment and brownfield development. The HafenCity waterfront project converting a former harbor area in Hamburg into a lively urban area with new housing and jobs is an example.

In a post-industrialized Germany, big cities have a specific urban climate: their extensive developed area impacts local climate reflected by cities more vulnerable to heat waves and flooding due to substantial impervious surface. However, already built-up areas such as brownfields can be recycled as open space. Green infrastructure, including urban gardens and green roofs, can mitigate the negative effects of severe precipitation since permeable surfaces absorb excess water, while green corridors provide better ventilation. The redevelopment of abandoned industrial sites into parks and urban greening will prevent inner-city floods and mitigate urban heat waves and thus increase resilience of a city (ICLEI- Local Governments for Sustainability 2017). Different planning strategies are employed for lagging eastern cities where abandoned housing and public infrastructure in the inner cities are demolished and adapted to changing demands. This is accompanied by green space creation on dilapidated land, public space renovation incorporating high-quality design. Yet another feature characteristic of German planning is a vast amount of former military land within many German states and cities, which allows their conversion to urban development Fig. 2.15.

Fig. 2.15 Climate-proof urban development of an inner-city quarter (University Duisburg-Essen) in Essen, Germany. This urban redevelopment aims at 50% reduction in CO2 by 2020 (Author 2010)

2.7 The US National Planning

While the political power is centralized in many European countries accompanied by a top-down planning approach, in the United States, many planning decisions and initiatives are taken at the state or local level. The federal government sets federal guidelines and incentives including federal money and federal land (Levy 2013). According to the Statistical Abstracts of the United States (2012), as of 2003, out of 1937.7 million acres, the federal land represents 20.7%, or 401.9 million acres.

2.7.1 Federal Legislation and US Territorial Expansion

The US federal government was involved in a series of actions that impacted the development of the settlement and urbanization of the country. The acts included acquisition of additional land in the post-revolutionary period west of the Mississippi River through the Louisiana Purchase in 1803, cession from Spain in 1819, the Oregon Compromise with England in 1846, cession of lands by treaty after the Mexican War in 1848, purchase from Texas in 1850, and Alaska purchase from Russia in 1867, among others, which greatly contributed to national expansion. The Continental Congress adopted the Ordinance of 1785, which set up a survey system for the public domain lands in the form of townships six mile-by-six mile large and sections one mile-by-one mile large.

Another federal action was later disposing of land and selling large pieces of land at a cheap price to investors and speculators who after new settlements were created would in turn resell farm-sized land plots to individual landholders. These and other federal actions contributed to fast settlement and development of the western states and the establishment of the scattered grid farm pattern typical of the rural American land.

2.7.1.1 1862 Legislation

Several pieces of legislation were passed by Congress in 1862 resulting in vast tracts of public land sold or granted by the federal government to settlers, veterans, towns, colleges, homesteaders, and railroad companies. For example, the Homestead Act of 1862 allowed settlers to claim 160-acre plots of land (and sometimes larger in places with drier climate) at no cost if they lived on it for five continuous years, effectively homesteading 125,000 square miles (about 80,000 acres) of land west of the Mississippi (Hofsatdter et al. 1957, p. 683).

Under another federal act of the same year, the Morrill Land Grant Act of 1862, each state was granted 30,000 acres (120 km^2) of federal land for each member of the state's congress as of the Census of 1860 with the purpose of establishment of agricultural and mechanical colleges and universities. This resulted in 76 land grant-based major state universities scattered across many US towns including Kansas State University, Cornell University, Louisiana State University and Agricultural and Mechanical College, Massachusetts Institute of Technology, Rutgers University, North Dakota State University of Agriculture and Applied Sciences, University of Tennessee, and Texas A&M University.

2.7.1.2 Western States Development

Between 1862 and 1873, about 160 million acres in total (250,000 square miles) of federally owned public land was granted to the railroads both directly via land grants and indirectly via states (Hofsatdter, Miler, and Aaron, p. 53). Federal lands granted to railroad companies resulted in tripling of the national rail mileage from less than 10,000 miles in 1850 to 30,000 within a single decade roughly completing the rail system east of the Mississippi and linking all major cities of the east by rail. The result of the First Pacific Railway Act of 1862, which authorized and provided financial incentives (including the right-of-way and public land for each mile of rail route constructed) granted to the railroads, as well as the issue of bonds, was the transcontinental railway spanning the entire continent and enabling coast-to-coast rail travel by 1869 (Levy 2013). However, the idea of the rail-served compact cities never had the mass support in the country whose citizens' have a love affair with the automobile.

The federal government helped further development of the Western states by helping to reclaim the desert lands for agriculture after the Reclamation Service established by the Reclamation Act of 1902 (later renamed the Bureau of Reclamation in 1923), which effectively subsidized water development including irrigation projects and dam construction during the era of reclamation until the 1980s when contributions from local state and municipalities to federal water projects have greatly increased.

2.7.2 Highway Legislation and Suburbanization

Since early twentieth century, the physical infrastructure including roads and transit supported development beyond the urban periphery. The interstate highway system represents a tremendous amount of planning that highly influenced the development of the entire nation and had important urban design implications reflected in the *beltway pattern* around many major cities. The Federal Aid Road Act of 1916 began the period of federal highway construction by providing federal funds for *intercity* highways construction. Federal legislation in 1934 authorized funds for state highway planning, which resulted in multiple traffic planning studies and led to a recommendation of a national single high-speed limited-access roadway system of about 40,000 miles planned by the federal government and the states who determined the exact highway routing within the states. The legislation needed for this vision, the Federal Aid Highway Act, was passed in 1956, which, among others, established the *Highway Trust Fund*, a mechanism of massive funding used only for highway construction from new vehicle excise taxes and motor fuel sale taxes. The entire system covering more than 42,000 miles was built to the same uniform design standards at the cost of about $129 billion connecting all cities and their hinterlands (Levy 2013).

2.7.3 Federal Legislation and Metropolitan Expansion

Legislation on financing and tax policy on housing passed by Congress after 1935 to increase homeownership and reduce Great Depression-era unemployment by constructing residential housing impacted the spatial form of metropolitan areas and contributed to suburbanization because of the location of land available for single-family house construction. The changes in mortgage lending included the introduction of federal mortgage insurance, which eliminated the risk of default offered by the FHA created in 1934. To buy housing, borrowers could borrow from the banks for a much longer time and with much less of a down payment. Other acts of the federal government included the creation of the FNMA and the VA, which encouraged home buying by creating a secondary mortgage market and insuring veteran's mortgages, respectively (Levy 2013) (refer to the Sect. 2.5.2.1 earlier in the chapter).

Changes in taxation consisted of favorable tax treatment of homeowners compared to renters. The homeowner tax deductions, including the deductibility of mortgage interest and property taxes on owner-occupied housing from taxable income, increase the demand for owner-occupied housing and influence people's decisions regarding housing: two thirds of all housing units in the United States are occupied by owners (Levy 2013).

The homeownership rate in the United States fluctuated between 64.2% in 1995 to 63.5% in October 2016, peaking at 69.1% in 2005. The 2016 homeownership rates were highest in the Midwest (68.6%) and lowest in the West (58.2%), while it is highest for those householders ages 65 years and older (79.0%) and lowest for those younger than 35 years of age (35.2%). Non-Hispanic White householders had the highest homeownership rate at 71.9%, followed by Asian or Native Hawaiian and Pacific Islander Alone householders at 55.6%. The lowest rate was for Black Alone householders at 41% (U.S. Census Bureau 2017).

2.7.4 US Regional Planning: Metropolitan and Regional Planning Entities

The forerunner of the planning for a whole city region is the 1909 plan for Chicago, drawn up by prominent designer Daniel Burnham and published on the Fourth of July. It is both citywide and regional in scope. Its transportation elements consist of radial and circumferential highways at a distance of 60 miles from the city center as far north as the Wisconsin shore and reaching Michigan City, IN, to the south, the distance covered by then-suburban electric lines and a two-hour drive by car (Burnham and Bennett 1909, pp. 36–37). The plan was commissioned and funded by business organizations (Levy 2013). The plan focused on six main points: improvement reclamation of the lakefront from industrial and commercial uses, including a 25-mile lakeshore development of parks, lagoons, and islands along the waterfront; creation of a regional highway system; relocation of railway terminals and development of a comprehensive freight handling system for land and water transport and passenger traction system (including elevated and subway loops surrounding a business core, and other transportation features); development of an outer chain of forest preserves forming the proposed open spaces and park system and parkway network; city-wide arrangement of internal streets to ease movement to and from the CBD; and establishment of civic centers and public structures, in accordance with the City Beautiful movement, to encourage social coherence and unity for the metropolitan area (Miller 1973, pp. 142–143). Although the plan had some shortcomings (e.g., it lacked model worker tenements or model neighborhoods and it did not mention the housing problem), no other city can compare with the amount of public improvements that would follow a specific plan (Scott 1995).

2.7.4.1 *Need for Comprehensive Regional Planning*

In 1920, for the first time in the nation's history, more people lived in urban areas of 2500 or more inhabitants than in rural areas (51.4% versus 49.6%). The increasing agricultural mechanization decreased the share of workers needed on the farm and technological advances created new employment opportunities in manufacturing and services in the cities luring more people from the rural areas and small towns. Alternatively, the larger urban centers had been producing a centrifugal movement of population fleeing congestion of the central cities and seeking more living space in the outlying territories. Suburbs and satellite cities of Atlanta, Buffalo, Chicago, Cleveland, Louisville, Pittsburg, Philadelphia, Seattle, and St. Louis enjoyed a greater population gain and accelerated development than the central cities between 1910 and 1920. Similarly, in southern California, the population grew 108% in outlying towns around Los Angeles attracting families of middle- and upper-income class in the thousands as well as blue-collar families; the population of Los Angeles itself increased 80%. However, in the latter case, population settled in the peripheral areas not due to congestion or bad housing of the central core but due to promotion of new tracts (Scott 1995).

2.7.4.2 *The 1920s: First Wave of Metropolitan Regional Planning in Los Angeles*

As built environment changed quickly from the compact city to vast urban regions sprawled across several administrative jurisdictions in the early twentieth century, there emerged the need for comprehensive regional planning for which the municipal boundary was no longer the proper boundary. The *regional planning agencies* were established in the 1920s to respond to the rapid widespread suburbanization greatly accelerated by increased automobile ownership.

The Los Angeles County Regional Planning Commission was the first regional planning organization in the nation established in 1922 to address the problems of 38 municipalities and 50 unincorporated towns within the mushrooming Los Angeles metropolitan area (Scott 1995). It was created two years after the Los Angeles City Planning Commission had been created and the members soon realized that to further the development of the city would be impossible unless an overall approach of regional planning for the municipalities and towns composing the urban complex of the Los Angeles Basin was applied. Neither Los Angeles nor a municipality in the surrounding area alone could solve effectively the area-wide problems of flood control, the provision of public beaches, the development of adequate water supplies, zoning, the development of a highway network and transit to serve all cities in the basin, the disposal of garbage and sewage, wild fire control, and other common problems that demanded coordinated effort and concerted action. During the 1920s, the entire area experienced an economic boom with manufacturing and other industries, including tire factories, oil refineries, automobile assembly plants, motion picture studios, offices, and retail, arrived in great numbers to both downtown Los Angeles and where open fields had recently been. Yet, it was felt that the region did not realize its chance of "superior planning and development" (Scott 1995, p. 210). The reasons included unprecedented immigration due to benign climate and job openings as well as the promises of free bus rides and free lunches by promoters, disproportionate real estate speculation with land in some areas grabbed by developers and industrialists and put under inappropriate use before a county zoning ordinance was adopted in 1927, and huge sums spent to provide essentials including highways, storm drains, schools, and so on, resulting in a low-density leapfrog development with more than 25% of a million vacant lots at the end of the decade.

2.7.4.3 Chicago Metropolitan Regional Planning

Another regional planning organization, founded in 1923, was the Chicago Regional Planning Association, which was established to implement various elements of the Burnham Plan in the three-state region including Illinois and parts of Wisconsin and Indiana. The outcomes included formation of an integrated circulation network with connected highways, the construction of the double-decked Wacker Drive, several bridges along the Chicago River, a system of lakefront parks and beaches, and cultural institutions and museums.

2.7.4.4 Other Examples of Metropolitan Regional Planning

Another example of a metropolitan-area planning effort is the Regional Plan Association in New York, the largest metropolitan region characterized by the big disparity between the extent of the administrative boundary lines of a city and the actual size of the city reflected by the economic and social interactions between the outlying areas and the core. The Plan Association was charged with the responsibility to define the region's boundaries, which spanned across numerous municipalities in a three-state region of New York, New Jersey, and Connecticut, and develop plans for transportation (including both highways and transit, with much of what was planned having been built), recreational areas, and other physical features.

Following these planning efforts for the Los Angeles, Chicago, and New York regions, the interest for metropolitan regional planning spread to other large urban clusters. Altogether, by the end of the 1920s, about 15 regional plans had been developed in other parts of the country. However, regional planning commissions mostly had no political status and therefore were limited in the ability to impact the physical development of urban regions. Other regional planning organizations created also in the 1920s were *public authorities*, quasi-governmental public organizations whose mission was to make region-wide planning decisions especially in public infrastructure, such as transportation, bridges, tunnels, utilities, and others. The first authority was the Port Authority of New York and New Jersey created in 1921 requiring regional cooperative actions by a large number of municipalities to implement a series of the river crossing projects and mass transit connecting both states into a unified region (Levy 2013).

The following decade witnessed an example of comprehensive regional planning in integrated development of the Tennessee River Valley made possible by the congressional passing of legislation in 1933 that created the Tennessee Valley Authority (TVA) to alleviate persistent poverty of the region, create jobs, and build on the already existing dam built in the mid-1920s for flood control, navigation improvement, and low-cost power generation (Levy 2013). Yet another example of regional development, however, involving less comprehensive efforts, implemented at a multi-state level, is the Appalachian Region Commission (ARC) established in the mid-1960s to improve economic underdevelopment by channeling federal funds on transportation construction, workforce training, and social services provision, including medical care.

2.7.4.5 Second Wave of Metropolitan Regional Planning: SANDAG

The second wave of regional planning occurred in the post–World War II period. To recognize continuing metropolitan growth and development of closely spaced urban areas in certain regions since the late 1950s, new regional governmental planning entities were established throughout the United States. Included are *councils of governments* (COGs), voluntary regional planning organizations of local governments responsible for developing plans covering a group of cities and counties. In the

1960s–1970s, 26 COGs were created (Cullingworth and Caves 2014) totaling about 450 today. An example of a COG is SANDAG (San Diego Association of Governments), which was established in the early 1970s and is responsible for regional comprehensive and transportation planning for 18 cities within San Diego County. It is also responsible for preparing the Regional Housing Needs Assessment (RHNA) and developing a long-term wide-ranging Regional Comprehensive Plan (RCP) for the San Diego region that integrates local land-use and transportation plans. Though the RCP may encourage implementation of certain objectives such as transportation demand reduction, habitat protection, or housing affordability, it has no regulatory power to implement these goals. Without such land-use authority, SANDAG faces problems with the development of efficient regional transportation because it depends on land use, density, and travel patterns.

Regarding regional growth, SANDAG applies a smart growth planning approach by promoting compact multi-use development with future growth occurring close to existing job centers and public facilities rather than in rural or other open areas. SANDAG is also responsible for resource allocation, and project development. Funding recently allocated among local agencies includes federal transportation funds for transportation improvement projects that use smart growth principles such as providing safety to pedestrians, transit users, and bike riders. Projects funded improved amenities including streetlights, paving, and landscaping. However, regarding other region-wide matters such as water quality across the metropolitan area, SANDAG has no regulatory authority, although recommendations for improving water quality have been developed in its RCP. Special districts composed of consortia of groups from municipalities and counties carry out control over air quality or water resources (Hill 2006).

2.7.4.6 Second Wave of Regional Planning: DRCOG

Yet another example is the Denver Regional Council of Governments (DRCOG), one of the nation's three oldest councils of governments (the Atlanta Regional Metropolitan Planning Commission and the Northern Virginia Regional Planning Commission were founded in 1947). It is a nonprofit membership organization of local governments in the Denver, CO. DRCOG initially consisted of four counties including Adams, Arapahoe, Boulder, and Jefferson, and currently serves an area comprising eight counties and a variety of cities and towns within its area (47 municipalities in total where each member has an equal voice). DRCOG was founded in 1955 as the region was undergoing a post–World War II growth to foster collaboration and cooperation and work together on common pressing issues and concerns to make life better as "working together to enhance our region's quality of life was a much better approach than going it alone" (https://drcog.org/about-drcog/about-drcog). DRCOG serves as a planning organization; it also provides technical assistance (e.g., an economic analysis of businesses and housing along rail lines) and a large-scale interactive public forum providing both education and engagement opportunities. DRCOG brings a regional perspective to discussions about the eight-county region's most important problems in three fields including transportation and personal mobility, growth and development, and aging and disability resources.

As a Regional Planning Commission, the DRCOG prepares the long-range plan for the physical development of the region (titled "Metro Vision"), and assists with local and regional planning efforts by producing travel models and a wide range of data and maps on various topics, including transportation, socioeconomics, land use, demographics, and employment. The first Metro Vision plan (Metro Vision 2020) was adopted by the DRCOG board of directors in 1997, and a revised Metro Vision plan was adopted in January 2017 (DRCOG 2017). As a comprehensive policy document, Metro Vision expresses a vision for growth, development, environmental quality, and transportation for the Denver region. It identifies the outcomes, objectives, and initiatives in individual communities within the eight-county region. Local government representatives of the region's counties, cities, and towns are guided by the Metro Vision regional growth and development plan to establish guidelines, set policy, and allocate funding (DRCOG 2014). As the federally designated Area Agency on Aging (AAA), it provides comprehensive services including transportation, home-delivered meals, senior centers, legal services, and other services to older adults and people with disabilities and their family members and caregivers to help them remain independent in the community. Required by the federal transportation law since 1973, a metropolitan planning organization (MPO) conducts regional transportation planning in urbanized areas (DRCOG 2017).

Functioning as the MPO for the Denver region, DRCOG is involved in *a continuing, cooperative, and comprehensive* transportation planning process for all modes of transportation in the region carried out by the three primary partner agencies including the DRCOG, the Regional

Transportation District (RTD), and the Colorado Department of Transportation (CDOT). DRCOG addresses both short-term needs presented in the Transportation Improvement Program (TIP) and the twenty-year transportation plan for the Denver region through the Regional Transportation Plan (RTP) and Metro Vision. The Metro Vision RTP is part of Metro Vision and includes two key components such as the Metro Vision transportation system (referred to as the "vision plan") and the air quality conforming fiscally constrained RTP. The other elements of Metro Vision are the comprehensive long-range land-use plan and development objectives for the Denver region. The TIP identifies the transportation projects using federal surface transportation funds to be implemented during a six-year period and is updated at least every four years (DRCOG 2017). The TIP implements the air quality conforming fiscally constrained RTP. The Denver region currently meets national air quality standards for CO and PM-10 and the Environmental Protection Agency considers the region to be attainment-maintenance for those pollutants. However, much of the Denver region was designated as a moderate nonattainment for ozone in 2016 under the 2008s 75 ppb eight-hour standard for this pollutant. Since then, a new eight-hour ozone standard of 70 ppb was set by the EPA in 2015, for which the region is now planning (DRCOG 2017).

DRCOG implements another important role: Denver aggressively pursues TOD. The DRCOG has been facilitating discussion about TODs and what TODs in different communities should include in the Denver region after FasTracks (the long-range, regional land-use and rail and bus rapid transit development program) was passed by voters in 2004. DRCOG compiles easily available resources for its members, holds meetings and workshops for developers and planners, and develops online forums.

2.7.5 US City Planning: Tools of Municipal Development

Local governments can influence their land-use patterns by using two groups of actions including *public capital investments* and *land-use controls* (Levy 2013). The former is the more powerful tool for the pattern of development of the municipality in the long term. Public capital can be invested into structures such as municipal buildings, roads, transit stops, bridges, water and sewer lines and treatment plants, schools, and the like that will stay in the municipality for decades. Public capital investments are financed by issuing general obligation and revenue bonds.

2.7.5.1 *Municipal Land-Use Controls*

Land-use control is one of the major powers of US local governments; in some cases, large metro areas are fragmented into hundreds of separate jurisdictions, such as incorporated cities or townships, and each can regulate land use within their borders, including rules on building safety, urban growth boundaries, and actual control over use (Lens and Monkkonen 2016). To control land-use development, municipalities have the power to regulate subdivisions and to zone. Most planning agencies develop and implement subdivision regulations and zoning ordinances (Levy 2013). The former represent an old form of land-use control enforced by the municipality as an exercise of its police power. *Subdivision regulations* are usually applied to residential development (Cullingworth and Caves 2014). Essentially, a community may regulate the process of dividing land into building lots for sale or property improvements by requiring the provision of roads, sidewalks, sewers, water mains, and other local infrastructure facilities as well as dedicate land for civic facilities, such as schools, fire stations, recreation facilities (or provide cash payments in lieu) and ensuring that these elements meet community standards (e.g., internal streets' width, quality, and safety). Although subdivision regulations have been in use since the early nineteenth century, they are less well known compared with zoning (described in the following section).

2.7.5.2 *Municipal Zoning*

Since the early twentieth century, land-use zoning has been applied to divide cities into areas by purpose: residential, industrial, and commercial. American cities increasingly took to zoning land use with the first zoning implemented in New York City in 1916 (Whittemore 2017). Conventional US planning uses zoning as a main tool of land-use control due to its effective capability to affect community land use while costing municipality very little, including only administrative and legal expenses (Hirt 2012; Levy 2013).

Modern zoning, as it is practiced in the United States presently, did not exist until the late nineteenth century, although there were housing, building, and fire codes in individual cities. Fire codes banned wooden structures in certain urban areas, and building codes controlled heights of buildings (e.g., height regulations in Washington, DC in 1899) (Silver 1997). Zoning differs from these urban codes in two important ways. First, it controls *urban functions* including residential, commercial, and industrial (that is, how the land and buildings are to be used). Therefore, zoning is often synonymous with land-use zoning. The focus on urban function goes back to the Industrial Revolution when industry development created pollution, congestion, and public health risks, which necessitated segregation of residential zones from incompatible uses. It was accompanied as much by a desire to segregate along racial and class lines (Silver 1997).

Other elements of zoning implemented within the US cities include *bulk and density*. Examples of the former include structure characteristics requirements including height, number of stores, and maximum floor area. Examples of the latter include requirements for site layout such as minimum lot area, minimum setbacks, site structure coverage (allowable percentage of site that a structure may cover), parking and signs requirements, and so on. Many U.S. cities implement their zoning based on these elements, with land-use zoning being the key focus area of comprehensive zoning.

The second distinction of zoning from urban codes relates to the fact that zoning divides a city into districts where different rules are applied. Private regulation through the use of deed restrictions existed before laws related to land use at the national scale were adopted. Neighborhood associations in upper-class areas in Boston, Chicago, and St. Louis have developed rules on the allowed building height and setback (Garvin 1996, 356–357). New York was the first in the country to enact comprehensive zoning law covering all five boroughs in 1916. Three types of districts that the city was divided into were based on function, bulk, and area (Bressi 1993). These *use districts* included residential, commercial, and mixed land-use types. Five height districts were overlaid on these use districts, where height was proportional to the street width. Additionally, the zoning law specified requirements for ground coverage including minimum lot sizes, and stepped-back structure design with mandatory setbacks from the street at higher stories rather than rising up from the lot lines (Fig. 2.16).

Edward M. Bassett designed this zoning law based on the privilege of a municipality to exercise the police power to regulate the use of land and the area and height of buildings to protect the general public welfare (Levy 2013). He is referred to as the father of zoning in the United States. Later, Bassett served as chairman to the special advisory committee appointed by Herbert Hoover in 1921 charged with drafting a standard state zoning-enabling act to allow municipalities of the states with enabling

Fig. 2.16 Stepped-back structure design in New York (Author 2013)

legislation to adopt zoning ordinances. Frederick Law Olmstead and J. Horace McFarland, among other members of the National City Planning Institute, served with him.

The standard zoning act encouraged the most appropriate use of land as well as overall city-wide zoning and in accordance with a comprehensive plan rather than haphazard zoning. A flood of zoning legislation followed: if in 1921 zoning was in effect in only 48 municipalities with less than 11 million residents, by the end of 1923, 218 cities and towns with 22 million inhabitants adopted zoning ordinances (Scott 1995, p. 194).

2.7.5.3 Eminent Domain Versus Zoning

Local governments can acquire and thus become the owner of private property by exercising the power of ***eminent domain*** for public purposes such as road building. If a "taking" has occurred, the initial property owner will be compensated for the imposed loss of value as condemnation award paid by the government. Just compensation for the property taken for public use is guaranteed by the taking clause of the Fifth Amendment of the US Constitution. When there is disagreement over government's regulatory and other actions towards the privately owned property, the judicial system of the courts will make a decision whether a taking has occurred and whether compensation stated by the Fifth Amendment to the property owner is due or need not be paid for any imposed loss of property value.

However, local governments can exercise their power of public regulation over the use of private property to protect public interests, including public health, community safety, and its welfare through *land-use zoning* with the loss imposed on the property owner not compensated. In this case, private activities are controlled under the concept of the *police power* of the local community with the considerations of air and sunlight reaching the sidewalk, and traffic congestion in the community. An example of zoning is the passage of legislation imposing a certain building height limit to prevent permanent shadows darkening the streets below, a prevention of some industrial or commercial activities or intense land development in a residential section to protect a residential environment from undue noise, dust, congestion, and so on.

2.7.5.4 Zoning and Land-Use Separation

The Federal Standard State Zoning Enabling Act was passed in 1924 and enabled states to adopt zoning. In the 1920s, the planners looked for the means of controlling density with debates on the ideal and maximum allowed density as well as density regulation (Scott 1995, p. 198). One consequence of early zoning was the increasing separation of uses resulting in the segregation of the different economic classes. However, city engineers and utility and transit companies, including public works, telephone, electric light, gas, and so on, found zoning helpful since it eliminated their guesswork regarding the type of future service needs and allowed for accurate calculations of future growth. For governments, zoning became a legitimate function after the 1926 Supreme Court case of *Euclid v. Ambler*, which claimed that separation of land uses prevents disorder. The case also compared an apartment building in a single-use residential area with a parasite. The effect of the *Euclid v. Ambler*, which gave name to conventional US practice—*Euclidean zoning* (Haar and Kayden 1989; Wolf 2008)—was tremendous: several hundred zoning ordinances were passed within a few years (Hirt 2012). During the following decades, each zone was divided into subcategories, increasing the number of zones. Multi-family homes were separated from single-family homes, and single-family homes on lots of different sizes were separated from each other (Abeles 1989).

The costs of land-use separation have long been recognized. Jane Jacobs, a prominent anti-zoning activist as early as the 1960s, criticized planners' excessive concern with land-use separation that results in "sterile" environments with a poor mix of uses and building types. Jacobs' ideal urban area was a desirable place to live with largely pedestrian access that contained a mix of structure types and both residential and non-residential uses in the same block with commercial uses being on the ground level while apartments are located above on the upper stories (Jacobs 1961).

Conventional (Euclidean) zoning has been criticized since the 1960s for favoring low-density patterns, which cause urban sprawl and infrastructure waste (Fischel 1999; Levine 2006). It supports segregation by class and race with the effect of exclusion of many from desirable areas (Haar and Kayden 1989; Frug 1996). Although there was a decline in racial segregation in the United States, segregation by income in cities increased in every decade from 1970 to 2010. In a recent examination of the link between land-use regulation and segregation by income based on the 95 biggest US cities in 2010, more restrictive density restrictions have been found, first, related to urban fragmentation; specifically, the segregation of the wealthy and middle-income classes, but not the poor. Inclusionary housing (that is, forcing wealthier parts of the city to build low-income housing) can reduce segregation. Second, local land-use regulation may result in higher segregation rates while more state control is linked to lower segregation at the metropolitan level (Lens and Monkkonen 2016). Zoning puts vulnerable individuals, including women, minorities, low-income citizens, the young, the elderly, and the disabled, at a disadvantage because of a need to use a car when homes and workplaces are separated by a long distance, and makes local communities less vibrant (Calthorpe 1993; Duany et al. 2000). Finally, it contributes to natural habitat degradation, waste of land, and pollution because of car driving and urban sprawl (Kleppel 2002).

2.7.5.5 *Environmental and Social Justice and Zoning*

Designed to protect public health, safety, and welfare, the most powerful tool of planning, land-use zoning, may not protect all residential neighborhoods equally, as noted earlier. When industrial land is zoned too close to residential areas, the outcomes may include degraded residential property values and increased health risks from household exposure to hazards. Low-income communities and communities of color bear most burdens by often locating proximate to polluting industries and activities (Whittemore 2017). In his quantitative analysis, Whittemore (2017) demonstrates that the city's neighborhoods of color were disproportionately *upzoned* to heavy commercial and industrial use, and White communities, regardless of income, were relatively protected from these upzonings. *Upzoning* is understood as a change from a less intense zone (regardless if it is a residential zone or a less intense commercial or industrial zone) to a heavy commercial or industrial zone. *Downzonings* represents the opposite, when land parcels are redesignated from heavy commercial and industrial to light commercial or residential categories. Downzonings may occur not due to health risk concerns because of proximate hazardous uses, but rather from the city's desire to redevelop disused industrial land (such as abandoned factories and warehouses) for residential and commercial uses, with gentrification occurring in tracts with significantly lower than average income within both African American and White areas (Whittemore 2017).

2.7.5.6 *Exclusionary Land-Use Regulation*

Some types of land use regulation appear to be related to exclusion. Local policies such as large lot zoning, restrictive single family definitions, minimum single-family home size (in square footage), housing location policies, and so on, may create barriers to housing opportunity and exclude certain groups from desirable neighborhoods; in fact, racial segregation is well documented, however, another outcome of such exclusionary policies is economic segregation (Lens and Monkkonen 2016). *Low-density only* zoning is a powerful exclusionary land-use control linked to both economic and racial exclusion through a reduction in multi-family housing, rental units, and affordability, leading, in turn, to limited growth in the minority population (Pendall 2000, 138). Further, density restrictions appear to be related to the segregation of the high- and middle-income households, that is, the concentration of affluence; however, the segregation of the lowest-income households was not affected (Lens and Monkkonen 2016).

Regulation that is most strongly related to housing market outcomes, that is, the number of approvals (or the number of independent reviews needed for project and zoning change approval) local governments require for new housing projects, is strongly associated with higher levels of economic segregation, more specifically with the low-income segregation. When local decision-makers (such as the incorporated city or township) have more power regarding land-use control, segregation is higher at the metropolitan level, and the effect is opposite when state government has stronger regulatory power; segregation levels are lower (Lens and Monkkonen 2016). Higher housing prices, demographic growth, and physical constraints (e.g., steep terrain or internal rivers, coastal location, etc.) result in more restrictive man-made land constraints or land-use regulations, with Miami, Los Angeles, San Francisco, Oakland, New York, San Diego, Boston, Chicago, and Seattle being the most inelastic cities in terms of housing supply. That is, population growth leads to housing demand and results in greater residential regulatory constraints in places with initially low land availability (physical constraints). The housing market then responds with increased land values and higher incentives for antigrowth regulations. Little change in future prices is expected in the large metro areas with high land availability, including Houston, Austin, Charlotte, Kansas City, and Indianapolis, which have a highly elastic housing supply (Saiz 2010).

2.7.6 Alternatives to Zoning

Several alternatives to Euclidean zoning have emerged in the United States to avoid rigidity of strict land-use separation. Newer techniques are more flexible, more mixed-use friendly, and more open to land-use negotiations between the two parties. For example, *bonus or incentive zoning* is a less frigid land-use control when in order to meet desirable public goals (e.g., within a residential area, a municipality may desire to reserve some affordable housing), the municipality provides a wide range of development incentives or bonuses to developers by including the option of re-zoning in exchange for the contribution of certain urban site features or amenities not legally required but deemed beneficial to the public. For example, if a developer provides something "extra" for public benefit, such as multi-family housing with some units earmarked for low-income tenants, builds a day-care center, provides landscaping, parks, or public plazas, or makes improvements to the nearby mass transit stations, and so on (Levy 2013). In return, the developer is granted an increase in density, building height, and the like.

Transfer of development rights (TDR) channels development into some areas (that is, the receiving area) from where it is not wanted (the sending area) by allowing property owners to sell their development rights if they had not fully developed their holdings as permitted by the law. By using this technique, a historic district may preserve its old buildings and not be re-developed at higher density, or an ecologically vulnerable area may limit development.

Other alternatives include *Planned Unit Developments* (PUDs), which might be entirely residential, commercial, mixed use, and industrial, with all types incorporating greater design flexibility than stipulated by the traditional zoning ordinance (Levy 2013; Cullingworth and Caves 2014). *Flexibility in design* means that items such as lot size, setbacks, building heights, parking requirements, and so on, are negotiable between the developer, the local officials, and the property owners. Overall, within the community (subject to the inflexible zoning ordinance requiring uniform conditions for the entire community), the property developed as a PUD is subject to the PUD ordinance with controls that may be the same or different; however, allowing more room for creativity and innovation, which results in less sterile areas. The PUD projects are discussed by a group who review the entire site plan at one time.

Performance zoning provides some flexibility into the conventional zoning by identifying the allowable intensity of land uses rather than the use of the land. It applies a rating/point system in the form of performance zoning codes to development proposals where the projects are evaluated for the created impacts on the environment or public services, including noise, traffic, light, odor, smoke, open space, and so on. (Kendig et al. 1980; Elliott 2008). The developer has to decide how to meet these performance standards and ensure the positive impacts for the development proposal to be approved.

Form-based zoning is an alternative to use-based Euclidian zoning and it is consistent with neotraditional design in that it promotes walkable and mixed-use communities. It offers more flexibility about permitted land uses, but like a traditional zoning, it imposes requirements on some items including building shape and style and their relationship to the street by specifying setbacks. In contrast to the conventional zoning approach, which does not specify what the development will look like, the form-based approach is concerned with the actual physical form and design of the area in question as well as scale of development such as community- or neighborhood-wide. It uses broader land categories such as "residential," "commercial," and so on, and fewer zones compared with the large number of zones in the zoning ordinance. The multiple zones found in the zoning ordinance are a result of subdividing each zone into distinct classes based on structure type and density. For example, the zoning ordinance may specify various classes of housing within residential districts including several standards of single-family detached, several standards of single-family semi-detached, apartments with different height requirements, and row houses. In contrast, form-based zoning identifies a few basic zones including "urban core," "urban center," "general urban," "suburban edge," "rural reserve," and "rural preserve" (Duany and Talen 2002a; Parolek et al. 2008). Form-based zoning focuses on physical form by regulating streets and public spaces, rather than just separate incompatible uses; that is, the emphasis is placed on the physical dimension such as the strict control of the size, placement, and street and building configuration versus densities or permitted land uses regulated in traditional zoning (Stone 2008). The *transect* concept substitutes form-based codes where various environments with various degrees of urban levels and intensity are arranged progressively from rural to densely urban (that is, natural zone, rural, sub-urban, general urban, urban center, and urban core) with different structures, streets, and civic space types situated within these six transect zones (Duany and Talen 2002b). Form-based zoning is relatively new (Hall 2007, p. 917). It was recently implemented in a very few unique places including Denver, Miami, and Flagstaff. One example is Seaside, FL, a resort town designed by Duany and Plater-Zyberk. The websites of the Congress of New Urbanism, the Form-Based Institute, and Smart Code Complete cite around 300 form-based codes making their impact rather modest in a country of more than 40,000 US locales (Hirt 2012) Fig. 2.17.

Fig. 2.17 Form-based zoning in Seaside, FL (Author 2012)

Bibliography

Abeles, P. 1989. Planning and Zoning. In *Zoning and the American Dream: Promises Still to Keep*, ed. C. Haar and J. Kayden, 122–153. Chicago: Planners Press.

Adams, J.S. 1970. Residential Structure of Midwestern Cities. *Annals of the Association of American Geographers* 60 (1): 37–62.

Alexander, E.R. 2017. Chance and Design: From Architecture to Institutional Design. *Journal of the American Planning Association* 83 (1): 93–102.

Anderson, W.P., P.S. Kanaroglou, and E.J. Miller. 1996. Urban Form, Energy and the Environment: A Review of Issues, Evidence and Policy. *Urban Studies* 33 (1): 7–35.

Angelini, F. 2017. *La Ticosa non sia più un simbolo di Como* (Ticosa is no longer a symbol of Como), January 28. At: http://www.laprovinciadicomo.it/stories/Editoriali/la-ticosa-non-sia-piu-un-simbolo-di-como_1221753_11/.

Antipova, A., L. Skryzhevska, and H. Kung. 2017. The Impact of Geography and Labor Upon the Size of Job Agglomerations in Memphis, Tennessee. *Data-Enabled Discovery and Applications* 1: 7–17.

Banai, R., and A. Antipova. 2016. Retail-Center Viability and Urban Form: A Micro Analysis. *The International Review of Retail, Distribution and Consumer Research* 26 (5): 521–540.

Bechle, M.J., D.B. Millet, and J.D. Marshall. 2011. Effects of Income and Urban Form on Urban NO2: Global Evidence from Satellites. *Environmental Science and Technology* 45 (11): 4914–4919.

Bereitschaft, B., and K. Debbage. 2013. Urban Form, Air Pollution, and CO2 Emissions in Large U.S. Metropolitan Areas. *The Professional Geographer* 65 (4): 612–635.

Berry, B. 1973. *The Human Consequences of Urbanization*. London: Macmillan.

Bhatta, B., S. Saraswati, and D. Bandopadhyay. 2010. Urban Sprawl Measurement from Remote Sensing Data. *Applied Geography* 30 (4): 731–740. https://doi.org/10.1016/j.apgeog.2010.02.002.

Booth, P. 2003. *Planning by Consent: The Origins and Nature of British Development Control*. London: Routledge. Crossref.

Bressi, T.W., ed. 1993. *Planning and Zoning New York City: Yesterday, Today, and Tomorrow*. New Brunswick: Rutgers University Press.

Burnham, D.H., and E.H. Bennett. 1909. Plan of Chicago (Chicago, 1909), pp. 36–37.

Calthorpe, P. 1993. *The Next American Metropolis: Ecology, Community, and the American Dream*. Princeton: Princeton Architectural.

Cervero, R. 1998. *The Transit Metropolis: A Global Inquiry*. Washington, DC: Island Press.

Cherry, G. 1996. *Town Planning in Britain Since 1900: The Rise and Fall of the Planning Ideal*. Oxford: Blackwell.

Cheshire, P., C.A.L. Hilber, and I. Kaplanis. 2011. *Evaluating the Effects of Planning Policies on the Retail Sector: Or Do Town Centre First Policies Deliver the Goods?* SERC Discussion Paper 66. http://www.spatialeconomics.ac.uk/textonly/SERC/publications/download/sercdp0066.pdf.

Chester, M., A. Fraser, J. Matute, C. Flower, and R. Pendyala. 2015. Parking Infrastructure: A Constraint on or Opportunity for Urban Redevelopment? A Study of Los Angeles County Parking Supply and Growth. *Journal of the American Planning Association* 81 (4): 268–286.

Conzen, M.P. 2001. The Study of Urban Form in the United States. *Urban Morphology* 5 (1): 3–14.

Conzen, M.P., K. Gu, and J.W.R. Whitehand. 2012. Comparing Traditional Urban Form in China and Europe: A Fringe-Belt Approach. *Urban Geography* 33: 22–45.

Cullingworth, J.B., and R. Caves. 2014. *Planning in the USA: Policies, Issues, and Processes*. 4th ed. New York: Routledge.

Dargay, J., and D. Gately. 1997. Income's Effect on Car and Vehicle Ownership, Worldwide: 1960–2015. *Transport Research* 33A (7–8): 101–138.

de Roo, G., and D. Miller. 2000. Introduction—Compact Cities and Sustainable Development. In *Compact Cities and Sustainable Urban Development; a Critical Assessment of Policies and Plans from an International Perspective*, ed. G. de Roo and D. Miller, 1–13. Aldershot: Ashgate Publishing.

Dear, M. 2005. Comparative Urbanism. *Urban Geography* 26 (3): 247–251. https://doi.org/10.2747/0272-3638.26.3.247.

Denver Regional Council of Governments (DRCOG). 2014. At: https://drcog.org/about-drcog/about-drcog. Accessed 22 Aug 2017.

———. 2017. *Transportation Planning in the Denver Region*. February. At: https://drcog.org/sites/drcog/files/resources/Transportation%20Planning%20in%20the%20Denver%20Region-v03-abl.pdf. Accessed 22 Aug 2017.

Duany, A., and E. Talen. 2002a. Making the Good Easy: The Smart Code Alternative. *Fordham Law Journal* 29: 1445–1468.

———. 2002b. Transect Planning. *Journal of the American Planning Association* 68: 245–267. Crossref.

Duany, A., E. Platter-Zyberk, and J. Speck. 2000. *Suburban Nation: The Rise of Sprawl and the Decline of the American Dream*. New York: North Point.

Elliott, D. 2008. *A Better Way to Zone: Ten Principles to Create More Livable Cities*. Washington, DC: Island.

European Environment Agency. 2016. *Urban Sprawl in Europe*, EEA Report No 11/2016. Joint EEA-FOEN report. Luxembourg: Publications Office of the European Union. https://doi.org/10.2800/143470.

Ewing, R. 1997. Is Los Angeles-Style Sprawl Desirable? American Planning Association. *Journal of the American Planning Association* 63 (1): 107–126.

Ewing, R., and R. Cervero. 2010. Travel and the Built Environment: A Meta-analysis. *Journal of the American Planning Association* 76 (3): 265–294.

———. 2017. "Does Compact Development Make People Drive Less?" The Answer Is Yes. *Journal of the American Planning Association* 83 (1): 19–25.

Ewing, R., R. Pendall, and D. Chen. 2002. Measuring Sprawl and Its Impacts. Smart Growth America Report, Smart Growth America, Washington, DC. At: https://www.smartgrowthamerica.org/app/legacy/documents/MeasuringSprawl.PDF.

———. 2003. Measuring Sprawl and Its Transportation Impacts. *Transportation Research Record* 1831: 175–183.

Fischel, W. 1999. Does the American Way of Zoning Cause the Suburbs of Metropolitan Areas to Be Too Spread Out? In *Governance and Opportunity in Metropolitan America*, ed. A. Altshuler, M. Morill, H. Wolman, and F. Mitchell, 151–191. Washington, DC: National Academy.

Fishman, R. 1987. *Bourgeois Utopias: The Rise and Fall of Suburbia*. New York: Basic Books.

Frug, J. 1996. The Geography of Community. *Stanford Law Review* 48: 1047–1108. Crossref.

Frumkin, H., L. Frank, and R. Jackson. 2004. *Urban Sprawl and Public Health*. Washington, DC: Island Press.

Garvin, A. 1996. *The American City: What Works, What Doesn't*. New York: McGraw Hill Professional.

Gipps, P., J. Brotchie, D. Hensher, P. Newton, and K. O'Connor. 1997. Journey to Work: Employment and the Changing Structure of Australian Cities, Australian Housing and Urban Research Institute (AHURI). Research Monograph 3, Melbourne.

Giuliano, G. 1999. Land Use and Transportation: Why We Won't Get There from Here. *Transportation Research Circular* 492: 179–198.

Giuliano, G., and D. Narayan. 2003. Another Look at Travel Patterns and Urban Form: The US and Great Britain. *Urban Studies* 40 (11): 2295–2312.

Grunfelder, J., and T.S. Nielsen. 2012. Commuting Behaviour and Urban Form: A Longitudinal Study of a Polycentric Urban Region in Denmark. *Geografisk Tidsskrift-Danish Journal of Geography* 112 (1): 2–14.

Haar, C., and J. Kayden. 1989. *Zoning and the American Dream: Promises Still to Keep*. Chicago: Planners Press.

Hall, P. 1997. Modeling the Post-industrial City. *Futures* 29 (4–5): 311–322.

Hall, E. 2007. Divide and Sprawl, Decline and Fall: A Comparative Critique of Euclidean Zoning. *University of Pittsburgh Law Review* 68: 915–952. Crossref.

Handy, S. 2017. Thoughts on the Meaning of Mark Stevens's Meta-Analysis. *Journal of the American Planning Association* 83 (1): 26–28.

Hawley, A.H. 1986. *Human Ecology: A Theoretical Essay*. Chicago: University of Chicago Press.

Heck, S., and M. Rogers. 2014. *Resource Revolution: How to Capture the Biggest Business Opportunity in a Century*. Boston: Houghton Mifflin Harcourt.

Herold, M., J. Hemphill, C. Dietzel, and K.C. Clarke. 2005. Remote Sensing Derived Mapping to Support Urban Growth Theory. *Proceedings of ISPRS vol. XXXVI-8/W27*. At: http://www.isprs.org/proceedings/XXXVI/8-W27/herold_hemphill_etal.pdf.

Hill, E.G. 2006. SANDAG: An Assessment of Its Role in the San Diego Region. http://www.lao.ca.gov/2006/sandag/sandag_033006.pdf. Accessed 23 Aug 2017.

Hirt, S. 2012. Mixed Use by Default. How the Europeans (Don't) Zone. *Journal of Planning Literature* 27 (4): 375–393.

HM Government. 2016. Enterprise Zones. Looking for a Place to Grow Your Business? At: http://enterprisezones.communities.gov.uk/ez-data-release-09-11-2016/. Accessed 13 Sept 2017.

Hoffmann-Martinot, V., and J. Sellers. 2005. *Metropolitanization and Political Change*. Wiesbaden: Verlag fuer Sozialwissenschaften.

Hofstadter, R., W. Miller, and A. Daniel. 1957. *The American Republic*. Vol. 1. Englewood Cliffs: Prentice-Hall, Inc.

Hordon, H.E. 1973. *Introduction to Urban Economics. Analysis and Policy.* Englewood Cliffs: Prentice Hall, Inc.

Howe, F.C. 1911. *The Municipal Real Estate Policies of German Cities.* Proceedings of the Fourth National Conference on City Planning.

Hoyt, H. 1939. *The Structure and Growth of Residential Neighborhoods in American Cities.* Washington, DC: Federal Housing Administration.

Huang, J., X.X. Lu, and J.M. Sellers. 2007. A Global Comparative Analysis of Urban Form: Applying Spatial Metrics and Remote Sensing. *Landscape and Urban Planning* 82: 184–197.

ICLEI- Local Governments for Sustainability. 2017. *Nature-Based Solutions for Sustainable Urban Development.* March, ICLEI Briefing Sheet—Nature-Based Solution. At: https://unfccc.int/files/parties_observers/submissions_from_observers/application/pdf/777.pdf. Accessed 25 Sept 2017.

Jackson, K. 1985. *Crabgrass Frontier: The Suburbanisation of the United States.* New York: Oxford University Press.

Jacobs, J. 1961. *The Death and Life of Great American Cities.* New York: Random House.

Kendig, L., S. Connor, C. Byrd, and J. Heyman. 1980. *Performance Zoning.* Washington, DC: Planners Press.

Kleppel, G. 2002. Urbanization and Environmental Quality: Implications for Alternative Development Scenarios. *Albany Law Environmental Outlook* 37: 37–64.

Knaap, G.-J., U. Avin, and L. Fang. 2017. Driving and Compact Growth: A Careful Look in the Rearview Mirror. *Journal of the American Planning Association* 83 (1): 32–35.

Lens, M.C., and P. Monkkonen. 2016. Do Strict Land Use Regulations Make Metropolitan Areas More Segregated by Income? *Journal of the American Planning Association* 82 (1): 6–21.

Levine, J. 2006. *Zoned Out: Regulation, Markets and Choices in Transportation and Metropolitan Land Use.* Washington, DC: Resources for the Future.

Levy, J.M. 2013. *Contemporary Urban Planning.* 10th ed. Upper Saddle River: Pearson Education.

Levy, J.M., S. Hirt, and J. Jessen. 2013. Planning in Other Nations. In *Contemporary Urban Planning*, ed. J.M. Levy, 10th ed. Upper Saddle River: Pearson Education.

London Chamber of Commerce and Industry (LCCI). 2012. Enterprise Zones: Only One Piece of the Economic Regeneration Puzzle. At: http://www.londonchamber.co.uk/docimages/10504.pdf. Accessed 13 Sept 2017.

Lowry, J.H., and M.B. Lowry. 2014. Comparing Spatial Metrics that Quantify Urban Form. *Computers, Environment and Urban Systems* 44: 59–67.

Luck, M., and J. Wu. 2002. A Gradient Analysis of Urban Landscape Pattern: A Case Study from the Phoenix Metropolitan Region, Arizona, USA. *Landscape Ecology* 17: 327–339.

Manville, M. 2017. Travel and the Built Environment: Time for Change. *Journal of the American Planning Association* 83 (1): 29–32.

Marshall, J.D. 2008. Energy-Efficient Urban Form. *Environmental Science & Technology* 42 (9): 3133–3137.

McCarthy, L.M., and C. Johnson. 2012. Cities of Europe. In *Cities of the World: World Regional Urban Development*, ed. S. Brunn, M. Hays-Mitchell, and D. Ziegler, 5th ed. Lanham: Rowman & Littlefield.

McKenzie, E. 1994. *Privatopia: Homeowners Association and the Rise of Residential Private Government*. Binghamton: Yale University.

Measuring Sprawl. 2014. Smart Growth America. April. At: https://smartgrowthamerica.org/app/uploads/2016/08/measuring-sprawl-2014.pdf. Accessed 7 Sept 2017.

Miller, Z.L. 1973. *The Urbanization of Modern America. A Brief History*. New York: Harcourt Brace Jovanovich, Inc.

Nelson, H.J. 1961. Walled Cities of the United States. *Annals of the Association of American Geographers* 51: 1–22.

Nelson, A.C., T.W. Sanchez, and C.J. Dawkins. 2004. The Effect of Urban Containment and Mandatory Housing Elements on Racial Segregation in US Metropolitan Areas, 1990–2000. *Journal of Urban Affairs* 26 (3): 339–350.

Nemes, V. 2011. The Favorability of Relief for Territorial Expansion of Marghita City. *Analele Universitatii din Oradea—Seria Geografie* (211114–526), 127–133.

Newman, P. 1996. Reducing Automobile Dependence. *Environment and Urbanization* 8 (1): 67–92.

Newman, P.W.G., and J.R. Kenworthy. 1989. *Cities and Automobile Dependence: An International Sourcebook*. Aldershot: Gower Technical. 388 pages.

Newman, P., and A. Thornley. 1996. *Urban Planning in Europe: International Competition, National Systems and Planning Projects*. London: Routledge. Crossref.

Nivola, P.S. 1999. *Laws of the Landscape: How Policies Shape Cities in Europe and America*. Washington, DC: Brookings Institution Press.

Organisation for Economic Cooperation and Development (OECD). 2000. *Managing Urban Growth*, 15. Paris: DT/TDPC.

Ou, J., X. Liu, X. Li, and Y. Chen. 2013. Quantifying the Relationship Between Urban Forms and Carbon Emissions Using Panel Data Analysis. *Landscape Ecology* 28: 1889–1907.

Parolek, D., K. Parolek, and P. Crawford. 2008. *Form-Based Codes*. Hoboken: Wiley.

Pendall, R. 2000. Local Land Use Regulation and the Chain of Exclusion. *Journal of the American Planning Association* 66 (2): 125–142.

Rickert, J.E. 1967. House Facades of the Northeastern United States; A Tool of Geographic Analysis. *Annals of the Association of American Geographers* 57 (2): 211–238.

Robinson, C.M. 1910. The Improvement of Fort Wayne, Indiana.
Rode, P., and C. Gipp. 2001. Dynamische Raeume: Die Nutzungsflexibilisierung urbaner Mobilitaetsraeume am Beispiel der Berliner Innenstadt, Technical University Berlin. At: http://www.dynamicspaces.net/Rode%20and%20Gipp%202001%20-%20Dynamic%20Spaces.pdf. Accessed 10 Sept 2017.
Rode, P., G. Floater, N. Thomopoulos, J. Docherty, P. Schwinger, A. Mahendra, and W. Fang 2014. Accessibility in Cities: Transport and Urban Form. NCE Cities Paper 03. LSE Cities. London School of Economics and Political Science.
Saiz, A. 2010. The Geographic Determinants of Housing Supply. *The Quarterly Journal of Economics* 125 (3): 1253–1296.
Sallis, J. F., C. Spoon, N. Cavill, J. K. Engelberg, K. Gebel, M. Parker, … D. Ding. 2015. Co-benefits of Designing Communities for Active Living: An Exploration of Literature. *International Journal of Behavioral Nutrition and Physical Activity* 12: 30.
Salomon, I., P. Bovy, and J.P. Orfeuil. 1993. *A Billion Trips a Day: Tradition and Transition in European Travel Patterns.* Dordrecht: Kluwer Academic Publishers.
Scholz, M., Å. Hedmark, and W. Hartley. 2012. Recent Advances in Sustainable Multifunctional Land and Urban Management in Europe: A Review. *Journal of Environmental Planning and Management* 55 (7): 833–854.
Scott, M. 1995. *American City Planning Since 1890.* Chicago: American Planning Association.
She, Q., X. Peng, Q. Xu, L. Long, N. Wei, M. Liu, W. Jia, T. Zhou, J. Han, and W. Xiang. 2017. Air Quality and Its Response to Satellite-Derived Urban Form in the Yangtze River Delta, China. *Ecological Indicators* 75: 297–306.
Silver, C. 1997. The Racial Origins of Zoning in American Cities. In *Urban Planning and the African American Community: In the Shadows*, ed. J.M. Thomas and M. Ritzdorf. Thousand Oaks: Sage Publications.
Siy, E.J. 2004. *Learning from Abroad: The European Approach to Smarter Growth and Sustainable Development.* Global Innovations Group. At: http://www.fundersnetwork.org/files/learn/Learning_from_Abroad-FINAL.pdf.
Song, Y., and G. Knaap. 2004. Measuring Urban Form: Is Portland Winning the War on Sprawl? *Journal of the American Planning Association* 70 (2): 210–225.
Statistical Abstracts of the United States. 2012. 131st Edition. At: http://www2.census.gov/library/publications/2011/compendia/statab/131ed/tables/geo.pdf. Accessed 10 Mar 2017.
Stevens, M.R. 2017. Does Compact Development Make People Drive Less? *Journal of the American Planning Association* 83 (1): 7–18.
Stone, B. 2008. Urban Sprawl and Air Quality in Large US Cities. *Journal of Environmental Management* 86: 688–698.
"The Sustainable Farming and Environment: Action towards 2020" report. 2007.

U.S. Census Bureau. 2011. Housing and Household Economic Statistics Division. Historical Census of Housing Tables. At: https://www.census.gov/hhes/www/housing/census/historic/units.html. Accessed 20 Mar 2017.

———. 2017. Quarterly Residential Vacancies and Homeownership, Second Quarter. Release Number: CB17-110. At: https://www.census.gov/housing/hvs/files/currenthvspress.pdf. Accessed 14 Sept 2017.

Van Ommeren, J., P. Rietveld, and P. Nijkamp. 1999. Job Moving, Residential Moving, and Commuting: A Search Perspective. *Journal of Urban Economics* 46: 230–253.

van de Coevering, P., and T. Schwanen. 2006. Re-evaluating the Impact of Urban Form on Travel Patterns in Europe and North-America. *Transport Policy* 13: 229–239.

Van der Schaar, J., and A. Hereijgers. 1991. *Volkshuisvesting: een zaak van beleid.* Utrecht: Spectrum.

Van der Vlist, A. J., C. Gorter, P. Nijkamp, and P. Rietveld. 2001. Residential Mobility and Local Housing Market Differences. TI 2002-003/3. Tinbergen Institute Discussion Paper.

Vance, J.E. 1990. *The Continuing City: Urban Morphology in Western Civilization.* Baltimore: Johns Hopkins University Press.

Warner, S.B. 1968. *Streetcar Suburbs: A Process of Growth in Boston.* New York: Antheneum.

Weber, M. 1921/1958. *The City.* New York: Free Press.

Wei, J., M. Jia, R.W. Twibell, and K. Underhill. 2006. Characterizing Urban Sprawl Using Multistage Remote Sensing Images and Landscape Metrics. *Computers Environment and Urban Systems* 30 (6): 861–879.

Welsh Assembly Government. 2013. *People, Places, Futures. The Wales Spatial Plan Update 2008.* At: http://gov.wales/docs/desh/publications/130701wales-spatial-plan-2008-update-en.pdf. Accessed 20 June 2017.

Whittemore, A.H. 2017. Racial and Class Bias in Zoning: Rezonings Involving Heavy Commercial and Industrial Land Use in Durham (NC), 1945–2014. *Journal of the American Planning Association* 83 (3): 235–248.

Wolf, M. 2008. *The Zoning of America: Euclid vs. Ambler.* Lawrence: University of Kansas.

Yen, H. 2012. *US Growth of Distant Suburbs Falls to Historic Low.* Accessed 5 April. http://news.yahoo.com/us-growth-distant-suburbs-falls-historic-low-0403466997html.

Yokohari, M., K. Takeuchi, T. Watanabe, and S. Yokota. 2000. Beyond Greenbelts and Zoning: A New Planning Concept for the Environment of Asian Mega-cities. *Landscape and Urban Planning* 47: 159–171.

CHAPTER 3

Local Amenities and Neighborhood Perception

3.1 Introduction

According to the definition provided by the Oxford dictionary, the term 'perception' is "a way of regarding, understanding, or interpreting something." People may form a subjective opinion or an attitude about different things, for example, importance of residential places. It is necessary to study people's perceptions and evaluations to avoid the mismatch between objective (i.e., spatial characteristics including urban form, transport infrastructure, socioeconomics) and subjective (non-urban form characteristics expressed by residents as attitudes and perceptions towards urban aspects) indicators of the aspects of urban environment (Klinger et al. 2013). Recent research on public transit quality and green space availability in Vienna, Austria found an overall high degree of agreement between the subjective responses and objective measurements (Haslauer et al. 2015).

Since residents of neighborhoods benefit when different local urban amenities are conveniently located nearby (Tiebout 1956; Glaesenera and Carusoa 2015), residents' perceptions of their neighborhoods, measured here by the degree of perceived importance, may be influenced by the availability of various urban features, including ample green areas, nearby shopping and services, access to work, recreation, education, and sports facilities, that enhance the quality of life. Our understanding is limited regarding which urban elements influence these perceptions. The aim of this study was to quantitatively examine the relationship between the perceived

A. Antipova, *Urban Environment, Travel Behavior, Health, and Resident Satisfaction*, https://doi.org/10.1007/978-3-319-74198-7_3

importance of and satisfaction with residential neighborhoods and different neighborhood amenities. In other words, which urban amenities enhanced residents' perceptions of neighborhood importance? Availability of a feature perceived important or desirable might not be realized, however. For example, a resident may deem living close to good public schools important, but might reside far away from this amenity, and therefore express dissatisfaction with their actual proximity to good schools. The second objective was to explore whether the residents are actually satisfied with these urban amenities and with their neighborhood overall. Therefore, the terms *resident perception*, *perceived importance*, *residential preferences*, *neighborhood attractiveness*, and *attitudes* are used interchangeably in the project and express the subjective degree of importance of various urban amenities, while *satisfaction* has been used to denote the degree of the stated satisfaction with the actual urban amenities (Fig. 3.1).

Ahlbrandt (1984) noted that different urban groups attach different degrees of importance for the neighborhood: the poor, the elderly, and people with children depend upon their neighborhoods more than other

Fig. 3.1 Nearby parks are an important amenity in Berlin (Author 2016)

residents do and therefore, for them, their neighborhood is very important. In developing a model of neighborhood perception, he related feelings that people have about their place of residence, including attachment, loyalty, and satisfaction, to the neighborhood's social fabric and its quality of life as well as characteristics of the residents.

Neighborhood features perceived by residents as desirable, and therefore important, include public services, such as schools, neighborhood life cycle, and physical maintenance; commercial facilities, including restaurants, grocery stores, and nearby parks; and culture, socioeconomic characteristics of its residents, ethnic make-up, and crime levels. Accessibility is yet another important aspect of a neighborhood's appeal, including physical access to work, schools, friends and relatives, shopping and leisure activities (Greene and Pick 2012, p. 179). In recent decades, leisure facilities, including recreation centers and golf centers, became the key local amenities attracting residents (Clark et al. 2002) (Fig. 3.2).

Fig. 3.2 Nearby education facilities enhance the quality of life: The music school in Berlin (Author 2016)

Approaches to study urban features can be either objective when secondary data such as census demographic or economic data are analyzed according to certain criteria or subjective when residents' perceptions of various aspects of living environment are used for modeling relationships between these subjective assessments of specific phenomena and characteristics of the urban environment (Marans 2012). The latter approach involves a data collection element with survey data shedding light on levels of satisfaction with different aspects of urban life.

As a part of the study on the connection between neighborhood importance and satisfaction, we conducted a survey of urban residents in Essen located in the Ruhr area in Germany. The purpose of the study was to evaluate, first, whether 12 different local urban amenities are preferred by the residents similarly as they express whether or not their neighborhood is important to them, and second, whether the dis–/satisfaction with these amenities impacts the stated dis–/satisfaction with a neighborhood (Table 3.1). To achieve this, we designed a detailed questionnaire and conducted a survey in 2010.

Table 3.1 Perceived importance of and satisfaction with the different urban amenities

	Coded variable	
Local urban amenities:	*Perception*	*Satisfaction*
Access:		
Access to work	pwork	satwork
Access with public transit	pacpub	sacpub
Access to downtown	pcenter	satcenter
Proximate/nearby:		
Public transit near	ppubnr	satpubnr
Shopping	pshop	satshop
Services	pserv	satserv
Facilities for adult leisure	padult	satadult
Schools and daycare	pschool	satschool
Playgrounds and children's facilities	pkid	satkid
Protection from:		
Crime	pcrime	satcrime
Environmental nuisances: noise, litter, pollution	penvt	satenvt
Availability in the neighborhood:		
Parks and green space	pgreen	satgreen
Car parking	pparkg	satparkg
Connection to a limited-access freeway	pinter	satinter

The chapter is structured as follows. The following section reflects on work that has been done on local urban amenities and how they relate to neighborhood perception and satisfaction. Then we introduce and describe the study area. Analysis of the data and the methodology come next. We conclude by discussing the revealed findings concerning the importance of different elements of urban environment for their inhabitants, especially with regard to the policy and planning implications.

3.2 Local Urban Amenities and Neighborhood Perception and Satisfaction

We review urban amenities shaping neighborhood perception and satisfaction. While the list of amenities that residents find important may be quite extensive, we have chosen to focus on the most pertinent amenities that affect residents' attitudes and satisfaction with their neighborhoods.

Urban green areas have been documented to carry an important role in urban societies regardless of social status or nationality, and especially in the neighborhoods, with accessibility often determining the preferences for specific types of nature, including gardens, parks, and greenways, in the pursuit of improved quality of life (Priego et al. 2008). Environmental aspects of their neighborhood may improve or decrease global well-being measures such as life satisfaction, especially in older residents (Forjaz et al. 2015). For example, nearby vegetation and natural environment substantially increase neighborhood satisfaction (Kaplan 2001). Therefore, the presence of urban green spaces in the residential areas is among the most crucial features of the urban form contributing to residents' appreciation of their neighborhoods (Crow et al. 2006). Prior studies have noted the importance of meeting recreation and play needs across all age and socioeconomic groups (Gobster 1995; Lindsey 1999; Fjørtoft and Sageie 2000; Oguz 2000; Shafer et al. 2000; Chiesura 2004; Jim and Chen 2006). Accessible natural areas influence human well-being (Matusoka and Kaplan 2008). Accordingly, the survey included questions on the importance of children's and adult leisure places within the residential neighborhoods (Fig. 3.3).

Work accessibility seems to be of less importance nowadays with residents' preferring certain social environments and trading shorter commuting trips with locating their homes close to high-quality schools (Giuliano 1995; Fan et al. 2011). However, Antipova et al. (2011) did not

Fig. 3.3 A residential zone in Berlin has an incorporated playground in the inner yard (Author 2016)

find that proximity to a high-performing school resulted in a statistically significant commuting increase. Public schools in general are being considered "a valuable repository of history and identity" (Good 2015) and school closures are rigorously contested by neighborhood residents.

Neighborhood satisfaction also relates to land uses (Ellis et al. 2006). Accessibility to commercial parcels within residential areas may support a greater sense of community through higher social interactions and decrease automobile dependency (Duany and Plater-Zyberk 1994; Katz 1994), while easily accessed public green spaces in many different forms (including parks, gardens, greenways, or urban forests) can increase both the sense of community (Stewart et al. 2004) and physical activity (Cohen et al. 2006, 2007; US Department of Health and Human Services 2010) thus improving life quality (Matusoka and Kaplan 2008). However, neighborhood satisfaction is negatively related to non-residential zoning (Jirovec et al. 1984; Kaplan 2001) where proximate commercial facilities may generate more garbage, which, combined with traffic-related odors, noise, and low visual quality, may decrease the quality of life of the residents. At the same time, well-landscaped areas tend to increase housing prices by making an area overall more attractive (Luttik 2000; Tyrväinen 2001).

Other factors contributing to neighborhood satisfaction include physical appearance of homes and yards and the neighborhood (Lansing and Marans 1969; Sirgy and Cornwell 2002; Parkes et al. 2002), and sense of safety (Cook 1988). Therefore, we were also interested whether protection from crime, noise, and pollution in the residential areas are important factors in the stated preferences and satisfaction of the residents. Table 3.1 describes variables we used in the study capturing urban form at a broader city level including accessibility to work and access to downtown area, as well as at a finer neighborhood scale including availability of the nearby children's and adult's leisure facilities, schools and daycare centers, services, shopping, parks, public transit, car parking, and freeway connections. These indicators have been analyzed to understand how their perceived importance and residential satisfaction with these amenities impact residents' perceived importance and satisfaction with their residential neighborhoods.

Socioeconomic and demographic variables that have been identified as important contributors to neighborhood satisfaction include age, life cycle stage, income level, family size, home ownership, and so on. (Francescato et al. 1987; Lu 1999). Respectively, covariates included age, gender, mobility status, occupation, marriage status, presence of children, and income.

3.3 Study Area

The study area is the city of Essen, which is located in the Ruhr region of Western Germany (Fig. 3.4). It is located in the federal state of North Rhine-Westphalia with a population density of 1140/sq. km. The Ruhr Regional Association (Regionalverband Ruhr, RVR) is the area's metropolitan planning association (MPO), which is responsible for the development of regional business and tourism, infrastructure projects, preservation of green open spaces, and land-use planning in the cities and communities for the enhancement of life quality. The region has undergone a massive structural economic, spatial, and social transformation and changed from an agricultural region to a highly productive industrial landscape, and finally to a service-, high-tech-, research-, and culture-oriented metropolis.

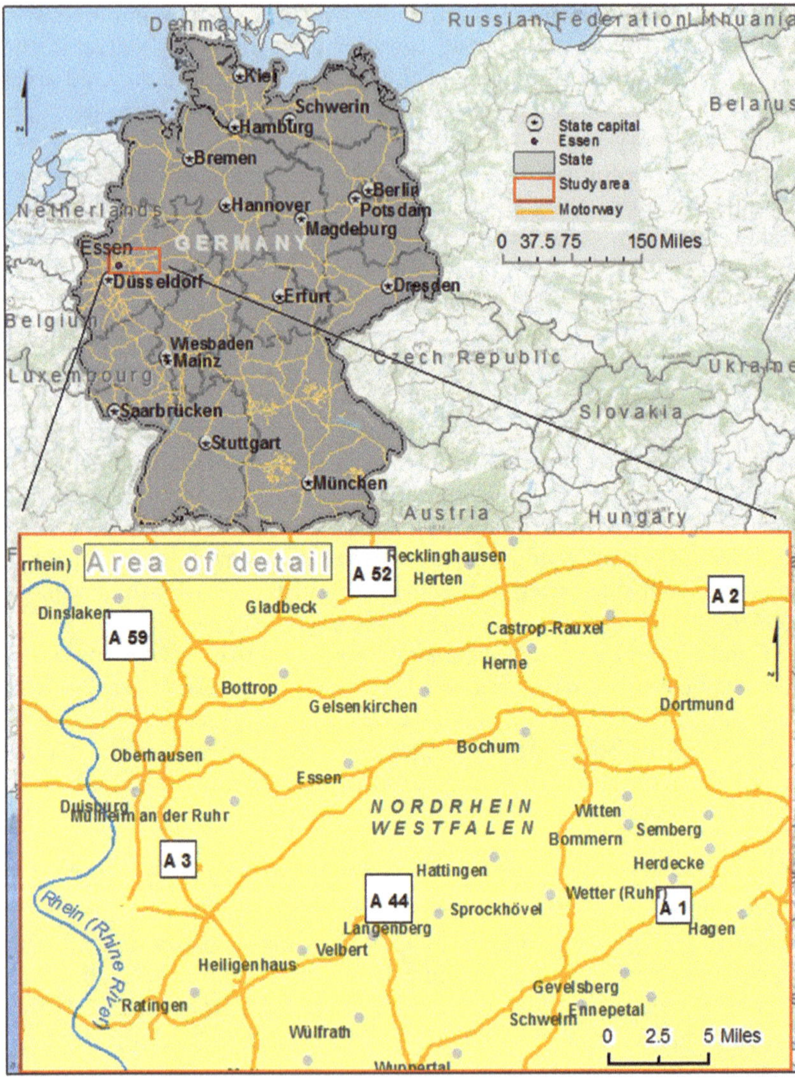

Fig. 3.4 Study area

The area's infrastructure includes a dense and efficient rail network, an extensive freeway (autobahn) and road grid-like system (one of the densest in Europe), international ports and airports. Driven by unparalleled growth in online retail with a demand for logistics and warehouse space, the Ruhr region is an important logistics hub ranking second (behind the Hamburg region) in terms of warehouse space with 3.8 million square meters (JLL 2015).

Demographically, the population has been declining since the 1960s due to de-industrialization and the resultant high unemployment. It is distributed equally throughout different parts of the region (Transforming City Regions 2015). Population composition has changed to become more international with the earlier immigrants coming from Southern and Eastern Europe and Asia, while more recently refugees came from African countries (Cameroon, Ghana, and Nigeria) due to political instability (Meierkord et al. 2012) as well as from the Middle East.

The polycentric area represents an excellent study site for the empirical investigation of planning-related problems in post-industrial areas regarding attitudes towards residential neighborhoods as the local residential areas consist of traditionally designed communities with many features of urban form incorporated into them as well as ample public parks, recreational areas, and open and green spaces shaping the urban landscape. Understanding the perceived preferences and stated satisfactions with such communities might help avoid future conflicts and contribute toward sustainable development where housing, access to jobs, convenience, and a safe and clean environment are the key ingredients for such development (Transforming City Regions 2015, p. 69).

3.4 Methodology

This study is a part of a larger study investigating the relationships between urban environment, travel behavior, and travelers' health and activity patterns (not analyzed here). The premise of the study is that residents' perceptions and satisfactions regarding their neighborhoods are shaped by the spatial arrangements of various urban features and are conditioned by socioeconomic and demographic characteristics. In order to confirm this hypothesis, we conducted an empirical study in Germany.

One of the aims of this project was to study attitude of population towards residential areas and urban amenities measured by degree of importance. One way to study related problems is by conducting survey campaigns (e.g., via household surveys), which can be very costly. To reduce costs and to ensure safety to survey conductors, the survey targeted travelers to generate a sufficiently representative sample related to the study goals. We based our data collection strategy on a selection of strategic survey places located along large shopping centers. Such data collection approach reduces costs incurred via household surveys while profiling travel behavior patterns (Monzon and Rodriguez-Dapena 2006).

We collected the data during the fall of 2010. Authors and a small group of students of the Geography Department at the University of Duisburg-Essen (Germany) conducted a survey and collected information of specific variables related to travelers and their stated preferences. The detailed questionnaire was designed and passers-by were approached, explained the survey's purpose, and requested to answer questions. Participation was voluntary. The questionnaire was conducted in person by the interviewers. A questionnaire had stated preference and satisfaction questions related where appropriate to the following categories: (1) individual-level; (2) local access to various amenities at the neighborhood- and city-level; (3) daily travel and activity pattern; and (4) negative travel outcomes and travel-related health impacts. The last two problems are not analyzed in the project, thus only data pertaining to the study will be described.

Guided by literature review, the questionnaire included socioeconomic and demographic items for respondents and their households as differences in sociodemographic factors impact perceptions of residential environment (Carp and Carp 1982; Fried 1982; Francescato et al. 1989; Bonaiuto et al. 1999). Individual-level characteristics contain information on gender, residential postcode and name of a home town/city, work/place of study postcode, marital status, presence of children, and residential mobility. We included questions on residential mobility into the survey as higher residential mobility might impact the attitudes towards neighborhoods. Residential mobility was captured by whether a resident moved house or not moved at all. Background data were used as independent variables in the analysis of preference and satisfaction.

Specifically, we collected the following demographic information: gender, residential mobility (moved house, not moved), residential postcode, name of residential urban place, marital status, and presence of children. Marital status was coded as single, divorced, having a partner, widowed, and married. Presence of children was coded as "No children," "One Child less than 5y.o.," "One Child 5–15y.o.," "One Child 15–18y.o.,"

Table 3.2 Demographic composition of travel survey respondents

Gender	*Male*	*Female*	*Total*
N (%)	27	25	52
Residential mobility	Moved house	Not moved	
N (%)	23 (44.5%)	29 (55.8%)	100%
Children	No children	Children	
N (%)	37 (71%)	12 (23%)	100%
Marital status	Single	Not-single	
N (%)	19 (36.3%)	33 (63.6)	100%

"Two Children less than 5y.o.," "Two Children 5–15y.o.," and "Two Children 15–18y.o." Table 3.2 reports the demographic composition of travel survey respondents.

To capture less concrete concepts including resident satisfaction or attractiveness, several Likert-type items are commonly created and grouped into a survey scale that subsequently is used in factor analysis to group items measuring the underlying latent variable (Sullivan and Artino 2013). Accordingly, the survey employed the four-point ordinal Likert scale developed in 1932 by Rensis Likert (1932) to measure respondents' attitudes by asking the extent of degree to which they agree or disagree with a particular statement. To analyze neighborhood attractiveness we asked survey takers how important is a residential neighborhood itself and how they perceived accessibility to various urban amenities (including easy access to work, nearby shops, services, entertainment for adults as well as children, schools and daycare, parks and green space, neighborhood travel infrastructure (public transit including bus or train nearby, interstate connection, and parking), and neighborhood safety and environmental features (protection from crime, and exposures to noise, pollution, and litter).

Ab initio stated preferences were measured as "Not important," "Less important," "Important," and "Very important." On a scale of 1 to 4, the survey participant evaluated what elements of the built environment were of importance. Analogously, we also asked travelers their actual satisfaction with urban amenities as well as with their neighborhood. On a scale of 1 to 4, satisfaction was coded as "Very unsatisfied," "Unsatisfied," ""Satisfied," and "Very satisfied." Table 3.3 presents the described Likert-type scale used in the questionnaire.

Table 3.3 Likert-type scale used in the questionnaire

	Attitudes (perceived importance)				*Stated satisfaction*			
Scale	Not important	Less important	Important	Very important	Very dissatisfied	Dissatisfied	Satisfied	Very satisfied
Code	1	2	3	4	1	2	3	4

The initial data contained 58 completed questionnaires, but after observing that some categories consisted of a few observations, we consolidated these. For example, we merged all observations with children into a single group: "Children present" versus "No children." The same logic dictated merging "Single," "Widowed," and "Divorced" into a "Single" group, and "Married" and "With a partner" into a "Married" group. We deleted three respondents who did not provide any data on their residential hometown or a zip code. The final sample contained 52 data entrances and was more balanced (Table 3.1). There are 27 males and 25 females in the survey. There are slightly more people who did not move residential address (56%), while those who moved account for 45% (refer to Table 3.3).

3.5 Factor Analysis

Prior to data analysis, we checked data consistency and obtained the value of the Cronbach coefficient alpha (raw-0.7, and standardized=0.68). A common technique for analyzing subjective perceptions from survey responses is factor analysis while the factors obtained from factor analysis could be used in place of the original variables in subsequent statistical analyses (Santos et al. 1998). Correspondingly, survey respondents' data was analyzed in two steps. First, to probe underlying interrelationships in Likert-type variables, a factor analysis was used (Santos et al. 1998). The purpose was grouping perception and satisfaction variables into uncorrelated factors to be used for the subsequent analysis. Our original variables included various detailed elements of urban form, including access to work, proximate public transit, nearby shopping and services, facilities for adult leisure, schools and daycare, playgrounds and children's facilities, importance of protection from crime and environmental nuisances including noise and pollution, availability of parks and green space in the neighborhood, availability of parking, and connection to a limited-access freeway (refer to Table 3.1).

The initial variables representing the attitudes towards local amenities coded "Not Important," "Less Important," "Important," and "Very Important" were assigned values 1 through 4. Analogously, the variables representing the actual satisfaction with these local amenities have been coded in ascending order with an increase in satisfaction from "Very dissatisfied" coded as 1, "Dissatisfied" as 2, and "Satisfied" and "Very Satisfied" as 3 and 4, respectively (refer to Table 3.3). Factor analysis explains the relationships among a set of observed dichotomous variables using continuous latent variables: the example is described in Muthen and Muthen (2000, pp. 885–886) who used the National Longitudinal Survey of Youth data in the study of anti-social behavior. After we factorized the initial 14 variables, the variable representing perceived importance of the access to downtown (*pcenter*) was removed due to its low communality value from the analysis of neighborhood perception.

Our sample size satisfies the recommended subjects-to-variables ratio of 4:1 or greater ($N = 52$), while high communality values of the remaining 13 variables (the average communality value was 0.77) ensured accurate and stable factor recovery (MacCallum et al. 2001, p. 636). Varimax rotation was used. We retained five factors observing the graphical scree plot pattern (Fig. 3.5a). Retained factors account for 77% of common variance in the data set (Fig. 3.5b). Factor analysis of the 13 urban amenities for neighborhood perception yielded five internally consistent categories: Education (Factor 1 includes schools and children's daycare, playgrounds and children's facilities nearby); Commercial facilities (Factor 2 includes nearby shops, services, facilities for adult leisure); Travel infrastructure (Factor 3 includes a convenient connection to a limited-access freeway, nearby parking, and access to work); Personal and environmental safety and green space (Factor 4 includes protection from crime and environmental nuisances, such as noise, litter, and pollution, as well as nearby parks and green space); and Public transit (Factor 5 includes access with public transit, nearby public transit stops).

To analyze satisfaction with urban amenities, two variables representing green spaces (*satgreen*) and environmental exposure (*satenvt*) had low communality values and were removed from the analysis of satisfaction. The remaining 12 variables have been factorized. Using the same criteria as for perception (i.e., an observation of a scree plot pattern and the variance accounted for), we retained six factors (Fig. 3.5c): Education (Factor 1 includes children's educational and leisure facilities); Public transit (Factor 2 includes access with public transportation and nearby public

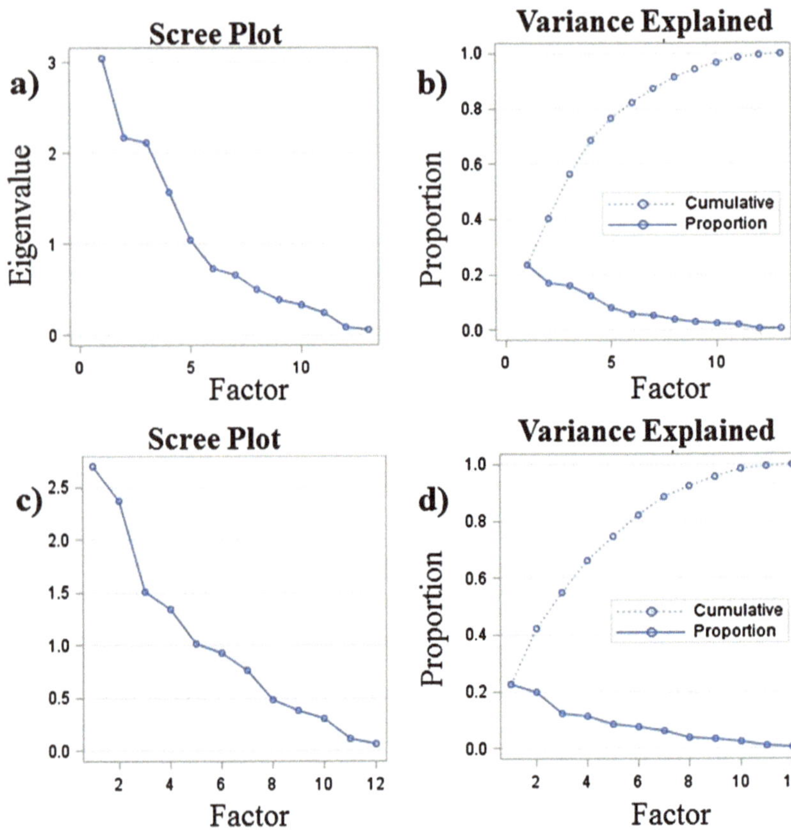

Fig. 3.5 Factors analysis of neighborhood perception (upper panel): (**a**) scree plot pattern of factors (**b**) amount of variance explained as a percent of common variance in the data set; and neighborhood satisfaction (lower panel): (**c**) scree plot pattern of factors, and (**d**) amount of variance explained

transportation stops); Travel infrastructure (Factor 3 includes a convenient location to a limited-access freeway, nearby parking, and services); Shopping (Factor 4 includes access to downtown and shopping); Safety and leisure (Factor 5 includes adult leisure activities and protection from crime); and Work access (Factor 6 includes access to work). These account for 82% of the total variance (Fig. 3.5d). Table 3.4 presents the combination of the local amenity variables into these factors. We used SAS® to perform the analyses.

Table 3.4 Factor analysis—rotated factor pattern

	Attitude						*Satisfaction*					
	Factor 1	*Factor 2*	*Factor 3*	*Factor 4*	*Factor 5*		*Factor 1*	*Factor 2*	*Factor 3*	*Factor 4*	*Factor 5*	*Factor 6*
Variable	*Education*	*Commercial*	*Road infrastr.*	*Safety/ Green*	*Transit*	*Variable*	*Education*	*Safety*	*Shopping*	*Road infrast.*	*Transit*	*Work access*
pkid	0.95	0.1	0.09	0.17	0.01	satkid	0.94	0.13	−0.09	0.06	0.05	−0.02
pschool	0.95	0.12	0.01	0.03	0.02	satschool	0.93	0.09	−0.02	0.12	0.09	0.03
padult	−0.06	0.72	0.16	−0.07	−0.11	satadult	−0.07	0.79	0.06	−0.17	0.01	0.28
pserv	0.12	0.82	−0.15	0.09	0.01	satcrime	0.26	0.81	0.05	0.07	−0.12	−0.16
pshop	0.15	0.89	−0.06	0.14	−0.03	satshop	0.14	0.11	0.88	0	−0.01	−0.15
pwork	−0.13	−0.11	0.68	−0.27	0.19	satcenter	0.05	−0.01	0.68	0.05	0.17	0.47
pinter	0.03	0.07	0.84	0.04	−0.22	satinter	0.27	0.17	−0.30	−0.58	−0.16	0.49
pparkg	0.30	−0.02	0.78	0.21	−0.16	satparkg	0.47	0.25	−0.24	0.53	−0.26	0.16
penvt	0.01	−0.01	0.06	0.88	−0.09	satserv	0.02	0.02	−0.09	0.92	−0.05	0.17
pgreen	0.46	−0.2	0.12	0.51	0.42	satacpub	0.11	−0.04	0.04	0.04	0.97	0.06
pcrime	0.17	0.22	−0.1	0.78	−0.13	satpubnr	0.03	−0.07	0.07	0.02	0.97	0.1
pacpub	−0.09	−0.1	−0.28	−0.02	0.82	satwork	−0.02	0.05	0.04	0.06	0.13	0.83
ppubnr	0.11	−0.01	0.04	−0.18	0.92							

3.6 Ordinal Logistic Regression

The factors obtained from factor analysis could be used as predictors in logistic regression (Meyers et al. 2013, p. 641). Correspondingly, in the second step, stable categories of items identified in factor analysis were used as explanatory variables in ordinal logistic regression.

First, we used the cumulative logit as an approach most extensively used for analyzing ordered categorical outcomes (Ingelmo et al. 2011; Faden and Graubard 2000; McCusker et al. 1994; Kumanyika et al. 1994; Ashby et al. 1989). Second, we used the *invariance* property of the model: the slope parameters (the coefficients of the covariates) and interpretations will not change after collapsing of contiguous response categories (Agresti 1996; Greenland 1994; Murad et al. 2003). To illustrate, our response variable "Perceived Neighborhood" represented neighborhood attractiveness. After observing counts of expected numbers in the frequency table, we collapsed two adjacent categories "Not Important" and "Less Important, which contained sparse counts, into a single category of "Not Important" to increase the number of observations in this category, yielding $k = 3$ ordered categories ("Very important," "Important," and "Not important"). The resulting proportions for neighborhood attractiveness in each category were 33%, 46%, and 21% (categories 1–3, respectively). In a similar way, another dependent variable, "Satisfaction with the Neighborhood," expressing stated neighborhood satisfaction, has been recoded into an ordinal three-level variable: "Very Satisfied" = 1, "Satisfied"= 2, and "Unsatisfied" = 3 (with the resulting proportions for neighborhood satisfaction in each category of 19.2%, 65.4%, and 13.5%, respectively).

The cumulative models model the probabilities that the response Y (here, the perceived importance of and the satisfaction with the one's residential neighborhood) falls in category j or below, that is, the lower levels of Y are contrasting with the higher levels of Y; with odds ratios interpreted as the effect of the variable on the odds of being in a "lower" category (Gameroff 2005). To illustrate, in our case, we modeled the probabilities that the ith observation falls into "Very important" category rather than in "Important" and "Not important").

Let the jth cumulative probability be denoted as:

$$F_{ij} = P(Y_i <= j) = \pi_1 + \cdots + \pi_{j,j} = 1,\ldots,J \tag{3.1}$$

where $\pi_{j,j}$ is the probability that the *i*-th observation falls in the *j*th ordinal category of the dependent variable.

The cumulative logit model models two cumulative logits for the three-level (1, 2, and 3) ordinal outcome, one for each of the following cut points: 1 vs. 2,3; and 1,2 vs. 3 (the numbers refer to the three levels of status of importance and satisfaction: "Very important" (Level 1), "Important" (Level 2), and "Not important" (Level 3), and respectively, "Very Satisfied" = 1, "Satisfied" = 2, and "Unsatisfied" = 3):

$\operatorname{logit}(F_{i1}) = \log\left(\frac{\pi i1}{\pi i2 + \pi i3}\right)$ (i.e., here: the log-odds of residents perceiving their neighborhood as very important versus important and not important) and

$$\operatorname{logit}(F_{i2}) = \log\left(\frac{\pi i1 + \pi i2}{\pi i3}\right)$$
$$\left(\begin{array}{l}\text{here : the log-odds of residents perceiving their}\\ \text{neighborhood as very important and important}\\ \text{versus not important}\end{array}\right) \tag{3.2}$$

Exponentiating the cumulative logits produces the cumulative odds, so that one can solve for the cumulative probabilities:

$$\operatorname{Logit}(F\,ij) = \alpha\, j + X'\beta; \tag{3.3}$$

where X' is the vector of explanatory variables and $\beta = (\beta_1, \beta_2, \ldots, \beta_n)$ are regression coefficients to be estimated, to examine the effect of different factors on perceived importance of one's residential neighborhood (Hosmer and Lemeshow 2000).

The model is predicting the log odds of reporting the perceived importance of neighborhood:

$$\frac{F\,ij}{1 - F\,ij} = \exp\left(\alpha\, j + \bar{X}'\beta\right); \tag{3.4}$$

$$F_{ij} = \frac{\exp\left(\bar{X}'\beta\right);}{1 + \exp\left(\bar{X}'\beta\right);} \tag{3.5}$$

From (3.2), (3.3), and (3.5), we can plug in our predictor variables for Model 1, which evaluates the relationship between various urban amenities (now captured by five factors) and the perceived importance of neighborhood *not controlling for sociodemographic characteristics* and get:

$$\log\left(\frac{\pi \text{ VI}}{\pi \text{ IM} + \pi \text{ NI}}\right) = \text{Intercept } 1 + \textit{Factor } 1 \times \beta\ 1 + \textit{Factor } 2 \times \beta\ 2 + \textit{Factor } 3 \times \beta\ 3 + \textit{Factor } 4 \times \beta\ 4 + \textit{Factor } 5 \times \beta\ 5 \quad (3.6)$$

and

$$\log\left(\frac{\pi \text{ VI} + \pi \text{ IM}}{\pi \text{ NI}}\right) = \text{Intercept } 2 + \textit{Factor } 1 \times \beta\ 1 + \textit{Factor } 2 \times \beta\ 2 + \textit{Factor } 3 \times \beta\ 3 + \textit{Factor } 4 \times \beta\ 4 + \textit{Factor } 5 \times \beta\ 5 \quad (3.7)$$

We were interested in testing the hypothesis Ho: $\beta = 0$ (there is no relationship between various urban amenities and the perceived importance of neighborhood) against the alternative H_1: β not =0 (there is a relationship between various urban amenities and the perceived importance of neighborhood).

In ordinal logistic regression model, predictors may be categorical and/or continuous. To explore the impact of various urban amenities on the attitudes towards neighborhoods measured by the various degree of "importance," we used factor scores from all five factors obtained from the factor analysis as explanatory variables, including Factor 1, which measures education (schools and children's daycare, playgrounds and children's facilities nearby); Factor 2, which captures commercial facilities (nearby shops, services, facilities for adult leisure); Factor 3, which reflects travel infrastructure (a convenient connection to a limited-access freeway, nearby parking, and access to work); Factor 4, which measures personal and environmental safety and green space (protection from crime and environmental nuisances such as noise, litter, and pollution as well as availability of nearby parks and green space); and Factor 5, which captures public transit (access with public transit, nearby public transit stops). Factor scores are standardized with a mean of zero and a standard deviation of one. Table 3.5 reports the results from the analysis of the relationship between various urban amenities and neighborhoods perceptions (Model 1).

Table 3.5 Results of the ordinal logistic models of perceived preference for and stated satisfaction with a neighborhood

Dependent V.: Perceived neighborhood importance	*Model 1*			*Model 2*			*Dependent V.: Stated neighborhood satisfaction*	*Model 3*			*Model 4*		
	Est.	*OR (p-value)*	*95% Wald CI*	*Est.*	*OR (p-value)*	*95% Wald CI*		*Est.*	*OR (p-value)*	*95% Wald CI*	*Est.*	*OR (p-value)*	*95% Wald CI*
Intercept 1	−0.93			−0.6			Intercept 1	−1.86			−1.05		
Intercept 2	1.84			2.12			Intercept 2	2.23			2.96		
Urban amenities:							*Urban amenities:*						
Education (Factor 1)	**0.84**	**2.32 (0.0059)**	**1.27,4.2**	**0.95**	**2.59 (0.007)**	**1.29, 5.22**	*Education (Factor 1)*	**0.94**	**2.57 (0.008)**	**1.28, 5.14**	**0.89**	**2.44 (0.02)**	**1.13, 5.26**
Commercial facilities (Factor 2)	0.008	1.008 (0.097)	0.59; 1.74				*Safety (Factor 2)*	**0.68**	**1.98 (0.04)**	**1.02, 3.84**	**0.69**	**1.99 (0.04)**	**1.02; 3.87**
Road Infrastructure (Factor 3)	−0.3	0.74(0.28)	0.43, 1.28				*Shopping (Factor 3)*	−0.23	0.8 (0.47)	0.42, 1.48			
Safety & Green (Factor 4)	**0.9**	**2.47 (0.003)**	**1.36; 4.5**	**0.88**	**2.41 (0.009)**	**1.24; 4.7**	*Road Infrastructure (Factor 4)*	−0.07	0.94(0.82)	0.51, 1.71			
Transit (Factor 5)	0.28	1.33(0.32)	0.76, 2.31				*Transit (Factor 5)*	−0.05	0.95 (0.87)	0.52; 1.76			
							Work access (Factor 6)	−0.42	0.67 (0.19)	0.35; 1.23			
SES & Dem.:							*SES & Dem.:*						
Gender: Female vs Male (ref)				0.11	1.12 (0.85)	0.33; 3.8	Gender: Female vs Male (ref)				−1.08	0.34 (0.11)	0.09; 1.28

(*continued*)

Table 3.5 (continued)

Dependent V.: Perceived neighborhood importance	*Model 1*			*Model 2*			*Dependent V.: Stated neighborhood satisfaction*	*Model 3*			*Model 4*		
	Est.	*OR (p-value)*	*95% Wald CI*	*Est.*	*OR (p-value)*	*95% Wald CI*		*Est.*	*OR (p-value)*	*95% Wald CI*	*Est.*	*OR (p-value)*	*95% Wald CI*
Single vs Partner and Married (ref)				0.51	1.67 (0.43)	0.46; 6.05	Single vs Partner and Married (ref)				−0.13	1.14 (0.85)	0.28;4.5
Resid.mob.: Moved vs Never Moved (ref)				−0.75	0.47 (0.22)	0.14;1.5	Resid.mob.: Moved vs Never Moved (ref)				−0.4	0.67 (0.57)	0.17;2.6
Children vs No children (ref)				−0.85	0.43 (0.29)	0.09; 2.1	Children vs No children (ref)				0.04	1.04 (0.97)	0.15;7.2
Score Test: p-value		0.21			0.062		Score test		0.001			0.025	
Deviance GOF: p-value		0.66			0.66		Deviance GOF: p-value		0.91			0.90	
Pearson GOF: p-value		0.12			0.73		Pearson GOF: p-value		0.22			0.45	
Global test: b=0		**0.002**			**0.003**		Global test: b=0		**0.039**			**0.028**	
−2 Log L		90.6			84.2		−2 Log L		74.7			71.3	
Max Rescaled R^2		0.34			0.45		Max Rescaled R^2		0.28			0.31	

The probability that *i*th observation will be in the "Very important" category rather than the "Important' and "Not important" categories can be fitted by:

$$\pi_{\mathrm{VI}} = \frac{1}{1+\exp\left(-\begin{pmatrix}\text{Intercept } 1 + \mathit{Factor}\,1 \times \beta\,1 + \mathit{Factor}\,2 \times \beta\,2 + \\ \mathit{Factor}\,3 \times \beta\,3 + \mathit{Factor}\,4 \times \beta\,4 + \mathit{Factor}\,5 \times \beta\,5\end{pmatrix}\right)} \quad (3.8)$$

while the probability that *i*th observation will be the "Very important" and "Important" categories rather than in the "Not important" category can be fitted by:

$$\pi_{\mathrm{VI}} + \pi_{\mathrm{IM}} = \frac{1}{1+\exp\left(-\begin{pmatrix}\text{Intercept } 2 + \mathit{Factor}\,1 \times \beta\,1 + \mathit{Factor}\,2 \times \\ \beta\,2 + \mathit{Factor}\,3 \times \beta\,3 + \mathit{Factor}\,4 \times \beta\,4 + \\ \mathit{Factor}\,5 \times \beta\,5\end{pmatrix}\right)} \quad (3.9)$$

and the probability that *i*th observation will be in the "Not important" category can be found from:

$$\pi_{\mathrm{NI}} = 1 - \left(\pi_{\mathrm{VI}} + \pi_{\mathrm{IM}}\right) \quad (3.10)$$

In Model 2, we kept significant factors from Model 1 as primary predictors and added several sociodemographic variables as covariates to control for socioeconomic and demographic characteristics (Table 3.5, Model 2). Thus, Model 2 includes the perceived importance of different urban features including schools and children's daycare, playgrounds and children's facilities nearby (captured in Factor 1, "Education"), protection from crime and environmental nuisances such as noise, litter, and pollution, as well as nearby parks and green space (captured in Factor 4, "Personal and environmental safety"), gender, residential mobility, marital status, and presence of children, used in the modeling as categorical dichotomous variables (female and male, moved house versus never moved, single versus married, and children and no children, respectively).

Similarly, we explored the impact of various urban amenities on the satisfaction with one's neighborhood measured by the various degree of "satisfaction" (Model 3). Model 3 contains six factors including Education

(Factor 1), Public transit (Factor 2), Travel infrastructure (Factor 3), Shopping (Factor 4), Safety and leisure (Factor 5), and Work access (Factor 6), while Model 4 contains significant factors from the previous model and includes the same modifiers as in the perception models to be consistent including gender, residential mobility, marital status, and presence of children.

The cumulative logit model is available in SAS software for modeling ordered categorical outcomes. SAS version 9.4 has been used for all models.

3.7 Results and Findings

Perception Model 1 captured the impact of five factors reflecting access to local amenities on residential perceptions of neighborhoods, while Model 2 kept the significant factors from Model 1 and included socioeconomic and demographic explanatory variables to control for gender, marital status, residential mobility, and presence of children. The value of the score test for the proportional odds assumption (0.21) informs that the test does not reject the null hypothesis that the assumption holds indicating that the cumulative logit model can be used. This is corroborated by the deviance and Pearson goodness-of-fit statistics (both of which have high p-values); both tests suggest an adequate fit.

Generally, the analysis of perceived importance of neighborhoods indicates that mostly residents render neighborhoods important and very important. For example, the first intercept in Model 1, −1.05 −0.93, is the predicted log-odds of being in category 1 ("Very Important") rather than in categories 2 or 3 ("Important" and "Not Important," respectively), when all the explanatory variables have values of 0 (Allison 2012). Exponentiating, it gives us 0.4, which is the predicted odds of being in category 1 versus the other two. Farther, the second intercept, 1.84, is the predicted log-odds of being in categories 1 or 2 ("Very Important" and "Important") rather than in category 3 ("Not Important"), when all the *X* variables are 0. Exponentiating, it gives us odds of 6.3. So, residents are more than six times as likely to be in categories 1 or 2 rather than category 3 if all the independent variables have values of 0. A similar pattern is observed across all models.

Factors 1 and 4 are consistently statistically significant in Models 1 and 2. Figure 3.6 displays the relationship of these factors to neighborhood

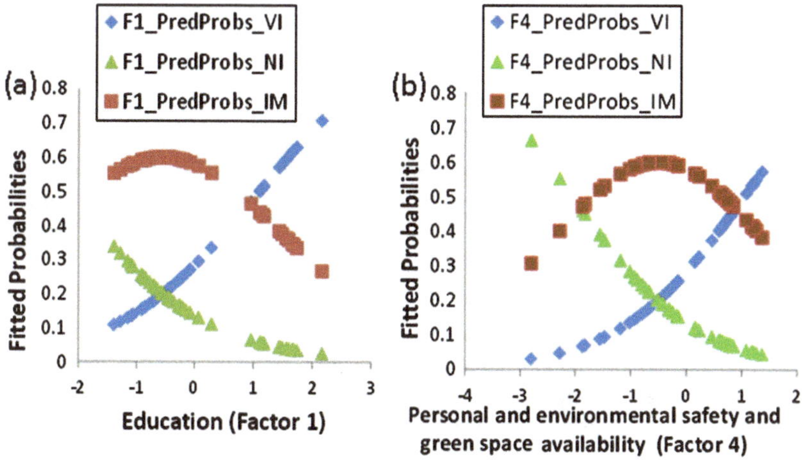

Fig. 3.6 Ordinal logistic model: fitted probabilities for three levels of neighborhood importance—*VI* very important, *IM* important, *NI* not important—as a function of (**a**) education perception (Factor 1); and (**b**) perception of personal and environmental safety and green space availability (Factor 4)

perception: cumulative probabilities are visualized over a range of perception levels. We observed that children's leisure (playgrounds, children's centers) and educational facilities (schools and daycare) and availability of parks in the residential areas (all captured in Factor 1) contribute to greater attachment towards neighborhoods (Fig. 3.6a). Incrementing Factor 1 (Children's facilities in Table 3.5) by 1 (that is, going from residents' perception of this urban amenity from unimportant to important) increases a resident's odds of expressing the importance of their neighborhood by a factor of 2.3 and 2.6, respectively (Models 1 and 2) (Table 3.5). Figure 3.6a fits probabilities for three levels of neighborhood importance: very important, important, and not important, as a function of education perception captured in Factor 1.

Personal and environmental safety and green space (Factor 4) contribute to a greater importance of residential places (Fig. 3.6b). Respondents were 2.5 (Model 3 with no sociodemographic characteristics) and 2.4 times (Model 4 with sociodemographic characteristics added) more likely to contribute this feature to the perceived preference. This finding was in accord with an earlier observation that a sense of personal safety is paramount as it contributes to a sense of well-being in an urban setting (Amerigo and Aragones 1997; Bonaiuto et al. 1999). Similarly, prior studies found access to open space with trees and vegetation both an important community feature and an important consideration in purchasing a house, with residents often voicing the need for more natural areas within their neighborhoods (Crow et al. 2006). Figure 3.6b illustrates how perception of personal and environmental safety and green space availability captured in Factor 4 influences neighborhood importance by category: very important, important, and not important.

Although Fujita and Thisse (2013) argue that consumers rather prefer short trips to retail and services, regarding residential areas, residents give low ranking to the business/commercial environment as an attribute of the residential community (Crow et al. 2006). Jakle and Mattson (1981) describe how when the implementation of the commercial zoning and the resulting accumulation of businesses resulted in the demise of the residential place in a medium-sized American city. Accordingly, proximity to shopping (Factor 2, which includes nearby shops and services as well as facilities for adult leisure) was not significant in our model. Figure 3.7, which shows flat fitted probabilities (i.e., no influence) for three levels of neighborhood importance as a function of commercial facilities captured by Factor 2. In another European country, the global effect for retail and service diversity is found to be non-significant in relationship to the price of residential land in Luxembourg (Glaesenera and Carusoa 2015).

Public transit has a negative sign in the factor loading (Table 3.5) demonstrating the relative more importance of individual transport and access to a freeway over public transport. However, access to this local amenity (Factor 3, "Road infrastructure," includes a convenient connection to a limited-access freeway, nearby parking, and public transit) is not significant in neighborhood perceptions.

Adding socioeconomic variables improves model performance as reflected in the greater value of the generalized R^2 and −2 Log L values (here, the lower is the value the better). However, sociodemographic variables were not significantly impacting the perception of residential communities. For example, no significant gender-based differences were found for perceived urban features; similarly, Crow et al. (2006) found no differences between men and women regarding their perception of recreation categories, green residential atmosphere, and business/commercial environment in Berwyn community. In Germany, social policies facilitate access by low-income people to green areas (Priego et al. 2008).

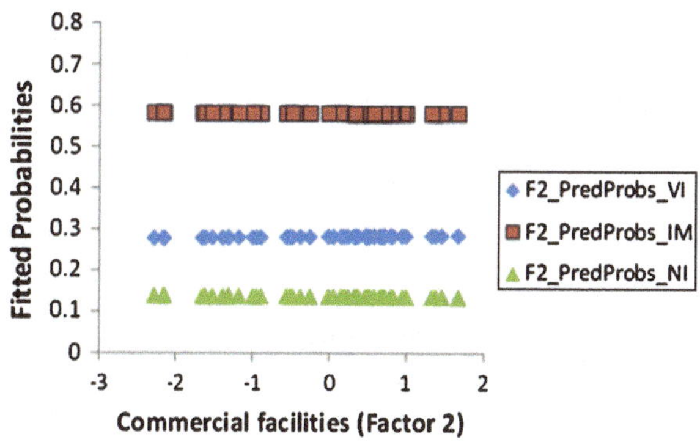

Fig. 3.7 The ordinal logistic model: fitted probabilities for three levels of neighborhood importance—*VI* very important, *IM* important, *NI* not important—as a function of commercial facilities (Factor 2)

We did not find the impact of residential mobility (the length of residential tenure) and the presence of children in the household contributing to neighborhood perceptions. The finding was in contrast to the prior documentation of differences in age, gender, and length of residency impacting perceptions of residential environment (Carp and Carp 1982; Fried 1982; Francescato et al. 1989; Bonaiuto et al. 1999). Respondents with resident children younger than 18 years of age as well as younger people value the recreational attributes of their community significantly more compared to respondents without resident children and older respondents, respectively; while longer residential tenure contributes to higher appreciation of green spaces and vegetation within their communities (Crow et al. 2006).

Satisfaction To keep consistent, Model 3 measured the impact of the main factors only on the stated satisfaction with neighborhoods, while Model 4 kept the significant factors from Model 3 and included socioeconomic and demographic explanatory variables as controls (Table 3.5).

Overall, higher endorsements of the various urban features shaping residential neighborhoods were measured in the analysis of perception compared to the actual satisfaction. However, both analyses agreed that availability of children's educational and recreational facilities and green spaces as having a positive effect on their neighborhood appreciation.

Respondents consider easy access to residential parks and natural areas beneficial due to a sense of tranquility and an escape from the stresses of work and city life they provide (Kaplan 1995, 2001; Crow et al. 2006).

3.8 Discussion and Conclusion

Frequently, the definition of a neighborhood is subjective, with residents not reflecting upon their neighborhoods as defined by administrative borders (e.g., zip codes), but instead often using individually determined ones. Nevertheless, residential neighborhoods play an important role in the urban area as they provide identity, support, and a sense of place to its inhabitants (Greene and Pick 2012). However, urban landscape is not perceived or used similarly by people of different social and cultural backgrounds (Priego et al. 2008; Dony et al. 2015).

A survey is often a useful research tool to capture abstract ideas or concepts that are more difficult to quantify or not directly observable, such as attitudes and perceived preferences or satisfactions (Rickards et al. 2012). Accordingly, to assess neighborhood attractiveness, survey data were collected. The point of departure for the study was that the importance residential inhabitants ascribe to their communities is shaped by the easy accessibility of places frequented by the residents. Preferences are also shaped by cultural differences, thus resulting in different groups of the population using amenities in a varying way. Therefore, qualitative interviews with the public is an appropriate data collection method for neighborhood preference-related investigations. We explored residents' attitudes to their residential community as conveyed through perceptions of and satisfactions with local urban features. Specifically, we were interested to see whether urban residents perceive differences in neighborhood characteristics and how much of their satisfaction can be attributed to these differences. These questions were investigated in the city of Essen located in the Ruhr region of Western Germany.

We show that some urban features are perceived as more important than others. Residents recognize the value of children's facilities, green spaces, and safe and pollution-free environments as they endorse residential places. Both models of perception and satisfaction found proximity to parks and green spaces as contributing to perceived importance and stated satisfaction with neighborhoods. Earlier studies documented having "nature right outside my door" contributing to their sense of well-being, satisfaction, and comfort (Crow et al. 2006). Access to parks and green spaces increases physical activity and thus plays an important role in the development of effective disease-prevention strategies (Dony et al. 2015). The study found support for this position: easy accessibility to children's recreational and educational places and green areas (Factor 1), the offer of environmental protection including from noise and pollution and safety from crime (Factor 2), and access to work (Factor 3) are *perceived* in the same way by individuals from different sociocultural backgrounds, with or without variables of gender, residential mobility, marital status, and presence of children. One explanation of socioeconomic characteristics being insignificant in our model may be that all inhabitants in Germany enjoy similarly easy access to public gardens and green areas regardless of their socioeconomic status.

Having neighborhood shopping and services seems less important as close-by retail and service outlets might deteriorate residential environment through increased traffic volume, congestion, and traffic-related accidents, which are undesirable from the perspective of the residents. To support, one study documents how the intrusion of gas filling stations, auto-catering service outlets, and other types of business into a once stable residential area forced residents to leave due to dirt and noise from more parking lots spilling over onto adjacent residential properties; the area's land use gradually changed from single-family owner-occupied to mostly commercial with off-street parking, few surviving residencies consisting of multiple-family rental housing mostly catering to the students of the nearby university, and increased amount of vacant land (Jakle and Mattson 1981).

In contrast to perception models, analysis of *satisfaction* found that access to children's recreational and educational places (Factor 1) and safety from crime (Factor 2) shape residents' satisfaction with their neighborhoods.

The significance of easily accessible children's recreational and educational places as well as environmental protection from noise and pollution offered by neighborhoods demonstrates that these elements are a fundamental pillar of individuals' perceptions and satisfactions, regardless of their socioeconomic background. Scholars recognize the importance for every individual across all racial and ethnic lines and at any income level to use public goods and services equally (Talen and Anselin 1998), which is often not realized, with minorities experiencing worse access, at least in the context of the United States (Dai 2011). Being limited by the scope of the project, we did not investigate the actual use of parks and green areas, and therefore leave it for future investigations. Although authors gathered detailed information in their questionnaire, a small sample size presents a limitation of the project, therefore leaving a collection of a larger database for future investigations of personal attitudes towards urban amenities.

Urban planners and developers are challenged with designing "development with zero growth" appealing urban environments to attract new and retain existing residents. Understanding these features is not possible without careful examination of residents' valuation of these urban features.

Bibliography

Agresti, A. 1984. *Analysis of Ordinal Categorical Data*. New York: Wiley.

———. 1996. *An Introduction to Categorical Data Analysis*. New York: Wiley and Sons, Inc.

Ahlbrandt, R.S. 1984. *Neighborhoods, People, and Community*. New York: Plenum.

Allison, P.D. 2012. *Logistic Regression Using SAS. Theory and Application*. 2nd ed. Cary: SAS Institute.

Amerigo, M., and J.I. Aragones. 1997. A Theoretical and Methodological Approach to the Study of Residential Satisfaction. *Journal of Environmental Psychology* 17: 47–57.

Antipova, A., F. Wang, and C. Wilmot. 2011. Urban Land Uses, Socio-Demographic Attributes and Commuting: A Multilevel Modeling Approach. *Applied Geography* 31 (3): 1010–1018.

Ashby, D., C.R. West, and D. Ames. 1989. The Ordered Logistic Regression Model in Psychiatry: Rising Prevalence of Dementia in Old People's Homes. *Statistics in Medicine* 8 (11): 1317–1326.

Bonaiuto, M., A. Aiello, M. Perugini, M. Bonnes, and A.P. Ercolani. 1999. Multidimensional Perception of Residential Environment Quality and Neighbourhood Attachment in the Urban Environment. *Journal of Environmental Psychology* 19: 331–352.

Carp, F.M., and A. Carp. 1982. Perceived Environmental Quality of Neighborhoods: Development of Assessment Scales and Their Relation to Age and Gender. *Journal of Environmental Psychology* 2: 245–312.

Chiesura, A. 2004. The Role of Urban Parks for the Sustainable City. *Landscape and Urban Planning* 68: 129–138. https://doi.org/10.1016/j.landurbplan.2003.08.003.

Clark, T.N., R. Lloyd, K. Wong, and P. Jain. 2002. Amenities Drive Urban Growth. *Journal of Urban Affaires* 24: 493–515.

Cohen, D.A., J.S. Ashwood, M.M. Scott, A. Overton, K.R. Evenson, L.K. Staten, et al. 2006. Public Parks and Physical Activity Among Adolescent Girls. *Pediatrics* 118 (5): e1381–e1389.

Cohen, D.A., T.L. McKenzie, A. Sehgal, S. Williamson, D. Golinelli, and N. Lurie. 2007. Contribution of Public Parks to Physical Activity. *American Journal of Public Health* 97 (3): 509–514.

Cook, C.C. 1988. Components of Neighborhood Satisfaction: Responses from Urban and Suburban Single-Parent Women. *Environment and Behavior* 20 (2): 115–149.

Crow, T., T. Brown, and R. DeYoung. 2006. The Riverside and Berwyn Experience: Contrasts in Landscape Structure, Perceptions of the Urban Landscape, and Their Effects on People. *Landscape and Urban Planning* 75: 282–299.

Dai, D. 2011. Racial/ethnic and Socioeconomic Disparities in Urban Green Space Accessibility: Where to Intervene? *Landscape and Urban Planning* 102 (4): 234–244. https://doi.org/10.1016/j.landurbplan.2011.05.002.

Dony, C., E. Delmelle, and E.C. Delmelle. 2015. Re-conceptualizing Accessibility to Public Parks in Multi-Modal Cities: A Variable-Width Floating Catchment Area (VFCA) Method. *Landscape and Urban Planning* 143: 90–99.

Duany, A., and E. Plater-Zyberk. 1994. The Neighborhood, the District and the Corridor. In *The New Urbanism: Toward an Architecture of Community*, ed. P. Katz, 17–20. New York: McGraw-Hill.

Ellis, C.D., S.W. Lee, and B. Kweon. 2006. Retail Land Use, Neighborhood Satisfaction and the Urban Forest: An Investigation into the Moderating and Mediating Effects of Trees and Shrubs. *Landscape and Urban Planning* 74: 70–78.

Faden, V.B., and B.I. Graubard. 2000. Maternal Substance Use During Pregnancy and Developmental Outcome at Age Three. *Journal of Substance Abuse* 12: 329–340.

Fan, Y., A. Khattak, and D. Rodriegez. 2011. Household Excess Travel and Neighborhood Characteristics: Associations and Trade-offs. *Urban Studies* 48 (6): 1235–1253. https://doi.org/10.1177/0042098010369712.

Fienberg, S.E. 1980. *The Analysis of Cross-Classified Categorical Data*. 2nd ed. Cambridge, MA: MIT Press. (reprinted by Springer, New York, 2007).

Fjørtoft, I., and J. Sageie. 2000. The Natural Environment as a Playground for Children: Landscape Description and Analyses of a Natural Playscape. *Landscape and Urban Planning* 48: 83–97.

Forjaz, M.J., P. Pedrero, M.-E. Prieto-Flores, F. Rojo-Perez, G. Fernandez-Mayoralas, and V. Rodriguez-Rodriguez. 2015. *Environmental Factors and Life Satisfaction of Community-Dwelling Older Adults*. Presented at 60th Annual Meeting of the Association of American Geographers (AAG), April 21–27, 2015, Chicago.

Francescato, G., S. Weidemann, and J.R. Anderson. 1987. Residential Satisfaction: Its Uses and Limitations in Housing Research. In *Housing and Neighborhoods: Theoretical and Empirical Contributions*, ed. W. van Vliet, H. Choldin, W. Michelson, and D. Popenoe, 43–57. New York: Greenwood Press.

———. 1989. Evaluating the Built Environment from the Users' Point of View: An Attitudinal Model of Residential Satisfaction. In *Building Evaluation*, ed. W.F.E. Preiser, 151–171. New York: Basic Books.

Fried, M. 1982. Residential Attachment: Sources of Residential and Community Satisfaction. *Journal of Social Issues* 38: 107–119.

Fujita, M., and J. Thisse. 2013. *Economics of Agglomeration: Cities, Industrial Location and Regional Growth*. 2nd ed. Cambridge: Cambridge University Press.

Gameroff, M.J. 2005. Using the Proportional Odds Model for Health-Related Outcomes: Why, When, and How with Various SAS Procedures. In *Proceedings of the 30th Annual SAS Users Group International Conference*. Cary: SAS Institute, April 10–13. At: http://www2.sas.com/proceedings/sugi30/205-30.pdf.

Giuliano, G. 1995. The Weakening Transportation-Land Use Connection. *Access* 6: 3–11.

Glaesenera, M.-L., and G. Carusoa. 2015. Neighborhood Green and Services Diversity Effects on Land Prices: Evidence from a Multilevel Hedonic Analysis in Luxembourg. *Landscape and Urban Planning* 143: 100–111.

Gobster, P.H. 1995. Perception and Use of a Metropolitan Greenway System for Recreation. *Landscape and Urban Planning* 33: 401–413.

Good, R.M. 2015. *Landmarks and Legacies: Neighborhood History, Claims to Space, and the Contestation of School Closures in Philadelphia*. Presented at 60th Annual Meeting of the Association of American Geographers (AAG), April 21–27, 2015, Chicago.

Greene, R.P., and J.B. Pick. 2012. *Exploring the Urban Community. A GIS Approach*. 2nd ed. Boston: Prentice Hall.

Greenland, S. 1994. Alternative Models for Ordinal Logistic Regression. *Statistics in Medicine* 13: 1665–1677.

Haslauer, E., E.C. Delmelle, A. Keul, T. Blaschke, and T. Prinz. 2015. Comparing Subjective and Objective Quality of Life Criteria: A Case Study of Green Space and Public Transport in Vienna, Austria. *Social Indicators Research*: 1–17. https://doi.org/10.1007/s11205-014-0810-8.

Hosmer, D.W., and S. Lemeshow. 2000. *Applied Logistic Regression*. 2nd ed. New York: John Wiley and Sons.

Ingelmo, F., M.J. Molina, J.M. de Paz, and F. Visconti. 2011. Soil Saturated Hydraulic Conductivity Assessment from Expert Evaluation of Field Characteristics Using an Ordered Logistic Regression Model. *Soil & Tillage Research* 115–116: 27–38.

Jakle, J.A., and R.L. Mattson. 1981. The Evolution of a Commercial Strip. *Journal of Cultural Geography* 1 (2): 12–25.

Jim, C.Y., and W.Y. Chen. 2006. Recreation-amenity Use and Contingent Valuation of Urban Greenspaces in Guangzhou. China. *Landscape and Urban Planning* 75: 81–96.

Jirovec, R.L., M.M. Jirovec, and R. Bosse. 1984. Environmental Determinants of Neighborhood Satisfaction Among Urban Elderly Men. *The Gerontologist* 24 (3): 261–265.

JLL. 2015. *Logistics Property Report*. At: http://www.jll.de/germany/en-gb/Research/Logistics-Property-Report-Germany-2015-JLL.pdf. Accessed Dec 2015.

Kaplan, S. 1995. The Restorative Benefits of Nature: Toward an Integrative Framework. *Journal of Environmental Psychology* 15: 169–182.

Kaplan, R. 2001. The Nature of the View from Home, Psychological Benefits. *Environment and Behavior* 33 (4): 507–542.

Katz, P. 1994. *The New Urbanism: Toward an Architecture of Community*. New York: McGraw-Hill.

Klinger, T., J.R. Kenworthy, and M. Lanzendorf. 2013. Dimensions of Urban Mobility Cultures—A Comparison of German Cities. *Journal of Transport Geography* 31: 18–29.

Kumanyika, S.K., J.R. Landis, Y.L. Matthews, S.L. Weaver, L.C. Harlan, and W.R. Harlan. 1994. Secular Trends in Blood Pressure Among Adult Blacks and Whites Aged 18–34 Years in Two Body Mass Index Strata, United States, 1960–1980. *American Journal of Epidemiology* 15: 141–154.

Lansing, J.B., and R.W. Marans. 1969. Evaluating Neighborhood Quality. *Journal of the American Institute of Planners* 35: 195–199.

Likert, R. 1932. A Technique for the Measurement of Attitudes. *Archives of Psychology* 22 (140): 55.

Lindsey, G. 1999. Use of Urban Greenways: Insights from Indianapolis. *Landscape and Urban Planning* 45: 145–157.

Lu, M. 1999. Determinants of Residential Satisfaction: Ordered Logit vs. Regression Models. *Growth and Change* 30: 264–287.

Luttik, J. 2000. The Value of Trees, Water and Open Space as Reflected by House Prices in the Netherlands. *Landscape and Urban Planning* 48: 161–167.

MacCallum, R.C., K.F. Widaman, K.J. Preacher, and S. Hong. 2001. Sample Size in Factor Analysis: The Role of Model Error. *Multivariate Behavioral Research 36* (4): 611–637.

Manor, O., S. Matthews, and C. Power. 2000. Dichotomous or Categorical Response? Analysing Self-Rated Health and Lifetime Social Class. *International Journal of Epidemiology* 29: 149–157.

Marans, R.W. 2012. Quality of Urban Life Studies: An Overview and Implications for Environment Behaviour Research. *Procedia—Social and Behavioral Sciences* 35: 9–22.

Matusoka, R.H., and R. Kaplan. 2008. People Needs in the Urban Landscape: Analysis of Landscape and Urban Planning Contributions. *Landscape and Urban Planning* 84: 7–19.

McCullagh, P. 1980. Regression Models for Ordinal Data. *Journal of the Royal Statistical Society, Series B* 42: 109–142.

McCusker, J., C. Bigelow, A.M. Stoddard, and M. Zorn. 1994. Human Immunodeficiency Virus Type 1 Antibody Status and Changes in Risk Behavior Among Drug Users. *Annals of Epidemiology* 4: 466–471.

Meierkord, C., B. Fonkeu, and E. Zumhasch. 2012. Diasporic Englishes in Translocal African Neighborhoods of the Ruhr Metropolis. In *Sociolinguistics Symposium 19*, Freie Universität Berlin, August 21–24, 2012.

Meyers, L.S., G. Gamst, and A.J. Guarino. 2013. *Applied Multivariate Research. Design and Interpretation*. Thousand Oaks: SAGE.

Monzon, A., and A. Rodriguez-Dapena. 2006. Choice of Mode of Transport for Long-Distance Trips: Solving the Problem of Sparse Data. *Transportation Research A* 40: 587–601.

Murad, H.A., S. Fleischman, O. Geyer Sadetzki, and L.S. Freedman. 2003. Small Samples and Ordered Logistic Regression: Does It Help to Collapse Categories of Outcome? *The American Statistician* 57 (3): 155–160.

Muthen, B., and L.K. Muthen. 2000. Integrating Person-Centered and Variable-Centered Analyses: Growth Mixture Modeling with Latent Trajectory Classes. *Alcoholism: Clinical and Experimental Research* 24 (6): 882–891.

Oguz, D. 2000. User Surveys of Ankara's Urban Parks. *Landscape and Urban Planning* 52: 165–171.

Parkes, A., A. Kearns, and R. Atkinson. 2002. What Makes People Dissatisfied with Their Neighbourhoods? *Urban Studies* 39: 2413–2438.

Peterson, B., and F.E. Harrell. 1990. Partial Proportional Odds Models for Ordinal Response Variables. *Applied Statistics* 39: 205–217.

Priego, C., J.-H. Breuste, and J. Rojas. 2008. Perception and Value of Nature in Urban Landscapes: A Comparative Analysis of Cities in Germany, Chile and Spain. *Landscape Online* 7: 1–22.

Rickards, G., C. Magee, and A.R. Artino. 2012. You Can't Fix by Analysis What You've Spoiled by Design: Developing Survey Instruments and Collecting Validity Evidence. *Journal of Graduate Medical Education* 4 (4): 407–410.

Santos, J.R.A., L. Lippke, and P. Pope. 1998. PROC FACTOR: A Tool for Extracting Hidden Gems from a Mountain of Variables. In *Proceedings of the 23rd Annual SAS Users Group International Conference*. Cary: SAS Institute Inc. At: http://www2.sas.com/proceedings/sugi23/Stats/p240.pdf. Accessed Dec 2015.

Shafer, C.S., B.K. Lee, and S. Turner. 2000. A Tale of Three Greenway Trails: User Perceptions Related to Quality of Life. *Landscape and Urban Planning* 49: 163–178.

Sirgy, M.J., and T. Cornwell. 2002. How Neighborhood Features Affect Quality of Life. *Social Indicators Research* 59: 79–114.

Sloane, D., and S.P. Morgan. 1996. An Introduction to Categorical Data Analysis. *Annual Review of Sociology* 22: 351–375.

Stewart, W.-P., D. Liebert, and K.-W. Larkin. 2004. Community Identities as Visions for Landscape Change. *Landscape and Urban Planning* 69: 315–334. https://doi.org/10.1016/j.landurbplan.2003.07.005.

Sullivan, G.M., and A.R. Artino. 2013. Analyzing and Interpreting Data from Likert-Type Scales. *Journal of Graduate Medical Education* 5 (4): 541–542. https://doi.org/10.4300/JGME-5-4-18. At: http://www.ncbi.nlm.nih.gov/pmc/articles/PMC3886444/.

Talen, E., and L. Anselin. 1998. Assessing Spatial Equity: An Evaluation of Measures of Accessibility to Public Playgrounds. *Environment and Planning A* 30 (4): 595–613.

Tiebout, C.M. 1956. A Pure Theory of Local Expenditures. *The Journal of Political Economy* 64 (5): 416–424.

Transforming City Regions. 2015. *Polycentric City Regions in Transformation—The Ruhr Agglomeration in International Perspective, 2015.* Dortmund: TU Dortmund University. At: http://transforming-city regions.com/files/150601_brochure_small.pdf. Accessed Dec 2015.

Tyrväinen, L. 2001. Economic Valuation of Urban Forest Benefits in Finland. *Journal of Environmental Management* 62: 75–92.

U.S. Department of Health and Human Services. (2010). *Healthy People 2020: Physical Activity.* Retrieved from www.healthypeople.gov/2020/TopicsObjectives2020/overview.aspx?topicid=33.

Velicer, W.F., and J.L. Fava. 1998. Effects of Variable and Subject Sampling on Factor Pattern Recovery. *Psychological Methods* 3 (2): 231–251.

CHAPTER 4

City Structure and Spatial Patterns

4.1 Introduction

Urban models have a short history of development due to a need to process large data quantities which only became possible after the computer had been developed (Foot 1981). Methods for analysis of city structure and growth date back to classical models developed by the University of Chicago scholars in the 1920s. With the coming of computer technology beginning in the 1950s and 1960s, complex mathematical models were implemented to understand the development of urban environment, provide forecast, and exercise control. Studying the patterns of land uses within a city is not merely an academic exercise. Thorough understanding of the spatial patterns formed by residential neighborhoods and other land uses of the urban landscape is needed to solve the various urban issues including slum clearance, planning and zoning, or market analysis. This chapter provides a review of urban growth and urban system modeling. It starts with a review of the classic internal city structure models developed by urban sociologists. The chapter will next review urban spatial structure models.

Population and employment suburbanization and improvements in transportation and communication have resulted in de-densification and horizontal expansion of urban areas (Rode et al. 2014) leading to a polycentric urban form in many cities (Gordon et al. 1998; Ingram 1998). Urban form and transport systems are interrelated (Smith 1984; Holtzclaw 2000; Rode et al. 2014), with high-density areas experiencing higher transport

A. Antipova, *Urban Environment, Travel Behavior, Health, and Resident Satisfaction*, https://doi.org/10.1007/978-3-319-74198-7_4

intensity including vehicle miles traveled (VMT) even when other factors are taken into account (Ewing et al. 2008). Other changes in urban structure such as the spatial distribution of activities also result in changes in travel patterns, especially those related to the journey to work. Therefore, studying urban form changes will help to understand changes in travel patterns including commuting.

The chapter concludes with an application of the spatial structure model to the contemporary medium-sized metropolitan area in the southern United States with the purpose of investigating the impact of spatial structure (employment locations) upon commuting. The objective of this study is twofold. First, urban form within the medium-sized Baton Rouge MSA region was observed for two time periods, 1990 and 2000, to determine any change in the spatial distribution of employment location. Second, to understand the impact of metropolitan structure on commuter travel, we investigated the commuting patterns under the monocentric and polycentric assumptions. We utilized two measures of the spatial separation between home and work locations including commute time and length. Commute times and lengths will be measured separately for 1990 and 2000 employment centers and compared for both time periods using the t-test of means.

4.2 Internal City Structure Models

Early urban models dealt primarily with city structure examining spatial forms or distributions (therefore, the general name of "internal city structure models").

The early generation of urban models includes concentric zone theory (Burgess 1967), sector theory (Hoyt 1939), and multi-nuclei theory (Harris and Ullman 1945). Urban system modeling, including residential and retail location model by Lowry (1964) and Harris and Wilson's (1978) retail model applicable to a wide range of systems of interest, uses the theories from regional science and urban economics to analyze urban structure, development, and decline (Klosterman 1994).

Internal structure of a city reveals "the physical body of the city in one sweeping view at one period of time" (Hoyt 1939, p. 96). Various components of the city include central and outlying business districts, residential areas, factories, blighted areas, and so on. At any given period of time they occupy specific locations within the city. Location of the business center, industrial areas, commercial districts, residential zones, parks, and other components of urban structure can be generalized though no city can be fit perfectly into an ideal spatial model due to various reasons such as rivers, transportation lines, and so on.

4.2.1 Concentric (Ring) Model

An investigation of the *internal structure* of urban areas defined as a "the present pattern of land use within the city" (Harris and Ullman 1945, p. 12) was seen "as not merely an academic matter," but as a practical way to solve vital urban problems including the selection of areas for slum clearance, mortgage lending policy determination, zoning decisions, and other problems (Hoyt 1939). Ernest Burgess (1925) was among the first sociologists who studied urban growth as a process in his "The Growth of the City."

Burgess developed his *concentric zone model* out of concern over an unprecedented expansion in area of the largest cities in the United States. He demonstrated the typical expansion of the city by a series of *numbered concentric circles* to designate the successive urban extension zones and the types of areas formed in the process. *Expansion* addresses both aspects of extension, the physical growth of the city and the extension of the technical services (such as communal public utilities: water system, electricity, and the like that make city life comfortable), while the aspect of expansion, *succession*, is defined as "the tendency of each inner zone to extend its area by the invasion of the next outer zone" (Burgess 1967, pp. 50–53). Any city tends to expand radially from the city center.

The five successive rings or zones include zone I: the Loop (Central Business District or CBD), zone II: zone in transition (residential deterioration), zone III: workingmen's homes, zone IV: residential zone, and zone V: commuters' zone. The rings are differentiated not only spatially but also by the type of areas. All outer and inner transportation converged in Zone I making it the most accessible. Apart from transportation, the CBD is a place of the concentration of commercial, social, and civic land uses containing department stores and shops, skyscraper office buildings, banks, entertainment facilities, hotels, the city hall, and organization headquarters. Zone II (industrial, or zone in transition) includes businesses and light manufacturing encroaching on the homes of the poorest immigrants. This is a zone of deterioration: poverty, disease, degradation and slums are concentrated in many American cities within the zones in transition. Hoyt described the zone of transition as "a breeding place for vice and crime" (1939, p. 107). Zone III is where workers in industries live; former residents of zone II would filter up to better houses and workers who had jobs in the industrial zone wanted to reside close to work. Often the second-generation immigrants who accumulated enough resources moved to zone III to escape from the blighted transition zone. High-class apartment buildings and single-family dwellings are found in zone IV. This zone of better residencies consists of exclusive "restricted districts." The most expensive zone V, "the best families" housing is located in the commuters' zone beyond the city limits. Suburban areas or satellite cities of the zone are within a 30- to 60-minute ride of the CBD; they consist of fragmented high-class housing development along rapid travel lines.

The general process of expansion in urban growth involves concentration (many city functions are concentrated and can be found in the CBD being the center of the economic, cultural, and political life) and decentralization (business sub-centers, "satellite loops," have grown up in outlying zones representing larger economic unities). A complex process re-organizes urban space into a centralized decentralized system where local communities merge into business sub-centers still dominated by the central business district. The process of centralized decentralization is reflected in the development of the chain store.

4.2.2 Limitations of Concentric Circle Theory

Concentric circle theory has certain limitations. The theory offers an ideal pattern of land uses. However, its pattern is never found in a real city where urban communities have many variations from this model. For example, the concentric circle theory stated that "In the zone of deterioration encircling the central business section are always to be found the so-called" "slums" and "bad lands" with their submerged regions of poverty, degradation, and disease, and their underworlds of crime and vice" (Burgess 1967, p. 54). However, the transition zone is not in every city a zone of deterioration with slums as described in the model (Harris and Ullman 1945); in fact, high-rise, high-class apartment development with upscale shopping and dining is found near the city core on the Gold Coast of Chicago and on Park Avenue in New York City, which includes the city's tallest apartment building completed in 2015 (Fig. 4.1).

The theory was soon questioned: the boundaries of the zones in a city are hard to demarcate with zones fading one into the other (Fisher 1930). This is especially true of the retail zone, which in the real world follows

Fig. 4.1 High-rise residential and commercial district at Lake Street, Chicago. (Author 2013)

population, and unless public or private restrictions exist, retail uses can be found in any zone. Advances in transportation and technology made industrial land-use pattern very different from the original concentric zone pattern, which implies that heavy industries encircle the CBD. Examination of the present pattern of heavy manufacturing reveals no support for the theory, with the industrial land uses often following transportation lines and occupying big tracts of land on the outskirts of urban areas while not forming a concentric circle around the CBD. Big variations in the patterns of industrial land uses among cities render the establishment of a general industrial pattern impossible. Another variation from the pattern suggested by the concentric model is the emergence in large cities of subcenters, which form as the neighborhood stores initiate the growth process around those similar to that around the main city's center (Fisher 1930, p. 126) and the independent formation of satellite business centers developing beyond the CBD, or on the city's periphery and located at or near transportation routes (Hoyt 1939). Nevertheless, despite its limitations, the concentric model of the internal city structure is a useful starting point for studying a pattern of land uses within a city.

4.2.3 Sector (Wedge) Model

Another model of urban land use explaining inner structure and growth of urban areas by *axial development* was developed by a land economist, Homer Hoyt. His work, *The Structure and Growth of Residential Neighborhoods in American Cities*, commissioned by the US Federal Housing Administration (FHA), was published in 1939. Using real property surveys available for 203 American cities (64 cities have been analyzed in the study), Dr. Hoyt and the officials at the Economics and Statistics Division of the FHA conducted a study to develop analysis tools and general principles to apply to examine the internal nature and growth of American cities to understand the causes of land-use arrangement and residential neighborhoods.

First, Hoyt examined the city structure and delineated the urban area *as a single unit*, that is, the extent and shape of the settled area as a whole (rectangular, circular, or star-shaped). Second, the component parts of a city were examined to formulate a generalization of land-use patterns and spatial relationships between commercial and industrial land uses as well as to segregate residential areas. At this stage, Hoyt studied whether dwelling units are segregated into sections or mixed with factories, office buildings, and stores into a "chaotic jumble of land uses." Third, using essential housing characteristics at the city-block level, Hoyt distinguished and analyzed the distribution of the several types of residential areas such as high, low, and intermediate residential rental areas within the city. To reveal housing conditions, factors deemed most pertinent to the quality of housing have been used, including the average rent for the block, percentage of aged residential structures (defined as structures 35 years old and older), owner-occupancy, structures in need of major repairs or unfit for occupancy, structures having no private bath or lacking central heat, persons living in the block who are non-White, and a percentage of *overcrowded dwelling units* (more than one person to a room).

4.2.3.1 Delineating Rent Areas

The main question to answer was whether buildings with certain characteristics mixed together in the same block or were segregated in certain parts of the city. To illustrate the idea, to delineate the areas where the most adverse housing conditions prevail, a series of special transparent maps have been produced and a series of patterns have been superimposed on each other. An area where several characteristics overlapped was designated the worst housing area. The technique is applicable in cities where the basic housing data is available. As an alternative to the described technique, Hoyt suggested using the average monthly home rent in a block as a reliable factor that represents other housing factors and thus can be used to analyze cities in which no real property surveys have been conducted and comprehensive data are lacking.

Rent was not used as an economic concept; it was determined for each unit as the price paid by tenants to use a structure for dwelling. Rental data is important to retailers to analyze the market for retail sales, mortgage investors to determine the risk in residential mortgages, as well as public welfare agencies and scholars of human ecology. Although many factors impact rental patterns, including topographical features, rates of urban growth, and industrial and transportation location within the city, a non-random pattern applicable to all cities could be observed that does not follow the successive concentric circle distribution or does not occur in the shape of abruptly changing rectangular areas. Instead, rent areas seem to conform to a sector pattern where areas of high rent form clusters with areas of lower rent surrounding the high-rent poles; the pattern repeats itself with a downward progression of rents in all directions, finally reaching the worst slums.

4.2.3.2 Rent Areas Are Generalizable

Although location, size, and shape of rental areas vary across the cities, it is possible to make some generalizations. The limited-extent high-rent zones occupy one or more sectors and are located on the one side of the city. The high-rent areas take the form of wedges radiating in sectors along radial lines from the center to the periphery. In many cities, high-rent poles are usually surrounded by the intermediate rental areas located in the same sectors as the high rental areas, while the low rent wedges have much larger extent and are located on the opposite side of the city extending from the center to the periphery.

Finally, *city growth* is examined, that is the movement of land uses and neighborhoods over a period of time in a city. For this purpose, factors influencing city growth have been considered, and the form of growth of the entire city has been studied observing the location and direction of growth and shape of settled area, that is, whether settlement occurs in concentric circles, axial lines, or independent nuclei. In his analysis of the growth of cities in order to trace the movement of each type of land use or occupancy, Hoyt used different periods of time (i.e., successive 15- or 20-year intervals) to understand whether business, commercial, and residential uses change their location in the process of urban growth. Using the data from a number of the largest cities, Hoyt made some generalizations and concluded that urban growth emanates from the city center (the CBD) and occurs along transportation routes as a series of sectors, not rings (therefore the name, "the sector model"). The land is used most intensively at the heart of the city where it has the highest value and therefore tall structures dominate the urban landscape at the center of the city. The sectors, or wedges, consist of similar land-use types such as industry, high-class residential, low-class residential, and middle-class residential. As sectors grow they move outward from the origin at the center toward the periphery.

Some land uses are seen as non-compatible with each other, for example, high-rent residential sector extends from the CBD to the edge and occupies the best high-lying well-drained land with natural amenities such as clean air, green areas, away from the industrial sector, along fast transportation lines, while low-quality residential housing of low-income class adjoins industrial land. Proximity of low-class residential land to industry can be explained by the desire to reduce transportation costs (since many residents work in nearby factories) as well as land made cheap due to pollution, traffic, noise, and other negative environmental conditions produced by industry. Middle class has residences in between but farther from industry and pollution.

4.2.4 *Monocentric Urban Structure*

The first two models (concentric circles and sector models) described previously represent monocentric structure of urban areas. According to the monocentric theory, land uses develop around a single center. At the core business center retail and financial uses have the highest land value. The introduction of new technology, such as steel-frame construction and steam and the electric elevator in the 1880s, enabled skyscraper construction and in turn led to a vertical expansion of the business center changing the skyline of the big cities. Structures located on land uses that can be used intensively grew vertically since a multistory building can yield higher rent. Hoyt (1939) observed land-use intensification of the downtown area in Chicago as the height of the structures increased from six stories in 1880 to 44 stories, with the large buildings having a capacity of housing thousands of workers in a single block (p. 107). Manhattan in New York had a similar vertical growth with the buildings having 80 and 100 stories (Figs. 4.2 and 4.3).

Urban environments are not static. As the city grows and new elements are introduced (e.g., new industries in turn attract more population), the structure does not remain rigid. With time, some areas in the city may retain their character, while other existing land uses may shift from a certain pattern established during the previous period transitioning to another type. For example, an established residential area may experience an influx of new arrivals and accordingly change in character from a fashionable area to a slum, and a new housing development may occur on the edge of the city. The filtering process allows newcomers to move in and occupy the vacant homes in old neighborhoods after the old residents abandon their homes and move out to new ones.

Writing in the late 1930s, Hoyt (1939), as Burgess (1925) did before him, explained the change in land-use patterns from the central point toward the edge of the city. Under the pressure of population increase, retail and financial land uses in the central business core encroach upon the land uses of less intensity, and these in turn push on to other land-use zones. For example, in Chicago in the early twentieth century, both the retail commercial and industrial uses expanded onto the area that used to be residential forcing residencies out of the business center of the city.

The speed with which high-grade neighborhoods move to new locations varies from place to place. Although many local factors determine the rate of neighborhood change and accordingly the rate of change in the internal structure of a city, including residential characteristics, type of

Fig. 4.2 Vertical expansion in New York. The High-Rise Crowne Plaza Hotel at Times Square. The iconic Hershey' Chocolate World closed its 15-year long location on Broadway in September 2017. (Author 2013)

Fig. 4.3 High-Rise Apartment Building (at the right) at Millennium Park, Chicago. (Author 2015)

buildings, and so on, the rate of population growth is one of the most important determinants. To illustrate, the 450% increase of the population of Detroit between 1900 and 1930, caused by the rise of the automobile industry, coincided with the relocation of the high-class residential neighborhoods initially located near the business center some 7–10 miles away from Detroit's downtown (Hoyt 1939, p. 82). Specifically, the population of Chicago grew from 3820 in 1836 to 93,000 in 1857, quadrupling from1857 to 1873, then tripling from 1874 to 1899 (Hoyt 1939, p. 82). Accompanying this population explosion was a sustained expansion and outward movement of high-grade fashionable neighborhoods farther from Chicago's downtown ("The Loop"). As homes are left behind by the previous owners, they deteriorate in quality while becoming affordable for lower-income newcomers.

4.2.5 Polycentric Urban Structure. Multiple Nuclei Model

Urban environment evolves with time. For example, as city population grows, residential neighborhoods may change, and the simple monocentric structure of a city (with a single business center) is replaced with a more complex pattern. An employer may relocate to a cheaper location. Stores may expand due to an increase in population and new business centers may form beyond the CBD. With rapid population growth on the periphery, satellite business sub-centers develop beyond the core of the original business center. To reflect the changing city structure, models have been developed acknowledging that urban areas may evolve around not a single but several centers. These models describe polycentric urban structure.

Multiple nuclei model was developed by Harris and Ullmann in 1945 in their *The Nature of Cities* to describe internal structure of cities. Both urban geographers, Edward L. Ullman and Chauncy D. Harris, were on active duty during the writing of their paper in 1945.

Despite uniqueness found in each city, their internal structure is generalizable, especially when city size, function, and regional setting are similar. The land uses in many cities develop not around a single node as the monocentric models described suggests, but several centers or nuclei exist, which in some cases date back to the city origin. For example, metropolitan London developed around at least two discrete centers, "The City" representing the focus of financial and commercial activity, and Westminster, the political center. These nuclei separated by open country existed as long as the city itself. Yet in other cities, multiple centers develop as the city is growing due to the processes of migration and specialization as certain land users may migrate out. For example, in Chicago, heavy industry initially occupied the heart of the city by the Chicago River but later moved southeast to the Calumet District around which extensive new urban development took place.

Depending on a city's function, the initial nucleus might start with a single retail district such as the CBD in a city serving as a central place, around the port or rail facility in a transportation city, or a resource-related facility such as mine or a beach in a specialized city, around which differentiated districts may form.

Over time, separate distinct centers evolve due to four reasons:

1. Some activities have specific location requirements and can only operate where particular facilities are available, such as transporta-

tion access to decrease transportation costs. Retail activities require ease of access to its customers so it is located at the accessible, easy to reach locations. Manufacturing, on the other hand, requires good connection via road, rail, or water network and large tracts of land, while the port needs access to water. The CBD locates at the point of maximum accessibility (based on 1945 Chicago, the CBD area was where all modes of transportation, including car, subway, elevated, and bus, converged making it the point of the most convenient access from any point with the highest land values.

2. Similar activities (not necessarily from the same industry) tend to co-locate. An example is clothing-related design and production facilities, department stores selling ready-made clothes, fashion advertising, showrooms, fashion schools, fabric suppliers, warehouses, zipper and button suppliers, textile manufacturing all benefitting from cohesion in the Garment District in New York City. Generally, similar shops tend to locate closely so that their potential customers can easily compare goods and prices. Buildings in a financial district need elevators due to their need to access other offices.
3. Some activities repel each other such as high-income residential and noise-, odor-, pollution-emitting heavy manufacturing with the former preferring well-drained land with clean air, and the latter requiring much space uninterrupted by residential streets making it locate away from the main city center such as the Calumet District in Chicago.
4. In some cases, economic activities locate in less desirable places due to inability to pay the high rents in the more desirable places of the city (e.g., storage facilities require a lot of cheap space accessible by motor trucks).

The number of nuclei depends on the size of the city. In the larger city, the nuclei are more numerous and specialized. The nuclei are also shaped by historical development and localization forces.

Most US cities can also be generalized in terms of the *land-use pattern* developing around their centers (nuclei). All cities tend to have a CBD (not necessarily at the city center, which contains retail, financial, and government activities), wholesale and light manufacturing activities (which are often located adjacent to the CBD), the heavy industrial district (it forms on the outer edge if the city), and residential areas (often including members of the same ethnicity or economic group, thus high-income and low-income housing tend to occupy opposite sides of the city).

4.3 Urban System Models

Urban models have traditionally been used in public and commercial planning due to their forecasting and optimization capabilities. Although long-time forecasting of structural variables may be impossible due to path dependence and phase transitions, the identification of desirable or undesirable phase transitions is an important capability so that appropriate actions can be taken either to encourage or avoid these transitions (Wilson 2009).

Previous urban models focus largely on *urban morphology* describing spatial forms or distributions (therefore, the general name "internal city structure models") while concentrating less on linkages between the elements of urban system, that is, socioeconomic interactions between population and economic activities (such as shopping trips between residencies and retail outlets, or journey-to-work movement between home and work places), or among economic activities (such as commodity flows). However, these spatial interactions are essential to proper city functioning.

During the 1950s, urban planners used computer technology rather than ad hoc planning to model urban land use and transportation to make planning more rational. The models wanted to understand how urban development occurs so that the future of cities could be forecasted and regulated (Wegener 1994).

The German economists developed *location theory*, that is, how geography affects economic development. Johann-Heinrich von Thünen, in "The Isolated State" (1826), developed the classical *agricultural location theory* based on an inverse relationship between land rent and transport cost, and used rings for different agricultural crops with different land intensity, where the ring producing the highest-value crops is located closer to the market. Alfred Weber (1909), in his "Theory of the Location of Industries," worked on the *manufacturing location* and used a concept of *least cost location* (the principle suggests that a manufacturing plant locates so as to minimize total transport costs, that is, assembling raw materials costs and distribution costs associated with shipping the finished product to the market), followed by Walter Christaller's "Central Places in Southern Germany" (originally published in 1933 and in English translation in 1966), with treatments of the size, spacing, and functions of service centers (cities). Another German scholar, August Loesch, developed a similar theory to Christaller's *theory of urban location* in "The Economics of Location" (1954), where he introduced a principle of *maximum market access.*

Economists in the United States wanted to measure economic activity (that is, employment) at a fine scale such as a region or city. For example, what factors shape business location decisions, why a certain industry is located in a certain place, and so on. Walter Isard, an American economist and a founder of the field of regional science, reviewed the principles of economic location theory (both least cost and maximum market costs were considered) in his "Location and Space Economy" (1956) related to the distribution of manufacturing activity, while William Allonso (1964) contributed by developing intrametropolitan location theory in his "Location and Land Use."

Ira Lowry initiated modeling of urban form represented by *location (land use) models* in the "Model of Metropolis" (1964) where he developed a *location model* of the spatial organization of human activities within a big metropolis using empirical data on Pittsburg, PA (described later in the chapter). A *location model* is guided by economic location theory and social physics. According to the economic location theory, business enterprises and households each make rational choices regarding location with the considerations of alternatives weighed against each other. The field of social physics studies empirical laws that govern social interactions and aggregate "mass" behavior while disregarding individual behavior. In a set of a group of actors (such as people, firms), the various relationships (termed "interactions") can be described using an equivalent to the law of gravity by Newton: the amount of interaction is directly proportional to the size of the groups of actors and inversely proportional to the distance between them. The gravity-based equivalent reliably estimates the aggregate outcome of individual actions; however, once the environment and various variables are modified, the gravity-based model may run out of date (Hoyt 1939). To illustrate, it cannot be used to test the impact of radical changes such as standards of living, since the parameters were fitted based on current data. To find the description of the work of social physicists, the reader is referred to George Kingsley Zipf (1949) and his publication, "Human Behavior and the Principle of Least Effort." However, a carefully constructed location model using a rational choice principle is less likely to become obsolescent. Lowry, using his words, "*I have followed the guidance of the social physicists more than that of the location theorists. ...a gravity model promised to be much easier to build and cheaper to operate*" (Lowry 1964, p. 23).

In general, urban models relate many essential activities, including urban land use, housing, population, urban networks (transportation, communications and utility), employment, workplaces, goods transport, and travel (Wegener 1994). However, despite the efforts made to explain how cities evolve, our understanding is still limited, including the question of what is the "optimal" size for a city (Batty 2008). A slowly developing theoretical foundation of urban evolution comes from several disciplines, including urban economics, transportation behavior, and recently, from network science, allometric growth, and fractal geometry. The recent theory of cities as complex hierarchical systems with human–natural interactions addresses important issues such as density, compactness, sprawl, and sustainability.

Contemporary urban models use computer technology, are comprehensive, and *operational* (that is, they can be applied towards a wide range of urban issues and understood by many professionals). A computer is used to run complex mathematical models that use economic theory and mathematical programming (Klosterman 1994) to analyze and predict the development of urban systems. For this purpose, the models need to be comprehensive, that is, they need to integrate the most basic processes of urban development. Comprehensive models include various types of urban land use (that is, residential, industrial, and commercial), and transportation. In contrast to comprehensive models, partial models address a certain part of the urban system, for example, transportation, housing, or retail. Contemporary spatial interaction models have been fully disaggregated and tested in different circumstances (Wilson 2009).

4.3.1 Urban Systems Modelling

In late 1960s, the powerful general methods of statistical mechanics were used in urban analysis to model transport flows in cities (Wilson 1967) deriving a family of spatial interaction models and the representation of flows in transport models (Wilson 1970). In late 1970s, the locational analysis was extended to model the evolution of structures enabling modeling of the evolution of cities, using retail outlets models as an archetypal model (Harris and Wilson 1978).

Besides theoretical contributions towards understanding of the distribution of towns and cities (locational order in the urban system), urban geographers, starting in 1970, have developed principles and theoretical foundations of interaction among urban components explaining many aspects of urban interconnectedness and thus explaining organizational order in the urban system. This section will focus on the latter, urban interactions.

An analytical approach to studying basic relationships within complex urban systems involves isolating individual components of urban system and its linkages and then analyzing each of these disaggregate constituent elements and linkages, and thus, is hugely impractical (Clark 1982). One focus of urban system approach involved understanding the factors guiding the spatial distribution of cities and towns and how they combined determine urban location (locational order in the system). Another focus of attention was on understanding of urban interactions (such as the various contacts among cities) and the constraints limiting urban interactions (organizational order in the system). Due to the impracticability of the analytical approach above, many urban geographers in the 1970s–1980s instead modeled urban systems at the macro (aggregate) scale of analysis, with the system as a whole becoming the focus of attention (Clark 1982), thus applying a *macroeconomic* approach.

In systems modeling, a *system* is understood as a set of components that can assume alternative states. For example, a city comprises a set of land parcels assigned to different purposes or uses, such as residential or industrial. The range of alternative states is determined by the number of options available, among other factors. For example, it can be small such as when the land can be used for only these two functions, however, as choices increase, the range does too, accompanied by the increasing complexity of the entire urban system. Given availability of a number of alternative states, one configuration of alternative states is likely to overwhelm all the others, that is, be dominant and represent *the most likely state* of the system. To demonstrate the idea, if a city consists of just two plots of land (residential and industrial), all possible combinations of states include three different ways (two plots are residential, one residential and another industrial, and two industrial), while within a more complex system such as the one consisting of four plots of land, the range of possible alternative states increases to 16, and when in a city with six plots of land there are 64 different land-use configurations, with a certain configuration of states being dominant reflected in the highest percentage. This important system property, that is, the most likely state of the urban configuration, is used in urban system modeling to identify this dominant configuration using the macro characteristics of the entire system, and not information on individual components and links.

The following sections review Lowry's residential and retail model as an example of location modeling. Then, Harris and Wilson's (1978) retail location model is illustrated as an archetypal model used to represent urban systems of interest. The retail model can be extended to model the evolution of urban structure. This is followed by description and examples of the urban systems modeling approach and applications, and finally, a classification of urban subsystems.

4.3.1.1 Lowry's Residential and Retail Location Model (1964)

Since the 1950s, the research focused on the simulation of complex systems including physical, social, and urban. Ira Lowry worked on the spatial structure of the metropolitan area and assumed there should be some consistency in the spatial arrangement of human activities. Due to the complex nature of the relationships between these activities, he attempted to make these relationships simplified and generalized through development of the quantitative model while making it sufficiently spatially detailed for its use in land-use and transportation planning.

The model was designed to simulate the real-world household and retail employment behavior given certain circumstances, some within and some outside their control. Lowry's model intended to serve as a tool for planners and provide practical assistance in comprehensive planning and decision-making, especially within the context of testing the impact of public decisions and changes in key variables (e.g., basic employment, transportation efficiency, population) on a metropolitan form. The examples of public decisions include urban renewal policies, taxation, control of land use, and transportation investments.

The *location model* was designed to assign urban activities occupying space (that is, basic and retail economic activities, and residential households, or population) to sub-zones (that is, basic building blocks of the analysis represented by one-mile square tracts) of a bounded region using the locational interdependence principles. *Basic activities* are treated as *exogenous* (that is, they are "given": the activities do not depend on the local economy and are not limited by labor access in local site-selection). Examples of *basic activities* include industrial (e.g., steel mills), business firms, and administrative industries. These are known as "export" activities as they are dealing with non-local clients. In contrast, *retail activities* consist of business and administrative establishments (e.g., local government agencies, schools, retail facilities and services such as department stores, gasoline dealers, services such as repair, professional, financial institutions, etc.) that are dealing with the local residential clients, and as such their site selection is impacted by labor access while employment is associated with local population growth. The locations and employment of the retail sector are assumed to be *endogenous*, that is, the model determines the values. Finally, household sector consists of residential population whose site selection is impacted by the location of resident's workplace, and therefore, the number and locations of households are determined within the model also. The entire region's population (the number of households) is proportional to total employment (in other words, it is determined by the number of jobs available at any given time in basic and retail sector), while at the tract level it is determined by that tract's accessibility to jobs (measured as a sum of total employment of the surrounding tracts weighted by distance to those tracts due to distance imposing costs on social interactions) constrained by land availability for residential uses. Residential population in the surrounding tracts directly impacts retail employment of each type within a tract (a number of persons employed in each line of retail sector). Distance term in the accessibility formula is measured by airline distance between pairs of points of interest (such as home and work, or home and retail store) ,which turned out to be accurate representations of time en route and over-the-road distance.

The model uses *inputs* such as assumed or actual data on the regional distribution of basic jobs, the amount of space occupied by basic activities, and land-use constraints due to the physical geography of sites, and legal constraints (e.g., zoning laws) for each one-mile tract to estimate retail employment, residential population (i.e., household number), and land use for sub-areas (mile-square tracts).

The model uses several *constraints* such as a minimum-size retail employment requirement imposed on the employment distribution for each type of retail represented as employment size (the constraint effectively controls density). If a location has the computed market potential below the minimum employment size, then the potential retail employees are assigned to another location. To illustrate, a tract is not allowed to have a three-man department store. Each tract's *market potential* is calculated as a linear combination of the number of employment and the number of population of the tract itself and surrounding tracts weighted by distance so that the contribution of a far-away household is much less significant to market potential than that of a household nearby. Clusters are allowed to be formed. So, facilities in the neighborhood cluster have a requirement of 30 minimum employees in total, and local and metropolitan clusters have 250 and 20,000 minimum employee size requirements (Lowry 1964, p. 72), respectively. High-accessibility locations in the vicinity of major employment centers may have very high population densities (a maximum number of households permitted per 1000 square feet of residential land), so a maximum density constraint is included to avoid extreme densities (surplus households are forced to settle elsewhere in the model). The value is set to 1.5 households per 1000 square feet, normally attained in a four- to six-unit apartment building on a residential land (to put this constraint into a context, in Pittsburg in the late 1950s, only the CBD area exceeded this constraint). The third constraint relates to retail land, which should not exceed the total amount of available land within each square-mile tract.

Mathematical equations were developed from traffic data (such as trip distribution for work and shop purposes) and relate residential places to workplaces and retail activity locations to the accessible consumer market. The model makes the calculations of the *potential* of a location to be a residential or retail site given the distribution of pre-existing basic workplaces. That is, given the distribution of basic activity such as steel mills and residential land available within a mile-square tract, a residential population can be allocated around the work clusters supplying it with labor. In those cases when all the allocated population cannot be accommodated within tracts surrounding major employment centers due to space restrictions (as most land is non-residential), other tracts are re-assigned the population overflow.

Through the iterative process, the residential locations of retail and service workers are identified whereby residential population is re-distributed (since retail land is prioritized over residential uses in the model) and market potentials are re-calculated given the restrictions of land availability, operational scales at which firms are most efficient, and maximum density for residential population. So, the residential population and retail employment are spatially distributed according to these potentials until a stable co-distribution of employment and population is attained given the constraints above (Lowry 1964). In other words, an equilibrium is achieved in terms of the land-use pattern and the distributions of population and employment.

The model has a series of simultaneous equations; an equilibrium is achieved in the land-use pattern and employment and retail distribution. Thus, given data on amounts and distribution of basic employment and basic land use, appropriate amount of residential population (the number of households) and retail employment are determined within the model. Residential population of an entire region is computed proportional to total employment, and residential population of a tract is computed proportional to that tract's accessibility to employment. The total population of the region must be equal to the sum of the tracts' population.

Next, retail workers and residential population are spatially distributed for each sub-area and land is assigned for residential and retail uses. Based on basic employment patterns, the model outputs residential population (that is, the number of households) and residential land use, retail employment, and retail land use spatially distributed. The model produced the spatial outputs that were highly similar to the existing residential and retail pattern. The model is also sensitive to variation in its parameters.

In generating *residential population* one issue is related to the *boundary problem*: not all employed workers reside in the study area, and some employed residents work outside the area. The number of residential population is a function of total employment and is determined inside the model with the function corresponding to a *labor force participation* rate (i.e., the average number of workers per household). Thus, to generate the appropriate number of households (that is, residential population), a labor force participation (LFP) rate is used. The limits for the LFP rate are calculated as a ratio of total jobs and resident households, or a ratio of employed residents and resident households. The chosen rate is applied to the existing total employment by dividing it by the LFP rate to arrive at the number of households. The number may or may not be the same as the actually located in the study area reported by the Census. In fact, the *net inflow of commuters* from surrounding areas is consistent with the model since workers can settle outside the boundaries. The locations of households are generated endogenously and not limited by the boundaries of the study area, noted previously.

Once residential population is allocated, the locations for local population-oriented activities are identified using the information on the market potential of each location (that is, access to customers), allowing to distribute retail employment including elementary schools, department stores, and so on, according to these potentials. Clusters of workplaces reveal a hierarchy of grouping of local population-serving activities: the neighborhood (the lowest level) cluster includes food, drug, gasoline retailers, and elementary schools, while the higher level "local" cluster has more retail firms, repair and professional services, and churches, and "metropolitan" clusters include mostly financial facilities, large department stores, producer services, and so on. Retail employment is allocated for each sub-area (mile-square tract) by type of cluster in proportion to the tract's market potential. To accommodate these retail establishments, space-use (not floor space) in each sub-area is also allocated for each retail cluster type, that is, neighborhood-level facilities have 1900 square feet per employee of site-area, and the local and metropolitan clusters have 1300 and 80 square feet, respectively (Lowry 1964, p. 79).

Retail behavior (that is, the choice of retail location) is guided by the *profit-maximization principle*: the individual firm attempts to locate in the place of the maximum attraction of patronage. The *shopping behavior* (the amount of customers attracted to a tract with at least one retail facility) is determined by the spatial distributions of population and employment with respect to both this tract and all other retail-containing tracts, thus including both competition and distance as factors affecting the probability of selecting a specific tract for shopping to the model. The model does not impose a capacity constraint on the locations of retail (that is, the number of retail and service facilities is not specified, only aggregate employment in each type of retail within tracts), thus allowing for a clustering of retail facilities even in locations already filled with basic establishments (as long as that location's market potential value is great enough) since the model does not impose a maximum-density constraint on the retail distribution. This implies mixed uses such as retail and wholesale/office, say, on the ground floor and upper stories, respectively, in multistory buildings, and no ceiling on the commercial building height is given.

Residential household behavior (the choice of residential location) is guided by factors of *commuting cost* (based on a particular job location, not necessarily proximity to the greatest job concentration) and *site rent* (adequate domestic space). To describe the relationship between residencies and workplaces, the workplace/residence co-distribution was estimated directly by using a single trip-distribution function capturing all occupations. The population (household) distribution generated by the model is converted to a map of residential density that has many potential applications, such as in social area analysis. To illustrate, Lowry argues for a high correlation between net residential density, income per capita, and distribution of population by age, that is, net density may be predicted by income and age distributions, so that *social characteristics* of the residents living in the various tracts of the density map can be specified.

A locational equilibrium emerges under the conditions of competition in the market for urban land. The *model fitting* in Lowry's words, "… involved a great deal of trial and error," with five major revisions between the first run and the final version, modifications of structural parameters, input data treatments, and so on. For example, colleges and hospitals were shifted between basic and retail activities (Lowry 1964, p. 55). Regarding the model's *validation*, there are the discrepancies between model and reality, for example, residential population generated endogenously, that is, it was determined by the model and differs significantly from the official reported numbers at least in some areas of the study area. Similarly, the model's retail estimates are close to the inventory numbers in some areas and divergent in others, likely due to the boundary problem and the imperfect model's measure of accessibility. Regarding land-use pattern classified as unusable, retail, basic, and residential, the model-generated residential densities are lower than the actual reported ones because the model does not allow for vacant land, with all available land within tracts assigned to the tract's households. The total retail space distributed by the model exceeds by 8% that listed by the inventory probably due to some institutional uses misclassified initially as "retail" when in fact they are "basic."

The residential location pattern generated by the model corresponds spatially to the workplace distribution much more closely than the actual residential pattern suggesting the possible future residential location pattern. Comparison of the actual inventory population numbers and those model-generated can be informative and serve a useful planning tool, especially in the core area. The model's output of residential population may identify areas with population deficiencies where appropriate corrective measures may be taken, making it all the more an important technique given the sprawl-containment ideal for the compact sustainable development. Similarly, the map of market potentials generated by the model can identify locations in terms of their accessibility for residentially based shopping trips. So, Pittsburgh's downtown area (the Golden Triangle) is not the most accessible in terms of its market potential values, at least at the time of this book's publication, with more retail facilities predicted to move to smaller shopping centers with local instead of metropolitan orientation.

In sum, Lowry's locational model uses the inputs of basic employment (given exogenously for each tract, and measured in number of persons) and basic land use (the amount of land used by basic firms, also given exogenously, measured in thousand square feet) and produces values for retail employment and residential population (in number of households) for each sub-zone (tract) within the metropolitan area, as well as assigns land for each use in each tract (that is, the amount of land available for basic sector, retail, and households). The model simulates household and firm locational behavior given that some factors are pre-determined and thus outside of their control, while others are within their control (Lowry 1964). Despite the model's shortcomings, designed as a tool for metropolitan and regional planning, it is a useful model to help understand issues of spatial organization of a metropolitan area and its temporal development.

4.3.2 Early Urban Models Limitations

Early urban models have been criticized due to the poor theoretical foundations the models used (Lee 1973) and unreliable and inefficient techniques used for calibrating the value of the parameters of the equations since the model results fluctuated from empirical observations. However, since then the models have improved regarding both the theories and empirical techniques (Klosterman 1994; Wegener 1994). More advanced theories explain spatial choice behavior while empirical techniques used to calibrate spatial choice models predict results that correspond well with real observations with models becoming more realistic (Klosterman 1994). These advances led to a broad agreement about state-of-the-art urban models (Wegener 1994).

State-of-the-art urban models explain the behavior of different urban actors or decision-makers (households, firms, travelers) with some models using random utility or discrete choice theory, while other models build upon the microeconomic theory, in particular on Alonso's (1964) theory of urban land markets or bid-rent theory, and still other models are hybrids of Alonso's bid-rent and random utility theory. Random utility or discrete choice models forecast choices between several alternatives as a function of their attributes, conditional on constraints that account for unobserved attributes, differences in taste between different actors, or lack of information. Regarding the level of analysis, state-of-the-art urban models apply both aggregate and disaggregate modeling techniques. The aggregate approach utilizes inputs and outputs at a meso level, that is, households and industries are aggregated at the level of medium-sized zones. Disaggregate modeling uses microsimulation techniques whereby input data are disaggregated. Inputs include samples of disaggregate data of individual households, or Geographic Information System (GIS)-generated detailed land information. Yet another criticism of large-scale urban models is that they were "data-hungry" (Lee 1973). This complaint on data dependency is still relevant today with the data collection for an urban model covering an area of a large metropolis being a major effort.

Applications of Large-Scale Urban Models Large-scale urban models are applied toward real-world problems, for research and policy analysis. For example, they are customarily used in metropolitan plan-making. Research and development in urban modeling is actively being conducted at different settings, including universities, public agencies, or private firms. Ohls and Hutchinson (1975) and Webb (1975) provide a review of the early urban and transportation models, while Wegener (1994) in his assessment of large-scale urban modeling addresses the questions of model specifications, reviews urban models of the mid-1970s–1990s, and potential urban issues that GIS and urban models can contribute to. Extensive urban modeling literature can be found in *Transportation Research*, the *Environment and Planning* series, the *Journal of Regional Science*, and other academic journals and manuscripts.

4.3.3 Harris and Wilson's (1978) Retail Location Model

Planners need to understand flow patterns in cities and depend on scientists' reliably modeling these flows both to explain an existing situation and to predict the future impact of the potential change (e.g., in population). A representation of flows to retail centers uses concepts similar to those of statistical mechanics such as thermodynamics (Wilson 2009). In a set of flows between origins and destinations in a spatial system, although each flow is the result of a certain individual decision made at the origins regarding the choice of a destination, aggregate destination-choice behavior modeling has been reasonably accurate based on the attractiveness of a destination (such as size, but not exclusively), and the separation between zones measured by distance between the origin and the destination, resulting in the well-known gravity models (Fotheringham 1985). Models forecasting spatial interaction need a third variable; the relative location of a destination with respect to all other destinations, besides the two factors mentioned, should be included as a basic component of spatial interaction models. The relative location variable can be measured as accessibility indicating the existence of spatial competition forces or agglomeration forces.

Besides underlying natural and man-made constraints, such as geographical factors or institutional regulations, different types of *agglomeration economies* promote the emergence of spatial variation (Osawa et al. 2017). Agglomeration economies occur when more production (in the retail modeling context, it is cost) is output within a specified geographical area (such as at a single location due to intra-firm economies of scale and scope, or due to close spatial proximity to other economic agents such as firms or retailers) leading to the decrease in average cost. The retail model is a spatial interaction model functioning also as a location model, that is, by using a single constraint of a fixed number of residents in the distribution of trips between origins and destinations the total flows into destinations can be calculated (Wilson 2009). Thus, Harris and Wilson (1978) developed a *retail location model* building on a static urban retail model of Huff (1963) and using the combination of "fast" and "slow" dynamics to describe the space economy.

The *fast dynamic* is used to describe the patterns of the short-run spatial interactions, such as flows (trips) between nodes (such as the residential origins and shopping destinations of zones), and the short-run payoff landscape (i.e., retailer profits), while the slow dynamic is used to describe the gradual evolution in the spatial distribution of mobile factors that guide the processes of flow generation (i.e., stocks at nodes, such as population at origins and destinations) (Osawa et al. 2017). The *slow dynamic* determines the long-run spatial distribution of firms.

Locations in urban systems can be represented by grids with one-mile intervals. According to the model, a city is divided into a set of zones each containing a continuum (i.e., there exists an infinite number of potential entrants to the retailing market) of retailers and a fixed number of consumers who patronize retail shops. The number of firms in the zone determines the relative attractiveness of each zone. In the short run, consumers' shopping behavior (i.e., consumer demand) is measured as a cash flow (that is, per capita money that residents of the zones spent on retail goods and services) expressed in the model by a series of origin-constrained gravity equations. Generalized transport costs (that is, a weighted sum of elements such as travel time, etc.) between the zones and consumer demand for shopping activity per capita are given exogenously. As travel cost increases (e.g., the distance to the store), consumer demand decreases. Revenue is assumed to be equally distributed for all firms in the zone, that is, after a certain zone is selected as the shopping destination, every shop in the zone has an equal probability to be visited by consumers. Every shop must only pay a fixed entry cost. The expected profit per floor space unit for retail goods and services moves toward an equilibrium of profit uniformity (with expected profits equal over all centers) (Harris and Wilson 1978).

Changes in the key parameters (known as a disturbance causing an ordered state to become disordered) such as attractiveness of retail centers, or travel costs (or any other exogenous variables) at a zonal level beyond some critical value may lead to system-wide changes characterized as phase transitions (Wilson 2009). The central focus is on the nature of the configurations: in urban system, a dispersed system with many small retail centers is thought less ordered versus a system with a small number of large centers (Wilson 2009). For example, in grocery retailing the small "corner store" has been replaced by the large supermarket (Fotheringham 1985). The catastrophe theory helps to explain how a slow and consistent change in consumer behavior can cause a fast change in the size of grocery retailing outlets. Harris and Wilson (1978) have further investigated this change in grocery retailing using bifurcation theory. The system returns to equilibrium after such a change slowly since developers require much longer time to construct a new retail center while individuals can adjust their transport routes more quickly.

In sum, Harris and Wilson (1978) developed a simple, dynamic model of agglomeration with spatial returns to scale. Instead of a static representation of the spatial pattern of retailers, spatial configuration of retail is allowed to dynamically evolve according to the payoff landscape.

4.3.3.1 Entropy Maximization

Wilson published his seminal work in the field where he developed *entropy maximizing techniques* in 1970. A geographical system of interest is described by a set of variables and parameters that constitute *a state space* (Wilson 2010, p. 384). Wilson (1970) used *entropy* as a measure of the degree of order in a system, so entropy-maximizing techniques can find the most ordered, that is, the most likely state of alternative configurations in a system (Clark 1982). In his work on gravity models, principles from information theory have been introduced to develop a typical trip table to create a series of spatial interaction models. Specifically, Wilson (1970) derived a journey-to-work model where he combined deterministic assumptions (i.e., origin and destination constraints) and the probabilistic entropy-maximizing method (Thomas and Huggett 1980, pp. 153–155). By using the entropy-maximizing approach the most probable state of a system of interest subject to constraints can be modeled; the models are "widely applicable" (Wilson 2010, p. 385). For example, trips are connected to structural variables such as number of workers at a residential location and number of jobs at employment locations. The models contribute to multiregional demographic and input–output models, to transport, and to location modeling such as retail.

Gould (1972) explained the mathematics behind the large equations involving the Lagrangian multipliers and applied Wilson's techniques to find the most likely journey-to-work trip configuration of the system in towns in Northwest England; additionally, he included several constraints such as fixed residential distribution (that is, a finite number of workers), job distribution (a finite number of employment), and overall cost constraints so that the modeled journey-to-work trip distributions by mode could be compared against the known existing patterns in the study area (Clark 1982). The model predicted the distribution of work movements highly accurately (0.95).

4.3.4 Spatial Structure and Spatial Interaction Models

Geographers wanted to understand spatial phenomena, specifically, patterns and their change, spatial structure, and spatial interaction. Various phenomena have their associated properties such as point, line, or area; examples of these entities include the population, land, physical infrastructure, organizations, goods and services, commodities, and so on. Also these entities generate the activities and flows (e.g., between population and economic activities such as journey to work, trade flows, migration flows, and so on).

Bennett et al. (1985) reviewed models of spatial structure, spatial interaction, and integrated location-interaction models. Starting in the 1970s, theories of spatial structure and theories of spatial interaction have been developed. Accordingly, the various alternative models contributed to the understanding of geographical patterns and change using the concepts of spatial structure and spatial interaction and their interrelationships. Geographical phenomena can be analyzed using the statistical analysis, especially in relation to structure. Models of spatial structure focus on the distribution of entities (e.g., retail stores) and their properties, with the entities often (but not necessarily) being subject to certain assumptions about the nature of underlying flows or interactions (either physical or informational). For example, a single homogeneous good might exhibit spatial variation of one of its attributes, say, urban retail outlets may sell the same good but price may vary across regions due to transport costs between markets (the movement behavior of buyers) and other reasons, such as the spatial price awareness of sellers vis-a-vis their competitors. Respectively, spatial price models were developed for the distribution of prices at urban retail outlets.

A new kind of location model is a *spatial interaction model*, where mathematical approaches (such as differential equations) are useful. Locational models are built explicitly on interaction concepts. Spatial interaction systems are the systems of flows. Spatial interactions between a pair of places, for example, between home and work or a retail location, are restricted by the costs of movement impacted by accessibility, which can be represented by distance between two locations. Distance is impacted in turn by the level of congestion, roadways pattern, public transit routes, and other attributes of the transportation network. In addition, the costs of movement are further impacted by traveler characteristics including auto ownership, time of travel, and income (Lowry 1964).

Models of spatial interaction explore specifically flows, given the spatial distribution of relevant geographical entities that generate them. For example, geographical entities, say, retail stores, generate consumer flows of patrons who travel between their homes and stores (i.e., flows) representing interactions between residential homes and retail stores. Shopping centers represent the spatial structure where these entities, shopping centers, have attributes, for example, size, while shopping trips (flows) represent spatial interaction. Geographers may be interested in understanding the distribution of shopping centers and shopping trips, thus models can be developed of the spatial structure of size and layout of shopping centers and spatial interaction of shopping trips. Thus, the shopping model models the flow of consumers (in other words, the flow of money) from residences to retail centers and simultaneously models the pattern of retail centers. Spatial systems of residential zones and retail zones; the two spatial systems may or may not coincide (Bennett et al. 1985). This spatial interaction model is essentially the model based on the Huff (1963) model and it explains an aggregate model of consumer flows (i.e., the flow of money from residents to shops) using several variables including the "size" of the retail center as a measure of attractiveness, the generalized cost of travel from residence to retail store, the per-capita income of residents of zones, and the population of zones.

4.3.5 Dynamic Models

Systems and subsystems evolve dynamically with variables and parameters describing the system having various speeds of change; some change relatively slowly (represented by "parameters") while others change quickly (represented by "variables"). For example, an urban system consists of the key subsystems (or structures) such as residential location, employment, industrial location, and so on, with spatial interactions represented as flows between them, say, work trips from residencies to workplaces. The relative rates of change of the two sets of processes are different: spatial interaction affects spatial structure; spatial structure affects other aspects of spatial structure; and spatial structure affects spatial interaction. Dynamic models can be built that explain change, focusing on models based on the interdependencies between structure and interaction. Another important example is journey-to-work and migration flows and these impact the spatial distribution of employment opportunities. Thus, the model is sensitive to the changes in its components.

The context of dynamic analysis is provided by Zeeman (1977). For example, interaction variables change "faster" in that they would return to equilibrium faster after a disturbance (e.g., prices could change quite rapidly) and structural variables change "slower" so that they can in effect be treated as constants even in a dynamic model. The retail model can illustrate the number of different speeds of change involved. "Retail" in the spatial interaction modeling can be interpreted very broadly (Wilson 2010): any system of interest where a population area generates a flow to some facility. To illustrate, spatial interaction modeling has been applied to the flow of students to universities (Wilson 1995). Other subsystems can be modeled using a spatial interaction modeling approach, such as residential location and housing [Clarke and Wilson (1983)], and industrial location (Wilson and Birkin 1983).

The retail centers structure remains in equilibrium given that flows are steady, that is, flows experience no changes. If flows are not steady, say, at the origin end there is a change (for example, there are changes in expenditure amounts or populations level), then the change will alter the flows, which will in turn change the retail centers structure. Sometimes changes may be "catastrophic": a small change in flow could trigger an extensive change in the structure. Similarly, the same analysis can be used to model work trips (journey-to-work flows), however, a set of journey-to-work flows is more slowly changing because migration flows affect both house and job moves, changing the structure and in turn changing the journey-to-work flows. In addition, when migration flows change the residential areas structure, these will in turn impact retail flows, which will impact retail structure.

4.3.5.1 Entropy Maximization and Spatial Interaction Models

Spatial interaction models have been enhanced by entropy-maximizing methods. Entropy-maximizing models represent a substantial subset of models of spatial interaction and location with widespread applications of models to problems of city functioning and evolution (Wilson 2010, pp. 377–378). Entropy in urban and regional modeling was used in transport modeling and in retailing (Wilson 2010, p. 365). Interaction and location are important to historical analysis (Wilson 2010, p. 378). Thus, the entropy-maximizing spatial interaction models have been applied in historical contexts. For example, Rihll and Wilson (1987a, b, 1991) used point data for settlement structures available from archaeological explorations for Greece in the ninth century BC, which were then used within a "retail" model to produce estimates of population sizes at different locations.

4.3.5.2 Location-Interaction Integrated Models

Theories about spatial structure and spatial interaction can be integrated. An example of the combined model is location-interaction models explaining the location of various entities and associated activities explicitly in relation to interaction; and models of structure-interaction adjustments addressing mutual adjustments that take place between spatial structure and spatial interaction where spatial structure and spatial interaction is, at least partly, a function of the other (Bennett et al. 1985).

Application of Location-Interaction Models Location-interaction models have a wide range of application. Besides the modeling of transport flows and retail flows, the entropy-maximizing models are usually applicable to nearly any situation that involves flows, and to locational structures, analogous to the retail case, that are based on flows (Wilson 2010).

Frequently, these models comprise components of other models. For example, models of migration are components of demographic models (Rogers 2008). Straussfogel (1991) examines the dynamics of suburbanization. The model-building principles still apply, and an integrated model can be built. The entropy-maximizing model can be applied in transport planning. Generally, in economics, the travel demand can be modeled using probabilities (e.g., the probability that a commuter will choose a certain travel mode, or a resident will select to live in a given area, using discrete choice models). The usual assumption includes a rational economic agent selecting an alternative that results in maximum satisfaction or utility (Wilson 2010, p. 375). If the random component is expressed as a Gumbel distributed variable, the probability of choice can be expressed as a multinomial logit model.

An article by Harris and Wilson (1978) incorporated the spatial interaction models within a framework that represents slow dynamics and describes the process of spontaneous spatial pattern formation in the economic landscape. Harris and Wilson (1978) offered a view of the characteristics of a network of retail interactions at equilibrium, so the Harris–Wilson model seeks for the equilibrium.

4.4 Urban Subsystems

For any model to be useful, relevant real-world features must be included while abstracting from those of small importance. For example, urban models include various important urban activities including housing, employment, commuting, and the like. Combined, these urban subsystems represent urban environment.

Wegener (1994) distinguishes eight types of major urban subsystem, which, by the speed at which they change, from slow to fast processes, can be grouped into four major categories.

1. First category includes urban subsystems that change slowly, including urban networks such as transportation, communications, and utility, as well as urban land use. Urban networks represent the most permanent components of the physical structure of the city. It takes a long period of time, a decade or longer, to implement a large infrastructure project. Their distribution is stable: once installed, the projects are maintained but rarely abandoned. Similarly, land use changes piecemeal, and compared with individual buildings it is more permanent.
2. Second category includes medium-speed change urban subsystems, such as workplaces and housing. Buildings require several years from planning to erection and exist for a long period of time (up to one hundred years). Workplaces include structures used for non-residential purposes. These include factories, warehouses, shops, offices, theaters, schools, and universities, which have a longer lifespan compared with the institutions that occupy them (e.g., firms, clubs, stores, etc.). Similarly, housing has a longer lifespan compared to the lifespan of individual households that occupy it.
3. Third category includes urban subsystems that change fast, including employment and population. Depending on market conditions and technologies, individual firms, once established, can close down, expand, or relocate. The process affects employment by creating new jobs or introducing new technology that makes workers surplus. The distribution of population is determined by household location. In the same way as firms, households are created, grow, decline, and dissolved. During their lifecycle they respond to their

changing needs by adjusting their housing consumption and location in each stage of the lifecycle.

4. Fourth category consists of urban subsystems with immediate change, including goods transport and travel. Since human activities occur in space, they necessitate spatial interaction such as transport of goods and travel. Spatial interactions have to respond to changing conditions such as congestion or changes in demand immediately, and therefore, these interactions represent the least permanent feature of spatial urban development.

All eight subsystems are interrelated. For example, the location of workplaces such as shopping centers is related to the location of other firms, clients, and employees. Additionally, it depends on access to goods transport and travel by customers and employees as well as the availability of land, utilities, and housing (Wegener 1994).

4.5 Contemporary Urban Modeling Development

At any spatial scale, contemporary scholars from various disciplines, including location theorists, geographers, and economists, observe spatial structures of population, firms, or shops and attempt to explain why and how these spatial, highly localized concentrations emerge and change over time.

Wegener (1994) addresses the issue of actual and potential application of the urban models. Despite a typical metropolitan area in industrialized countries having multiple planning problems, most urban models consider a rather narrow range of planning problems. Traditionally, they ask very important questions, including how land-use development and transportation will be impacted by land-use regulations or housing programs. Another issue includes the link between transportation improvements or changes in travel costs and the distribution of activities in an urban area. However, the future may bring other important problems such as equity and environmental issues, with urban models having a capacity to address those in an appropriate way. Similarly, urban models need to be sensitive to changes in employment, housing, or transportation, so that the effects of policy measures can be correctly evaluated.

With increasing environmental awareness and requirements by the existing laws, urban models should become greener. The amendments to the Clean Air Act (CAA) of 1990 requires that consistent air-quality forecasting be included into land-use transport (LUT) models studying the interaction between transportation and land use. According to the

Intermodal Surface Transportation Efficiency Act (ISTEA) of 1991, air quality must not be get worse by the implementation of new transport projects in non-attainment areas. Urban models can be used for environmental analysis.

On another note, GIS has a lot to contribute to urban models due to the continuing advances and developments. One way is by linking a GIS to the analytical capabilities of urban models.

4.6 Recent Applications of Spatial Interaction Models

In a recent application of Harris and Wilson's (1978) spatial interaction retail location model, Osawa et al. (2017) want to understand agglomeration behaviors in a multi-locational (more than two) setting. To overcome the model's incapability to explain polycentric spatial concentrations of economic activities often encountered in the real world, Osawa et al. (2017) study the evolution of the spatial structure and apply Harris and Wilson's (1978) model in a multi-location space. The structure has been found to have a remarkable previously unknown property, called a "spatial period-doubling cascade," which is not observed in the standard two-location setting. Essentially, the property refers to the finding that changes in a key parameter (here, decreasing transport costs) cause destabilization of the equilibrium in the system, or bifurcation, and each bifurcation halves the number of zones with positive numbers of retailers (market centers), while the average distance between market centers doubles, and the process proceeds until achieving a monocentric pattern. Bifurcations are defined as catastrophic phase transitions occurring in spatial patterns when some critical parameter values change. They also wanted to study sustainability of a given equilibrium, that is, test the impact of increasing transportation costs after an agglomeration is established. However, increasing transport costs leads to no destabilization of the equilibrium. The authors offer the theory to better understand agglomeration and spatial structure evolution.

On one hand, the spatial interaction modeling approach has been traditionally used to simulate the dynamics and evolution of urban spatial structure, such as the spatial relationships between retailers and their customers, including Harris and Wilson's (1978) study of equilibrium urban dynamics, the models have been limited by the difficulty of representing behavioral mechanisms and processes (Birkin and Heppenstall 2011). On the other hand, the *agent-based modeling* approach is used to model spatial behavior. Thus, to overcome the limitation, evolutionary models that use the spatial interaction principles can be enhanced with the more novel agent-based modeling approach allowing for a more flexible behavioral representation that makes models more "realistic." An agent-based model of retail market dynamics is driven by a process of price competition between retail agents. A model that replaces price competition with a mechanism of floor space adjustment reveals the equilibrium-seeking behavior (Birkin and Heppenstall 2011).

4.7 Case Study: Urban Structure and Commuting Patterns in the Baton Rouge Metropolitan Statistical Area

4.7.1 Introduction

According to the classic economic monocentric urban model, a city has a single, most accessible center with the highest land prices. All jobs are located in the central business district (CBD), so households living close to a CBD travel shorter distances to their workplaces. However, over time, many cities developed a polycentric urban form due to various reasons, including innovations in transportation, communication, employment suburbanization, and so on (Gordon et al. 1998; Ingram 1998). Some studies claim that peripheral cities appear more dominant than traditional downtowns urban cores (Dear 2005), while others argue that the CBD maintains its role as a major employment center (Greene 2008). Studying urban structure is crucial for understanding commuting trends.

In a centralized city all workers travel into the central city. Polycentric structure distributes the direction of commuting flows: it is the suburbs that attract commuters (Levine 1992). Polycentrism is believed to reduce commuting by car mainly due to a closer location to the employment centers (co-location theory) as well as reduced congestion, which intensifies within the monocentric concentration of activities (Gordon and Wong 1985).

The research was motivated by a lack of research regarding analysis of intraurban commuting and land-use patterns in small- to medium-sized metropolitan areas within the US South (Horner 2007). The objectives include (1) comparing urban form within the medium-sized Baton Rouge region between two time periods, 1990 and 2000 and (2) investigating the impact of metropolitan structure on commuter travel under the monocentric and polycentric assumptions.

Most commuting research utilizes a single measure of the spatial separation between home and work locations, such as travel time, because it is directly available from the travel data (Sultana and Weber 2007). Distances indicate commute lengths more consistently (Sultana and Weber 2007). Therefore, to explore commuting from the multiple perspectives, both spatial separation measures—the time and distance traveled'were incorporated into the study. Further, the commuting consequences of job decentralization were examined indicated by the commuting times and distances from the disaggregate individual-level 1997 Baton Rouge Personal Transportation Survey (BRPTS).

In a polycentric urban structure, as jobs move to the suburbs, commuters are thought to save travel time. However, a study focused on the effect of job decentralization on the average commute in the San Francisco Bay Area found that on average a one-way commuting distance grew by 12% and travel time by 5% to reach the sub-centers in 1990 compared to 1980 (Cervero and Wu 1998). Other studies, in contrast, point that people commute less to multiple centers than to the monocenter in a CBD.

There is empirical evidence that over time households relocate closer to workplaces and shorten commuting distances caused by job suburbanization (Sultana and Weber 2007). In an Atlanta-based study, Sultana (2000) found that people working in sub-centers, on average, have shorter commuting times than those working in the central city. There is more difference between distances than between times in monocentric and polycentric urban structures. This has the following important implications: as the cities grow outward, the overall commuting between homes and work places may grow too (Crane and Chatman 2004), however, the time spent traveling to job locations may not be very much different for the monocentric and polycentric settings, principally due to better roads with more capacities and faster speeds (Sultana and Weber 2007).

Contrary results were obtained by France- and Netherlands-based studies of the consequences of the employment decentralization revealing that the average commute is longer in a decentralized spatial structure. The finding may be due to spatial planning policies in those countries as well as cultural factors such a greater attachment to the place of residence and the regulated nature of housing provision resulting in reduced household mobility (Aguilera 2005). Both the theoretical and empirical research found that socioeconomic variables such as income, home ownership, racial membership, transportation mode, and time leaving for work, among others, influence in a significant way the length of the working trips (Sultana and Weber 2007). Other explanations regarding improving commuting include residential self-selection or time-lags, among others (Handy et al. 2006; Sarzynski et al. 2006).

4.7.2 *Study Area and Data*

The study area covers the Baton Rouge urbanized area, one of the fastest growing areas in the US South. It lies along the eastern banks of the Mississippi River. Boundaries of the Baton Rouge urbanized area provide, therefore, effective boundaries for the current study (Fig. 4.4). Its area is 3567 square kilometers (455 square miles). The study area encompasses a diverse range of land-use types across urban, suburban, and rural settings. In 2000, 526,248 people lived within the study area. There is almost twice as many Whites as African Americans in the study area (63% vs. 33%). There are 238,522 workers in total, with males comprising 52% of all resident workers. The average jobs-to-resident workers ratio (JWR), defined before, is 1.05, which is consistent with other metropolitan areas.

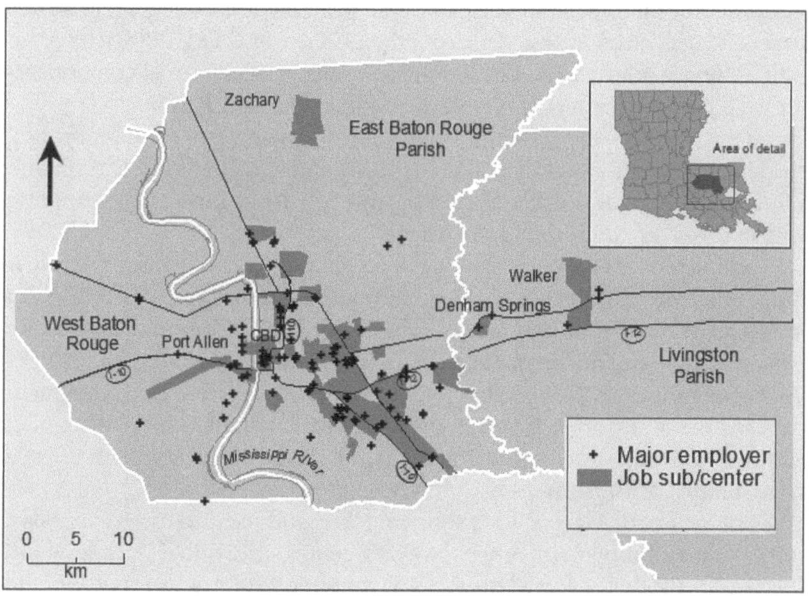

Fig. 4.4 Employment Sub/centers and the Major Employers in Baton Rouge in 2000

The first data source is the disaggregate BRPTS carried out in 1997. The BRPTS sample originally included 1395 surveyed households or 3069 persons, of whom 2934 reported travel-day data. Those with missing residential or work locations were excluded from the travel database. The final travel dataset provides detailed and disaggregate information on 1104 individuals whose work and home locations were geographically referenced therefore allowing for the detailed match of the land-use characteristics. Ethnically, they are predominantly White (76%), followed by African American (22%), Asian (<1%), and Other (<1%). Of the respondents, 45% are male and 55% are female. To define job concentrations within the study area we used the Census 2000 Transportation Planning Package (CTPP) parts 1 and 2 at the Traffic Analysis Zone (TAZ) level. For the identification of job locations in 1990, the urban element of CTPP 1990 for TAZs was used.

4.7.3 *Identification of the Employment Centers: Methodology*

To identify the employment centers, we obtained and geocoded major employers (at least 1000 workers) from the local area chamber (BRAC 2009). Addresses for the geocoding were retrieved from the Yellow Pages for the area studied. Then, employment data from the US Census CTPP 2000 part 3 was used to find workplaces. TAZs were used as the smallest spatial components for which employment data were available from the CTPP 2000.

This study uses the concept of JWR, emphasizing a net in-commuting to determine job concentrations (Forstall and Greene 1997). A higher value of JWR value indicates a better job provision for workers residing within this TAZ, and higher job access.

The following criteria were set for an area/s to qualify as an employment concentration: (1) The zones should be contiguous, (2) JWR of at least 1.0, and (3) total number of jobs per single TAZ should represent a peak in employment for the surrounding zones. A value of 750 jobs per TAZ was chosen as the threshold value based on a review of employment data for the study area. Contiguous zones with a total employment of at least 10,000 jobs were given the status of a job center, and sub-centers were defined as those areas with lower employment.

Zone properties layers were superimposed and used to assign an area a status of an employment center or a sub-center. Individual knowledge of the region also helped us identify employment concentrations (Weber and Kwan 2002). Finally, job concentrations were identified and given names reflecting an economic activity taking place or by the name of nearby roads (Table 4.1). They roughly follow the main road network structure. Figure below (Fig. 4.5) illustrates the location of these employment centers.

Table 4.1 2000 and 1990 employment concentrations (Ranked according to number of jobs)

2000						1990					
Center (rank)	*Name*	*Jobs*	*Workers*	*Pop.*	*E/R*	*Center (rank)*	*Name*	*Jobs*	*Workers*	*Pop.*	*E/R*
1	Essen Medical	24675	4925	9550	5.01	7	Essen Medical	6949	2017	2312	3.45
2	CBD	23905	1693	3733	14.1	1	CBD	24740	3727	5083	6.64
3	Cortana Mall Plaza	20540	6225	14595	3.3	2	Cortana Mall Plaza	21498	11312	14422	1.90
4	Harrells/S.Sherwood Commercial	14370	8165	14340	1.76	n/a	n/a	n/a	n/a	n/a	n/a
5	Florida Blvd Medical	13555	4665	12540	2.91	4	Florida Blvd Medical	14526	12561	15911	1.16
6	Hammond Aire Plaza	12950	3245	5930	3.99	6	Hammond Aire Plaza	6975	4513	5432	1.55
7	Siegen/Industriplex Commercial	12740	4905	9130	2.6	n/a	n/a	n/a	n/a	n/a	n/a
8	Louisiana State University	11425	3330	7325	3.43	3	Louisiana State University	16479	6106	6228	2.70
9	Corporate Blvd Commercial	8940	3030	5620	2.95	10	Corporate Blvd Commercial	4625	2173	2616	2.13
10	Industrial 1	8165	665	1755	12.3	5	Industrial 1	13375	1360	2088	9.83
11	EKL Memorial Hospital	2575	2085	5625	1.24	n/a	n/a	n/a	n/a	n/a	n/a
12	Denham Springs downtown	2360	655	1685	3.6	n/a	n/a	n/a	n/a	n/a	n/a
13	Zachary downtown	2180	820	2155	2.66	12	Zachary downtown	930	791	1120	1.18
14	Industrial WBRP	1880	160	325	11.8	n/a	n/a	n/a	n/a	n/a	n/a

(*continued*)

Table 4.1 (continued)

		2000						1990			
Center (rank)	*Name*	*Jobs*	*Workers*	*Pop.*	*E/R*	*Center (rank)*	*Name*	*Jobs*	*Workers*	*Pop.*	*E/R*
15	BR Metro Ryan Field Airport	1800	340	3165	5.29	13	BR Metro Ryan Field Airport	823	675	925	1.22
16	Southern University	1690	660	2820	2.56	9	Southern University	3002	2628	2832	1.14
17	O'Neal medical	1295	1010	2195	1.28	14	O'Neal medical	783	710	1079	1.10
18	Port Allen downtown	1190	475	1220	2.51	n/a	n/a	n/a	n/a	n/a	n/a
19	Industrial 2	1005	60	170	16.8	8	Industrial 2	3794	300	385	12.7
20	Walker downtown	800	555	1220	1.44	n/a	n/a	n/a	n/a	n/a	n/a
11		Central		1175		714		1023		1.65	
Total in centers	168,040 (67.1%)	47,668 (19.9%)		105,098 (19.9%)	3.53	119,674 (58%)		49,587 (24%)	61,456 (13%)	2	
Total within study area	250,516	238,522		526,248	1.05	207,464		206,357	470,050	1.01	

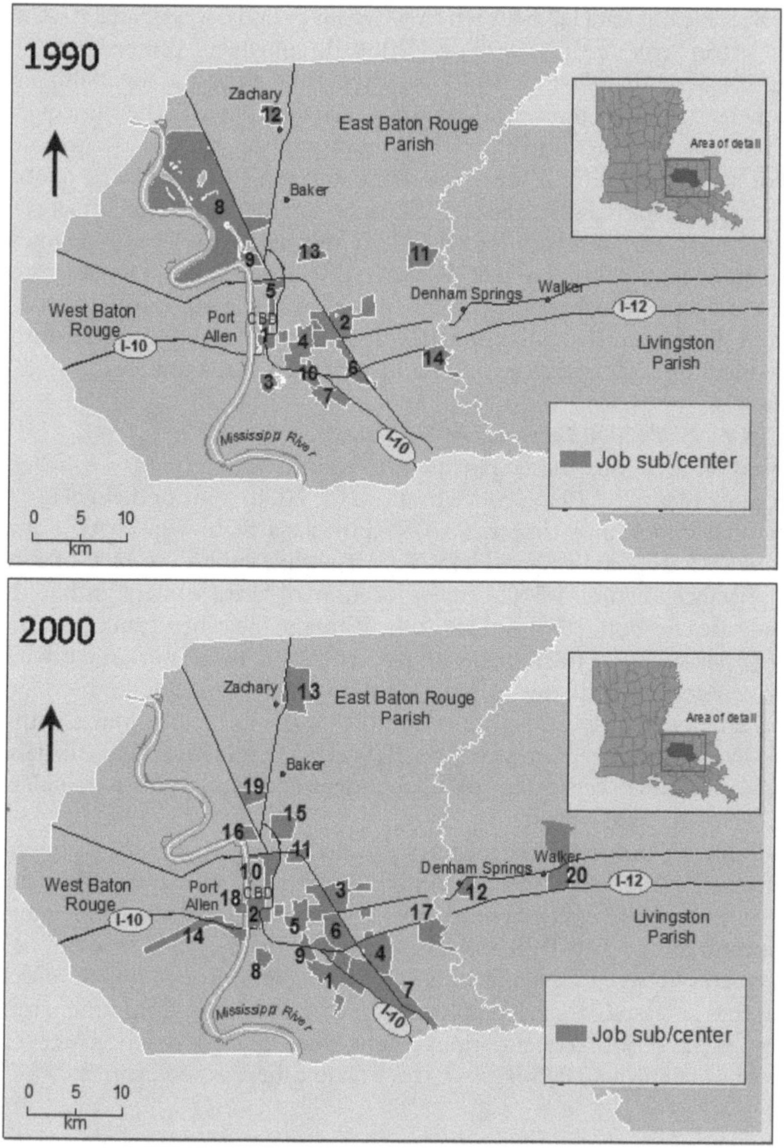

Fig. 4.5 Employment Subcenters in 1990 and 2000 IN Baton Rouge. (Note: center numbers can be found in Table 4.1)

Because our data fall between two Census periods, we decided to study the urban structure for 1990 and 2000. Employment centers and sub-centers were identified for 1990 using the same criteria as for 2000. The numerous discussions of the downtown revitalization with the subsequent development of the different revitalization projects, such as a downtown pedestrian mall, civic center, entertainment and gaming places, among others, have started since the late 1950s serving the purpose to attract residents and businesses back into the CBD area, and therefore, to augment the monocentric urban system (Wheeler 1994). The hypothesis was tested that polycentric urban structure is more favorable to journey-to-work travel, both spatially and temporally, then the monocentric system, using the medium-sized metro area of Baton Rouge, LA as a test area.

4.7.3.1 1990–2000 Employment Concentrations

Based on total jobs and employment to resident workers ratio (E/R) criteria, we identified 20 job concentrations for 2000, some of them being of regional importance (like the CBD, Louisiana State University, Baton Rouge Metropolitan Ryan Field Airport), while others are only of local importance. Figure 4.5 illustrates location of these centers within the study area for both 1990 and 2000. In 2000 the job centers rank from one to 20 according to total jobs, with the center one being the largest, while job centers rank from one to 14 in 1990 (Fig. 4.5). In 1990, the 14 identified employment locations contained 60.4% of all the jobs found within the Baton Rouge region with the CBD certainly standing out as the largest employment center. In 2000, the identified job locations contained 67.1% of all employment.

Table 4.1 provides a detailed description of these employment concentrations for 2000 and 1990. In 2000, 20% of population and workers and two thirds of all employment (67%) was concentrated in the employment concentrations. The JWR (jobs-to-resident workers ratio) of each TAZ fluctuates between 1.28 and 14.1 within the Baton Rouge region reflecting that at least 28% of all employees are in-commuting from other areas. Among these job concentrations, eight were defined as job centers (as there were at least 10,000 jobs in total) and others as sub-centers.

4.7.4 *Examining Impacts of Urban Structure on Commuting Patterns*

This research utilized both times and distances as measures of spatial separation between home and work (Sultana and Weber 2007). The role of job concentrations in affecting commuting outcomes was investigated under monocentric and polycentric assumptions of urban structure. The idea of a possible revitalization of a significant part of downtown area motivated a "what-if" approach.

4.7.4.1 *Commuting and Monocentric Versus Polycentric Structure: Hypothetical Scenario*

We tested the co-location hypothesis that a polycentric urban structure is conducive to shorter commutes compared to a monocentric structure, both in terms of time and distance. To test this hypothesis we proceeded in two steps. First, the commuting consequences of hypothetically concentrating employment at either the CBD or at employment centers were estimated. In a monocentric urban structure case, all the residents kept the same home locations, but their work places were assigned to the Baton Rouge CBD. The advantage of having the geographic coordinates for the places of residence within the Baton Rouge region contained in the data let us compute the commuting free-flow times (in minutes) and distances (in miles) using the real-world road network. To get more realistic results, the road type and varying travel speeds in different urban settings was taken into account (Luo and Wang 2003).

The polycentric scenario was investigated likewise using existing home locations and assuming workers worked at the nearest identified employment centers. The simulated monocentric commuting times and distances were compared to the simulated polycentric travel times and distances. An investigation of the commuting impacts of the urban structure supports the co-location theory as the experiment clearly demonstrated that monocentric structure does not contribute to a reduction in commuting. A paired t-test of means of the simulated commuting times and distances compared to the true times and distances resulted in significant differences. Workers commuting into the CBD where all the jobs are concentrated need to travel longer distances and spend more time commuting. For example, according to the experiment, the monocentric commuting distances were almost two miles longer (significant at less than 0.001 level) than distances required to reach multiple job concentrations. Similar

Table 4.2 T-test of mean commuting times and distances (Monocentric- and Polycentric-simulated)

Polycentric (Job centers)		*Monocentric (CBD)*	
Measure of spatial separation	*Mean*	*Mean*	*t Stat*
Distance, miles	8	9.85	8.87***
N	1104	1104	
Time, minutes	12.96	14.15	4.38***
N	1104	1104	

***Significant at 0.05 level of significance, Significant at 0.01 level of significance, Significant at 0.001 level of significance

results were obtained for the travel times, which were also longer for the monocentric conditions. Table 4.2 provides the paired t-test of means results of the experiment.

4.7.4.2 Commuting and Monocentric Versus Polycentric Structure: Observed Data

For the second step, we compared the observed commuting times (in minutes) and distances (in miles) as reported by the survey respondents who work within the CBD (the monocentric commutes) with those who work in all other employment concentrations (the polycentric commutes). The relative spatial stability of the job sub-centers between 1990 and 2000, as well as temporal proximity of data to 2000, allowed us to use the boundaries for the employment concentrations for the year 2000. Accordingly, for the second step, to determine whether real data support the findings from the experiment, we tested the data empirically using the self-reported commuting times and distances to work contained in the BRPTS survey. For this purpose, the workers working within the CBD as well as other employment concentrations were identified. The former were defined as the "monocentric group" while the latter workers were defined as the "polycentric group."

To check whether observed data follow the theoretical expectations, reported distances and time to work were examined among the CBD workers and those employed at other work destinations. Empirical data from the survey support the findings of the preceding experiment. Within the survey data, 107 respondents worked in the CBD, and 503 worked in the other 19 centers, making a total of 610 workers who worked in

Table 4.3 T-test of the mean commuting times and distances (Observed)

Polycentric (Job centers)		*Monocentric (CBD)*	
Measure of spatial separation	*Mean*	*Mean*	*t Stat*
Distance, miles	8.41	9.73	−2.11**
N	503	107	
Time, minutes	18.36	21.43	−2.49**
N	503	107	

**Significant at 0.05 level of significance, Significant at 0.01 level of significance, Significant at 0.001 level of significance

employment centers. The remaining 497 workers worked elsewhere outside of the identified job concentrations. Table 4.3 reports the findings from the empirical analysis together with the statistical results of the t-test.

A t-test of means conducted to determine whether commuting lengths were different between the commuters to the CBD and to the multiple job concentrations showed that commuting to multiple job concentrations results in significantly shorter times and distances (Table 4.3). For example, 610 workers identified from the 1997 Baton Rouge survey that commuted to 20 employment centers and sub-centers (including the CBD as one of the job centers), generally traveled about two miles less and spent about one minute less time on their journey to work than did 107 commuters working just in the CBD area (Table 4.3). The numbers of two miles and one minute may seem minor, however, with commuting being a vital activity for American households (Horner 2004), these numbers multiplied by the number of working days in a year and years over which the impact is being measured, can translate into much bigger numbers that could otherwise be spent with a family or other activities (Clark and Wang 2005).

Figure 4.6 visualizes the observed travel time for each TAZ as reported by the respondents to a workplace. Generally, individuals benefit from living closer to job-rich places as is demonstrated by the lower commuting times (which are less than 20 minutes), while those living farther away spend longer time commuting.

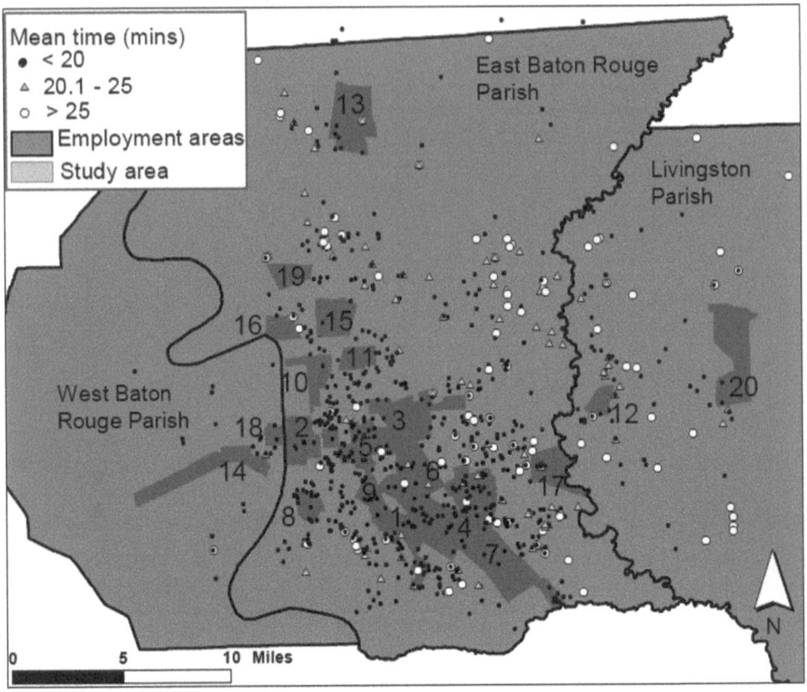

Fig. 4.6 1997 Baton Rouge Transportation Survey Respondents' Observed Commuting Times and Employment Centers Locations

4.8 Conclusions

In agreement with the co-location theory, our findings indicate that the polycentric system appears to reduce commuting lengths measured both by the distance and time. A paired t-test of means of the monocentric commuting times and distances veruss the polycentric times and distances, both simulated and observed, resulted in significant differences in time and distance. Under the monocentric assumption, the travel times and distances required for workers to reach a single center (CBD) were longer than for the decentralized structure. This finding is consistent with other research concluding that job suburbanization is associated with shorter commuting distances (Crane and Chatman 2004). Additionally, the travel times patterns were found strikingly similar to these revealed by the distances. This trend of similar behavior between travel times and mileage, although representing distinct commuting measures, were also found by Sultana and Weber (2007) in their Alabama-based study of two midsize southeastern metropolitan areas.

Bibliography

Aguilera, A. 2005. Growth in Commuting Distances in French Polycentric Metropolitan Areas: Paris, Lyon and Marseille. *Urban Studies* 42: 1537–1547.

Allonso, W. 1964. *Location and Land Use. Toward a General Theory of Land Rent.* Cambridge: Harvard University Press.

Antipova, A., and E. Ozdenerol. 2013. Using Longitudinal Employer Dynamics (LED) Data for the Analysis of Memphis Aerotropolis Tennessee. *Applied Geography* 42: 48–62.

Baton Rouge Area Chamber (BRAC). 2009. *Regional Overview, East Baton Rouge Parish.* At: http://www.brac.org.

Batty, M. 2008. The Size, Scale, and Shape of Cities. *Science* 319: 769–771.

Bennett, R.J., R.P. Haining, and A.G. Wilson. 1985. Spatial Structure, Spatial Interaction, and Their Integration: A Review of Alternative Models. *Environment and Planning A* 17: 625–645.

Birkin, M., and A. Heppenstall. 2011. Extending Spatial Interaction Models with Agents for Understanding Relationships in a Dynamic Retail Market. *Urban Studies Research* 2011, Article ID 403969, 12 pps. https://doi.org/10.1155/2011/403969. At: http://www.oalib.com/paper/51905#.WcMVncZyOpo.

Bronfman, N.C., L.A. Cifuentes, M.L. deKay, and H.H. Willis. 2007. Accounting for the Variation in the Explanatory Power of the Psychometric Paradigm: The Effects of Aggregation and Focus. *Journal of Risk Research* 10: 527–554.

Brueckner, J.K. 2003. Airline Traffic and Urban Economic Development. *Urban Studies* 40: 1455–1469.

Burgess, E.W. 1925. The Growth of the City: An Introduction to a Research Project. In *The City*, ed. R. Park, E.W. Burgess, and R.D. McKenzie. Chicago: University of Chicago Press.

Burgess, E.W. 1967. The Growth of the City: An Introduction to a Research Project. In *The City. Suggestions for Investigation of Human Behavior in the Urban Environment*, ed. R.E. Park and E.W. Burgess, 2nd ed. Chicago: The University of Chicago Press.

Cairncross, F. 1997. *The Death of Distance.* London: Orion Business Books.

Cervero, R., and K.-L. Wu. 1998. Sub-Centring and Commuting: Evidence from the San Francisco Bay Area. *Urban Studies* 35: 1059–1076.

Chinitz, B. 1960. *Freight and the Metropolis: The Impact of America's Transport Revolutions on the New York Region.* Cambridge: Harvard University Press.

Clark, D. 1982. *Urban Geography: An Introductory Guide.* Baltimore: The Johns Hopkins University Press.

Clark, W.V.W., and W.W. Wang. 2005. Job Access and Commute Patterns: Balancing Work and Residence in Los Angeles. *Urban Geography* 26: 610–626.

Clarke, M., and A.G. Wilson. 1983. *Exploring the Dynamics of Urban Housing Structure in a 56- Parameter Residential Location and Housing Model*, WP-363. Leeds: School of Geography, University of Leeds.
Cody, R.P., and J.K. Smith. 2006. *Applied Statistics and the SAS Programming Language*. 5th ed. Upper Saddle River: Pearson Prentice Hall.
Coleman, M.G. 2002. Contesting the Magic of the Market-Place: Black Employment and Business Concentration in the Urban Context. *Urban Studies* 39: 1793–1818.
Crane, R., and D.G. Chatman. 2004. Traffic and Sprawl: Evidence from U.S. Commuting 1985–1987. In *Urban Sprawl in Western Europe and the United States*, ed. H.W. Richardson and C.C. Bae, 311–325. Aldershot: Ashgate.
Dalton, S., Friesenhahn, E., Spletzer, J., and Talan, D. 2011. Employment Growth by Size Class: Firm and Establishment Data. *Monthly Labor Review*, December: 3–12.
Dear, M. 2005. Comparative Urbanism. *Urban Geography* 26: 247–251.
Economic Census. 2013. *Economic Census: Local Business Snapshot*. Economy Wide Key Statistics File.
Enright, M.J. 2003. Regional Clusters: What We Know and What We Should Know. In *Innovation Clusters and Interregional Competition*, ed. J. Bröcker, D. Dohse, and R. Soltwedel, 99–129. Berlin: Springer.
Ewing, R., K. Bartholomew, S. Winkelman, J. Walters, and G. Anderson. 2008. Urban Development and Climate Change. *Journal of Urbanism* 1 (3): 201–216.
Fisher, Ernest M. 1930. *Advanced Principles of Real Estate Practice*, 126. New York: The Macmillan Co.
Foot, D. 1981. *Operational Urban Models. An Introduction*. New York: Methuen & Co.
Forstall, R.L., and R.P. Greene. 1997. Defining Job Concentrations: The Los Angeles Case. *Urban Geography* 18: 705–739.
Fotheringham, A.S. 1985. Spatial Competition and Agglomeration in Urban Modelling. *Environment and Planning A* 17: 213–230.
Fox, J. 1997. *Applied Regression Analysis, Linear Models, and Related Methods*. Thousand Oaks: SAGE Publications, Inc.
Glaeser, E.L., and J. Kohlhase. 2004. Cities, Regions and the Decline of Transport Costs. *Papers in Regional Science* 83: 197–228.
Gordon, P., and H.L. Wong. 1985. The Cost of the Sprawl: Some New Evidence. *Environment and Planning A* 17: 661–666.
Gordon, P., H.L. Richardson, and G. Yu. 1998. Metropolitan and Non-Metropolitan Employment Trends in the U.S.: Recent Evidence and Implications. *Urban Studies* 35 (7): 1037–1057.
Gould, P.R. 1972. Pedagogic Review. *Annals of the Association of American Geographers* 62 (4): 689–700.
Greene, R. 2008. Urban Peripheries as Organizers of What Remains of the Center: Examining the Evidence from Los Angeles and Chicago. *Urban Geography* 29: 138–153.
Greene, R.P., and J.B. Pick. 2012. *Exploring the Urban Community. A GIS Approach*. 2nd ed. Upper Saddle River: Pearson Prentice Hall.

Handy, S., X. Cao, and P.L. Mokhtarian. 2006. Does Self-selection Explain the Relationship Between Built Environment and Walking Behavior? Empirical Evidence from Northern California. *Journal of the American Planning Association* 72 (1): 55–74.

Harkins, J.E. 1982. *Metropolis of the American Nile: An Illustrated History of Memphis and Shelby County*. Oxford: The Guild Bindery Press.

Harrington, L. 2011. Is US Manufacturing Coming Back? *Inbound Logistics* 31 (8): 38–46.

Harris, C.D., and E.L. Ullman. 1945. The Nature of Cities. *Annals of the American Academy of Political and Social Science* 242: 7–17.

Harris, B., and A.G. Wilson. 1978. Equilibrium Values and Dynamics of Attractiveness Terms in Production-Constrained Spatial Interaction Models. *Environment and Planning A* 10: 371–388.

Henderson, R. 2012. Employment Outlook: 2010–2020. Industry Employment and Output Projections to 2020. *Monthly Labor Review* 135: 65–83.

Hess, D.B. 2005. Access to Employment for Adults in Poverty in the Buffalo–Niagara Region. *Urban Studies* 42: 1177–1200.

Hillner, J., and B. Weiners. 1998. Silicon Envy. *Wired* 6: 136–137.

Holtzclaw, J.W. 2000. Smart Growth—As Seen from the Air: Convenient Neighborhood, Skip the Car. In *Air and Waste Management Association's 93rd Annual Meeting and Exhibition*.

Holzer, H.J. 1991. The Spatial Mismatch Hypothesis: What Has the Evidence Shown? *Urban Studies* 28: 105–122.

Hoover, E.M. 1948. *Location of Economic Activity*. New York: McGraw-Hill.

Horner, M.W. 2004. Spatial Dimensions of Urban Commuting: A Review of Major Issues and Their Implications for Future Geographic Research. *The Professional Geographer* 56: 160–173.

———. 2007. A Multi-Scale Analysis of Urban Form and Commuting Change in a Small Metropolitan Area (1990–2000). *Annals Regional Science* 41: 315–332.

Hoyt, H. 1939. *The Structure and Growth of Residential Neighborhoods in American Cities*. Washington, DC: Federal Housing Administration.

Huff, D.L. 1963. A Probabilistic Analysis of Shopping Center Trade Areas. *Land Economics* 39 (1): 81–90.

Ingram, G.K. 1998. Patterns of Metropolitan Development: What Have We Learned? *Urban Studies* 35: 1019–1035.

Isard, W. 1956. *Location and Space Economy*. Cambridge: Technology Press.

Klosterman, Richard E. 1994. An Introduction to the Literature on Large-Scale Urban Models. *Journal of the American Planning Association* 60 (Winter): 41–44.

Krugman, P. 1991. *Geography and Trade*. Cambridge: MIT Press.

Lee, Douglass B., Jr. 1973. Requiem for Large-Scale Models. *Journal of the American Institute of Planners* 39: 163–178.

Levine, J.C. 1992. Decentralization of Jobs and Emerging Suburban Commute. *Transportation Research Record* 1364: 71–80.

Lowry, I.S. 1964. *A Model of Metropolis*. RM-4035-RC. Santa Monica: The Rand Corporation.

Luo, W., and F. Wang. 2003. Measures of Spatial Accessibility to Health Care in a GIS Environment: Synthesis and a Case Study in the Chicago Region. *Environment and Planning B: Planning and Design* 30 (6): 865–884.

Marlay, M.C., and T.K. Gardner. 2010. *Identifying Concentrations of Employment in Metropolitan Areas.* Paper Presented at the Annual Meeting of the American Sociological Association Annual Meeting, Hilton Atlanta and Atlanta Marriott Marquis, Atlanta, GA. At: http://www.census.gov/population/www/cen2000/EmploymentClusters-Methods.pdf.

Memphis Regional Freight Infrastructure Plan 2010. March 2010.

Mills, E.S. 1967. An Aggregative Model of Resource Allocation in a Metropolitan Area. *American Economic Review* 57: 197–210.

Moses, L., and H.F. Williamson. 1967. The Location of Economic Activities in Cities. *The American Economic Review* 57: 211–222.

Muth, R. 1968. *Cities and Housing.* Chicago: University of Chicago Press.

Nelson, A.C., and T.W. Sanchez. 1997. Influence of the Metropolitan Atlanta Rapid Transit Authority on Population *and* Employment Location. *Transportation Research Board* 1604: 18–25.

Office of Management and Budget. 2010. Standards for Delineating Metropolitan and Micropolitan Statistical Areas. *Federal Register* 75 (123), June 28, 2010/ Notices'.

Ohls, James C., and Peter Hutchinson. 1975. Models in Urban Development. In *A Guide to Models in Governmental Planning and Operations*, ed. Saul I. Gass and Roger L. Sisson. Potomac: Sauger Books.

Osawa, M., T. Akamatsu, and Y. Takayama. 2017. Harris and Wilson (1978) Model Revisited: The Spatial Period-Doubling Cascade in an Urban Retail Model. *Journal of Regional Science* 57 (3): 442–466.

Porter, M. 1998. Clusters and the New Economics of Competition. *Harvard Business Review* 76 (6): 77–90.

Porter, M.E. 2000. Location, Competition, and Economic Development: Local Clusters in a Global Economy. *Economic Development Quarterly* 14: 15–34.

Pratt, A.C. 2000. New Media, the New Economy and New Spaces. *Geoforum* 31: 425–436.

———. 2007. *An Economic Geography of the Cultural Industries.* LSE Research Online.

Quah, D.T. 1999a. The Weightless Economy in Growth. *The Business Economist* 30: 40–53.

———. 1999b. *The Weightless Economy in Economic Development*, Discussion Paper No. 417. Center for Economic Performance.

Rihll, T.E., and A.G. Wilson. 1987a. Spatial Interaction and Structural Models in Historical Analysis: Some Possibilities and an Example. *Histoire et Me'sure* II (1): 5–32.

———. 1987b. Model-Based Approaches to the Analysis of Regional Settlement Structures: The Case of Ancient Greece. In *History and Computing*, ed. P. Denley and D. Hopkin, 10–20. Manchester: Manchester University Press.

———. 1991. Settlement Structures in Ancient Greece: New Approaches to the Polis. In *City and Country in the Ancient World*, ed. J. Rich and A. Wallace-Hadrill, 58–95. London: Croom Helm.

Rode, P., G. Floater, N. Thomopoulos, J. Docherty, P. Schwinger, A. Mahendra, and W. Fang. 2014. Accessibility in Cities: Transport and Urban Form. In *NCE Cities Paper 03. LSE Cities*. London School of Economics and Political Science.

Rodrigue, J.P., C. Comtois, and B. Slack. 2009. *The Geography of Transport Systems*. 2nd ed. New York: Routledge.

Rogers, A. 2008. Demographic Modelling of Migration and Population: A Multiregional Perspective. *Geographical Analysis* 40: 276–296.

Rossi-Hansberg, E., P. Sarte, and R. Owen. 2005. *Firm Fragmentation and Urban Patterns*, Working Paper 11839. Cambridge, MA: National Bureau of Economic Research.

Sarzynski, A., H.L. Wolman, G. Galster, and R. Hanson. 2006. Testing the Conventional Wisdom About Land Use and Traffic Congestion: The More We Sprawl, the Less We Move? *Urban Studies* 43: 601–626.

Smith, W.S. 1984. Mass Transport for High-Rise High-Density Living. *Journal of Transportation Engineering* 110 (6): 521–535.

Speck, J. 2013. *Memphis Riverfront Analysis and Recommendations*. AICP, CNU-A, LEED-AP, Hon. ASLA, Speck and Associates LLC.

Straussfogel, D. 1991. Modeling Suburbanization as an Evolutionary System Dynamic. *Geographical Analysis* 23: 1–23.

Stutz, F.P., and B. Barf. 2012. *The World Economy*. 6th ed. Upper Saddle River: Pearson Prentice Hall.

Sultana, S. 2000. Some Effects of Employment Centers on Commuting Times in the Atlanta Metropolitan Area, 1990. *Southeastern Geographer* 41 (2): 225–233.

Sultana, S., and J. Weber. 2007. Journey-to-Work Patterns in the Age of Sprawl: Evidence from Two Midsize Southern Metropolitan Areas. *The Professional Geographer* 59 (2): 193–208.

Tennessee Department of Labor and Workforce. 2010. *The Labor Market Report*, August 2010.

Thomas, R.W., and R.J. Huggett. 1980. *Modeling in Geography: A Mathematical Approach*. London: Harper & Row.

U.S. Census Bureau. 2012. *OnTheMap Application. Longitudinal-Employer Household Dynamics Program*. http://onthemap.ces.census.gov/.

U.S. Department of Commerce Economics and Statistics Administration. 1999. *Petrochemical Manufacturing. 1997 Economic Census*, Manufacturing Industry Series. EC97M-3251A.

U.S. Department of Transportation, Federal Highway Administration, Office of Freight Management and Operations, Freight Analysis Framework, version 3.1.

Van Arsdale, R., D. Arellano, K.C. Stevens, A.A. Hill, J.D. Lester, A.G. Parks, R.M. Csontos, M.A. Rapino, T.S. Deen, E.W. Wolery, and J.B. Harris. 2012. Geology, Geotechnical Engineering, and Natural Hazards of Memphis, Tennessee, USA. *Environmental and Engineering Geoscience* 18: 113–158.

van der Linde, C.M. 2002. *Findings from the Cluster Meta-Study*. Boston: Harvard Business School.

Vernon, R. 1959. *The Changing Economic Function of the Central City*. New York: Committee for Economic Development.

Webb, K.W. 1975. Models in Transportation. In *Models in Urban Development. A Guide to Models in Governmental Planning and Operations*, ed. Saul I. Gass and Roger L. Sisson. Potomac: Sauger Books.

Webber, M.J. 1984. *Industrial Location*. Beverly Hills: Sage.

Weber, A. 1909. *Über den Standort der Industrie*. Tubingen: Mohr JCB. English Translation by Friedrich C: *The Theory of the Location of Industries*. Chicago: Chicago University Press, 1929, reprinted 1957.

Weber, J., and M.-P. Kwan. 2002. Bringing Time Back In: A Study on the Influence of Travel Time Variations and Facility Opening Hours on Individual Accessibility. *The Professional Geographer* 54 (2): 226–240.

Wegener, M. 1994. Operational Urban Models State of the Art. *Journal of American Planning Association* 60: 17–29.

Wheeler, M. 1994. *Urban Revitalization Methods in Downtown Baton Rouge*. Unpublished Masters Thesis. Louisiana State University.

Willis, H.H., M.L. DeKay, B. Fischhoff, and M.G. Morgan. 2005. Aggregate and Disaggregate Analyses of Ecological Risk Perceptions. *Risk Analysis 25*: 405–428.

Wilson, A. 2009. Chapter 2. The "Thermodynamics" of the City. Evolution and Complexity Science in Urban Modelling. In *Complexity and Spatial Networks: In Search of Simplicity*, ed. A. Reggiani and P. Nijkamp. Berlin, Heidelberg: Springer-Verlag.

———. 2010. Entropy in Urban and Regional Modelling: Retrospect and Prospect. *Geographical Analysis* 42 (2010): 364–394.

Wilson, A.G. 1967. A Statistical Theory of Spatial Distribution Models. *Transportation Research* 1 (3): 253–269.

———. 1970. *Entropy in Urban and Regional Modelling*. London: Pion.

———. 1995. Simplicity, Complexity and Generality: Dreams of a Final Theory in Locational Analysis. In *Diffusing Geography: Essays for Peter Haggett*, ed. A.D. Cliff, P.R. Gould, A.G. Hoare, and N.J. Thrift, 342–352. Oxford: Blackwell.

Wilson A. G., and M. Birkin. 1983. *Industrial Location Theory: Exploration of a New Approach*, WP-361. Leeds: School of Geography, University of Leeds.

Zeeman, E.C. 1977. *Catastrophe Theory*. Reading: Addison-Wesley.

Zipf, G.K. 1949. *Human Behavior and the Principle of Least Effort*. Cambridge, MA: Addison Wesley Press.

CHAPTER 5

The Relationship Between Urban Environment and Travel Behavior

5.1 Urbanization and Travel Growth

According to the *Passenger Travel Facts and Figures* 2016, a report by the US Department of Transportation, during 2010–2014, the US population grew by 3.1%, from 309 million to 319 million with the most overall growth experienced by metropolitan areas. The expansion was accompanied by an increase in the number of people who travel as well as the distance traveled per person contributing to the demand for passenger travel and environmental pressure on urban areas. Most people travel using personal motorized vehicles: in 2009, 83.4% of trips were by personal vehicle. The average number of *person-trips* was 3.8 trips per day in 2009 (person-trips are trips by one person in any mode of transportation). Each day in 2009, Americans traveled an average of 36.1 miles with daily passenger travel patterns shaped by many demographic factors such as working age (those aged 36 to 65 years traveled the most), gender (men traveled more), and household income. Most workers (85.7%) got to work by driving in a personal vehicle in 2014. Over recent decades, workers and their workplaces have grown farther apart, in fact, commuting is the longest among trips for all trip-purpose categories and accounts for 25.3% of total miles traveled but only for 18.7% of total trips. Only slightly more than 3% of workers used non-motorized transportation to get to work (among them, 2.7% walked and 0.6% biked) (*Passenger Travel Facts and Figures* 2016).

A. Antipova, *Urban Environment, Travel Behavior, Health, and Resident Satisfaction*, https://doi.org/10.1007/978-3-319-74198-7_5

According to *Passenger Travel Facts and Figures* 2016, in general, road congestion worsened nationwide in urban areas of all sizes with measures of congestion delay and cost increasing during post-recession years, from 2013 to 2014. To illustrate, using *Planning Time Index* data (PTI can be defined as a highway reliability measure that evaluates the extra time one should plan for a trip, or in other words, it is extra time that should be budgeted for a trip compared with the time for this trip that would typically occur in free-flow conditions) for 52 metropolitan statistical areas (MSAs), travelers had to budget more than double the standard free-flow travel time to be "on-time" for 19 out of 20 trips. Washington, DC–VA–MD was one of the most congested urban areas by annual hours of delay per auto commuter in 2014 among the very large urban areas (with populations of 3 million or more). San Jose, CA represented the most congested large urban area (1 million to less than 3 million population), Honolulu, HI was the most congested medium urban area (500,000 to less than 1 million population), and Jackson, MS was the most congested small urban area (less than 500,000 population).

5.2 Active Travel: Walking and Cycling as an Alternative to Vehicle Travel

Rather than studying health benefits, transportation and urban planning researchers want to better understand how neighborhood environment affects physical activity including walking and cycling as an alternative to vehicle travel. These non-motorized (active) travel modes are regarded as the green modes of transportation due to their environmental, ecological, and sustainability benefits.

The National Household Travel Survey (NHTS) and its predecessor, the National Personal Transportation Surveys (NPTS), are the only travel sources for all trip purposes in the United States. The 2009 NHTS all-purpose data do not support an effect of MSA size on bike mode share or trip purpose: in the United States, city size does not affect cycling while in Europe small and medium-sized cities have the highest cycling rates and the biggest cities have lower levels. Regarding trends in trip purpose, the 2009 NHTS data compared with the previous NHTS of 2001 indicate a substantial increase in *utilitarian cycling* (that is, other than for recreation or exercise, with purposes such as for work, shopping, errands, visiting friends, and the like), rising from 43% of all bike trips in 2001 to 52% of bike trips in 2009. The portion of bike trips made for work grew from 8% to 12%, and that for shopping increased from 8% to 10%. Trips made by bike to and from public transport stops also grew over the same period, from only 1% to 3%. In 2009, a higher bike mode share was represented by social and recreational trips at 2.5% in the United States (among all trips in the United States, cycling accounts for just 1%). In contrast, there are countries with much greater shares of urban travel that is accounted for by cycling and walking. In Europe, for example, the Netherlands, Denmark, and Germany, cycling is mostly done for daily, utilitarian purposes including going to work, school, shopping, or visiting friends (Fig. 5.1).

Since the 1990s, the US federal government significantly increased financial support for creating, improving, and maintaining bicycle and pedestrian infrastructure and facility-related projects such as replacement or addition of sidewalks, accessible trails and parks, installation of bicycle storage facilities, lighting, as well as safety and education programs, and preservation of abandoned rail trails with the passage of three transportation bills that established flexible funding for multiple types of surface transportation including bicycle and pedestrian improvements (Cradock et al. 2009). The bills include the Intermodal Surface Transportation Efficiency Act (ISTEA) enacted in 1991, the Transportation Equity Act for the 21st Century (TEA21), and the Safe, Accountable, Flexible, Efficient, Transportation Equity Act (SAFETEA-LU) passed in 1998 and in 2005, respectively, and the added stimulus funds during the 2007–2009 economic recession (Pucher et al. 2011). Federal transportation funding contributes importantly toward achieving goals set by the Federal Highway Administration (FHWA) to double the share of non-motorized trips (i.e., walking and bicycling trips), and at the same time to make pedestrian and cycling safer in the country. Several FHWA programs incorporate funding for walking and cycling with the largest being the Surface Transportation Program (Cradock et al. 2009). Federal funding for walking and cycling projects increased dramatically from about $38 million in 1992 to a peak of more than $450 million in 2002. Federal funding for pedestrian and cycling projects has encouraged local and state governments to build new and improved cycling facilities (Pucher et al. 2011) (Fig. 5.2).

When walking/cycling is taken for other than recreational or health-related purpose, such as for basic transport for shopping or going to work, then walking and cycling are more subject to physical environmental influence and particularly, neighborhood context. To assess the extent of walking in specific neighborhoods, studies use the term "walkability,"

Fig. 5.1 Bike facilities in Rotterdam, Netherlands. (Author 2010)

Fig. 5.2 Bike storage facilities in Washington, DC. (Author 2010)

which has several definitions. In the smart growth study on the contribution of site design to walking, walkability is defined as "...the opportunity for continuous movement across some distance..." (Zook et al. 2011, p. 216).

5.2.1 Design Features Favorable to Walking/Cycling

Compact community design with grid-pattern streets, short block lengths, and close destination places can encourage active transportation. On the contrary, a sprawling community makes active transportation less safe and convenient (Green and Klein 2011). Land arrangement influences the choice between motorized or non-motorized transport with distance to destinations and connectivity (i.e., the configuration of the street network, travel directness) being key factors, though other factors such as travel costs, environmental quality, and access may also influence transport-related choice (Saelens et al. 2003). Yet other local-scale design features favorable to pedestrians include lighting, trees, street furniture, well maintained sidewalks, signage, and crosswalk markings at intersections.

Still other community design features linked with increased walking and biking include parks and open space, aesthetically pleasing places, multi-use paths, and bike racks (Green and Klein 2011). However, in most cities, planning for roads is done prior to planning for pedestrian facilities while planning for sidewalks does not include long-term maintenance, ease of access, or integrating the sidewalks with other neighborhoods and modes of travel, thus taking away from the quality of pedestrian paths and discouraging pedestrians from walking/cycling (Lawlor 2015) (Fig. 5.3).

Distance between origins (where trips start, e.g., homes) and destinations (where trips end, e.g., work places or shopping centers) is an important factor influencing walking/cycling. Distance can be measured as straight-line distance and is shaped by land use such as land density and land-use mix. High-density areas with apartment buildings are more conducive to walking and cycling because of convenience, with more potential activities (e.g., facilities or friends to visit) within walking or cycling distance. In fact, about 83% of all non-motorized trips (*trips* can be broadly

Fig. 5.3 In Washington DC, local-scale design features including trees, aesthetically pleasing places, and well-maintained sidewalks are favorable to pedestrians. (Author 2010)

defined as instances of moving from a point of origin to a destination) are short-distance, close to home, and for non-work purpose (Ross and Dunning 1997). In a low-density area with single-family housing such activities within convenient distance are less likely. *Land-use mix* is shaped by the distance between different land-use types including commercial, industrial, residential, office, public, and so on, where diverse land uses within a small area mean high mixed use, while a single land use within an area (which characterizes much of modern conventional suburban development) defines a place with poor land-use mix with widely separated land uses. In many European cities or older North American cities with a traditional layout, street-level shops and residencies above make it easy for neighborhood residents to walk from their residences to shops or places of work, while many suburban areas make it hard for neighborhood residents to walk to the closest store or get to work (Fig. 5.4).

Connectivity refers to how transportation systems (e.g., street network) connect destinations (Frumkin et al. 2004). It determines the ease of moving between origins and destinations within the existing street and

Fig. 5.4 Street-level commercial uses (a bakery is on the right and barber's salon is on the left) and residencies above in Essen, Germany. (Author 2010)

sidewalk network (Saelens et al. 2003). Connectivity has tremendous importance for travel since even nearby places can be made far apart within a poorly connected transportation system. In terms of active travel, connectivity is determined by the availability of continuous sidewalks and other paths and absence of significant barriers to pedestrians (Lawlor 2015). Connectivity also includes the elements of safety and accessibility beyond the neighborhood itself in connecting individuals with various activity sites including businesses, residences, commerce, and recreation within city and region via convenient and accessible links to other modes of transportation such as bus, streetcar, subway, or train within a reasonable time-distance (Southworth 2005). Streets laid out in a grid pattern and with few barriers (e.g., walls and freeways) to direct travel between origins and destinations result in high connectivity when route distances are direct and resemble straight-line distance. A simple gridded, interconnected road network is favorable to walking/cycling. A gridded pattern similarly offers shorter pathways and provides more than a single route to the same destination. In contrast, the *dendritic* (resembling the lifelines in the leaf on a tree) street pattern refers to the street arrangement based on the traffic volume with fewer streets organized into the hierarchy consisting of major arterials designed to carry cross-regional traffic, residential streets for local traffic, and few connectors. This street design forms the "loop and lollipop" neighborhoods where trips become more circuitous and lengthier (Frumkin et al. 2004).

Michael Lawlor (2015), in his analysis of the attributes of walkability, makes a case for the role of connectivity in city life on the example of the French city of Marseille, which has undergone urban decline, downtown disintegration, and population loss in the 1980s and 1990s due to deindustrialization leaving the city with just 6% of the metropolitan area's economic activity by the 1990s (Savitch 2011). To change the situation, one strategy was to focus on connectivity and promote walking as a source of transportation for both city residents and tourists while increasing pedestrian access, improving commuter experience, and diversifying land use. Marseille's connectivity within city limits and with the surrounding communities improved by integrating and expanding public transportation into neighboring communities, thus creating a more unified metropolitan area. Residents benefited during the daytime from getting good work access via mass transit and sidewalks while at night pedestrians can use the same facilities to reach various places of entertainment and travel to and from restaurants, cinemas, and other leisure places without having to use automobiles and taxis. More pedestrians appeared on the streets relying on travel by foot from public transit stations (rail and bus) to reach businesses, work places, and entertainment sites. The downtown area was restored where formerly abandoned warehouses and port buildings were turned into residential and commercial buildings; as a result, the population grew from 797,000 in the 1990s to 852,000 in 2009 and jobs were increasing at a yearly rate of 1.4% (Savitch 2011).

The importance of urban connectivity in making a place more walkable is illustrated in the pedestrian study of the Atlantic Station mixed-use residential and commercial development in Atlanta, Georgia. The Atlantic Station was conceived as a smart growth, sustainable development where its connectivity to the surrounding communities was an important feature. *Smart growth* is a generic term used by various organizations to promote development that attempts to ensure environmental sustainability and economic growth. To understand how site design including connectivity promotes walking beyond the immediate, local environment, walking habits and patterns of workers and residents of the Atlantic Station development have been analyzed using pedestrian counts as a measure of walkability and implementing behavior mapping to visualize movement patterns (Zook et al. 2011). The place's connectivity and midtown location that potentially extends walking beyond the boundaries of the development, allow residents and workers to use automobiles less frequently and thus reduce vehicle miles traveled compared with other residents of the metropolitan area.

Traditional communities have a good mixture of land uses as they integrate more and various land-use types within small areas. In contrast, modern suburbs possess low pedestrian connectivity, a low density of intersections (e.g., long block size), barriers to direct travel (e.g., cul-de-sacs), and thus offer few route choices (Randall and Baetz 2001). Conventional suburbs are also characterized by lower density of land use per unit area and a lack of land-use mix (Fig. 5.5).

5.2.2 *Safety and Active Travel*

Generally, the safest cycling is in cities with the highest bike mode shares since safer cycling promotes more cycling, and more cycling encourages greater safety, while cities with the lowest bike mode shares have the most dangerous cycling (Pucher et al. 2011). To illustrate, among Chicago, Minneapolis, Portland, San Francisco, and Washington, New York is the

Fig. 5.5 One-family house in a conventional US suburb in Memphis. (Author 2017)

largest city and it has the lowest cycling level at 0.6% in 2009 as well as a relatively high cyclist fatality rate of 8.58 fatalities per 10,000 daily commuter cyclists. The density of motor vehicle traffic may affect cycling safety. The greater percentage of pedestrian and cyclist fatalities occurs due to collisions with motor vehicles, thus making roadways the most dangerous environment for these activities. Further, motorists pay most attention on the center of the road focusing on other motor vehicles, and away from the road periphery where pedestrians and bicyclists are more likely to be found, therefore reducing car speed at intersections helps motorists to see bicyclists, motorcyclists, and particularly pedestrians better (Jacobsen et al. 2015).

The main strategy to increasing cycling and making it safer in countries of both Europe and North America has been the expansion of the supply and improvement of bike paths and lanes. The strategy is supported by the scientific evidence of the benefits of the approach revealing a positive correlation between cycling rates and the provision of bike paths and lanes, even after other explanatory factors such as city size, climate, topography, automobile ownership rates, income, and percentage of college students have been accounted for (Vandenbulcke et al. 2011).

Walking and cycling are safer when done on completely separate off-road facilities, (e.g., mixed-use recreational paths), in auto-free zones, on streets with traffic-calming road design features (e.g., narrow lanes and traffic circles that cause drivers to reduce speed), in residential areas (where speeds are slower and traffic is less), and physically separated on-street facilities (e.g., cycle tracks). For example, cyclists express a strong preference for separate paths and lanes rather than cycling in traffic (Menghini et al. 2010), while cyclists from the vulnerable populations including women, seniors, and inexperienced cyclists, specifically give preference to cycling on bicycle paths and lanes rather than riding on streets without bikeway facilities (Larson and El-Geneidy 2010). These findings provide scientific evidence that making cycling facilities better and safer is important for increasing cycling.

Traffic-calmed residential streets represent convenient, comfortable, and safe bike environments even without installation of special bike facilities (Pucher et al. 2011). Many cities in the Netherlands, Germany, and Denmark limit speeds to 30 km/hr (19 mph) or less on most residential streets. These legal speed limits are often reinforced through street redesign and infrastructure improvement used to decrease motor vehicle speeds and physically block through traffic with calming devices including street narrowing, traffic circles (used to lower speeds at minor uncontrolled intersections), speed humps, median islands, curb extensions (used to decrease the overall width of the roadway and reduce the crossing distance for pedestrians), raised intersections and crosswalks, and mid-block street closures with pass-throughs for bikes (Buehler and Pucher 2011).

Installing a roundabout at an intersection has a benefit of decreasing injury crash rates by as much as 50% at least (Transportation Research Board 2011). Accordingly, roundabouts have greatly increased in number in European countries. For example, in France, the total number of roundabouts grew from 10,000 to 30,000 adding to the reduction in France's fatality rate. In Sweden, where the *Vision Zero* strategy encompasses road design with safety being the main design criterion, road design discourages drivers from risky behavior and inattention. For example, to minimize conflicts between motorized and non-motorized traffic, intersection improvements can be implemented (e.g., when simple intersections with traffic lights are replaced with intersection roundabouts), barriers that separate pedestrians and cyclists from motor vehicles can be installed, and special features to keep drivers alert (e.g., instead of long straightway gentle curves) can be introduced and incorporated. To improving overall walking and cycling safety, cities need to provide more and better quality separate facilities. The provision of safety facilities is particularly important for certain populations including children and the elderly who are most vulnerable in terms of fatality or serious injury in case of a collision with a motor vehicle (Buehler and Pucher 2017) (Figs. 5.6 and 5.7).

5.2.3 *Active Travel in Germany and the United States*

Public policies have contributed to the impressive success of active travel such as walking and cycling in Germany, which is similar to the United States in terms of affluent economy, high per-capita incomes, high auto ownership rates, similar proportion of licensed drivers, vast road net-

Fig. 5.6 Traffic calming in residential areas: 30 km posted speed in Essen, Germany. (Author 2010)

Fig. 5.7 Higher than in Germany posted speeds of 35 miles/hour (>56 km/hour) in residential areas in USA. (Author 2017)

works, and advanced emergency medical care. Rapid motorization took place similarly after World War II coinciding with fast suburbanization (Buehler et al. 2011). Suburbanization provided relief to the housing shortage of the 1950s. German refugees from the former Eastern territories and the Soviet-controlled counties (Baron (1995) put the number at 12 million) had to be accommodated within the three western occupation zones in 1945. By the end of the war, in all major German cities, the central areas were ruined and the transport system, including all major railway hubs, bridges, airports, ports, and urban public transport, was severely damaged (Baron 1995). Urban areas were crowded and housing space was rationed with several parties sharing an apartment resulting in a housing shortage in cities. Most refugees were distributed in the rural areas, which mostly had not been damaged. To lessen the urban housing shortage, in the early 1950s the federal government offered substantial tax relief for persons willing to construct single- and two-family homes at the urban fringe (and essentially subsidized suburbanization), forcing local authorities to provide land for residential development. Land was developed for residential use far from the main urban public transport corridors resulting in the post-war suburbanization supported by the federal government's housing policy in Germany. The newly developed housing areas had issues similar to those of the US suburbs: they were not served by public transportation as the low residential density did not justify provision of regular services. Later, the federal government subsidized the automobile by providing low-interest rates and tax benefits for the purchase and use of private motorized vehicles for home-to-work trips in excess of six kilometers with most employment historically concentrated in the city center (Baron 1995) (Fig. 5.8).

Germany differs from the United States with regard to government policies toward walking and cycling (Buehler and Pucher 2017). For example, federal, state, and local transport policies made the costs of driving higher and limited car use and parking in cities, and at the same time, improved the safety of walking and cycling since the 1970s by installing pedestrian car-free zones in city centers, traffic-calming features on residential streets, and

Fig. 5.8 A residential suburb by Essen, Germany. (Author 2010)

providing extensive networks of footpaths and bikeways (Buehler et al. 2011). Mandatory cycling training in schools where schoolchildren are taught traffic education and learn safe walking and cycling habits is complementary to the aforementioned infrastructure improvements, and driver training emphasizes avoidance of the endangerment of pedestrians and cyclists, and traffic regulations are strictly enforced on both motorists and non-motorists. German land-use policies promote mixed-use, compact development where many short trips are generated, including in new suburbs, while restricting low-density sprawl. When trips are short people are more likely to walk or cycle. Transport policies also can indirectly influence the relative attractiveness of alternative modes of travel. For example, in post-war Germany, rail transport was promoted while private motor vehicles began to be subsidized later resulting in a massive increase in vehicle ownership and road traffic volumes in the early 1960s. The US transport policies create a strong incentive to drive a car due to the low cost of car use and absence of the policies limiting its use in the United States thus indirectly discouraging active travel (Buehler et al. 2011) (Fig. 5.9).

Fig. 5.9 Open air farmers market in pedestrian car-free zones in Essen center, Germany. (Author 2010)

Differences in active travel in terms of daily walking and cycling between the United States and Europe, using Germany as an example, and active travel trends in both countries have been examined using the 2001 and 2009 National Household Travel Surveys (NHTS) data for the United States and the similar 2002 and 2008 Mobilität in Deutschland (MiD) (the Mobility in Germany) surveys for Germany (Buehler et al. 2011). The two surveys have similar design and timing and thus enable a comparison of public transport demand in the two countries. The entire sample size for the United States was 308,901 individuals in 2009 and 58,882 individuals in Germany in 2008; both samples excluded children aged younger than five years. Time periods slightly differed due to data publication differences in both countries, and constituted the period 2002–2008 for Germany and 2001–2009 for the United States (Buehler et al. 2011). Active travel including walking and cycling is far more widespread in Germany compared to the United States. The population share who meets the 30 minutes per day goal of physical activity by walking and cycling is more than three times as high in Germany as in the United States (29% vs 8.7%). Trips by walking account for close to 24% of all trips taken in Germany in 2008 and only 10.5% in the United States in 2009; analogously, cycling trips comprise 10% of all German trips while cycling accounts for just 1% of all trips in the United States. The increases in active travel from 2002 to 2008 were much larger in Germany than in the United States from 2001 to 2009. To illustrate, the rise in the mean annual hours spent walking and cycling per person during 2002–2008 is more than twice in Germany compared with that in the United States during 2001–2009 (20.2 vs 8.4 hours), and the rise in the mean annual distance walked and cycled per person from 2002 to 2008 is four times greater in Germany than in the United States from 2001 to 2009 (109.6 km vs 25.6 km).

Increased active travel in Germany is largely due to the many more people who engaged in walking and cycling from 2002 to 2008, while there was the same proportion of the US population who walked and cycled over the period 2001 to 2009. Buehler et al. (2011) conclude that the main issue in the United States is the low percentage of the population engaging in active travel. In Germany, all socioeconomic groups engage in walking and cycling, and remarkably there are many more people from vulnerable population segments who walk and bike in

Germany. For example, German seniors (people who are at least 65 years old) who achieve 30 minutes of physical activity on a daily basis, more than five times exceed the number of their American counterparts (35.1% vs 6.4%), while the number of children (between 5 and 15 years old) in Germany meeting the 30-minutes threshold is more than three times as high (30.3% vs 8.6%), and women who exercise for at least 30 minutes daily exhibit similar differences in the two countries (29.3% vs 8.2%). The steep discrepancy in active travel between Germany and the United States indicates that policies need to be developed aiming at increasing walking and cycling among elders, children, and women. To increase active travel in the United States overall, a multifaceted approach is needed encompassing transport policies, land-use planning, and promotional programs because of the existing evidence of their positive impact on walking and cycling (Buehler et al. 2011). Prior research documents influence of population density, mixed land use, street connectivity, walking and cycling facilities, and overall urban design on active travel (Fig. 5.10).

Fig. 5.10 German senior citizens at a light rail stop near a residential neighborhood in Essen, Germany. (Author 2010)

5.3 Evidence of Impact of Environmental Factors on Active Travel

Studies examine environmental factors related to actual or reported rates of walking and/or cycling. Traditional neighborhoods are deemed walkable and bikeable due to high population density, a high land-use mix, high connectivity, and continuous sidewalks favorable for walking and biking, while communities with lower population density, less diverse land use (e.g., a single residential land use), lower street connectivity, and inadequate street design (i.e., with insufficient sidewalks, bike lanes, or stop signs at intersections), differences in travel behavior of residents, especially walking/cycling rates, can be compared between dwellers of neighborhoods that differ in neighborhood built design. Based on the results indicated by the studies, differences in the amount of walking/cycling done by neighborhood residents can be estimated between neighborhoods with different built-environment features (Saelens et al. 2003).

Bikeway facilities include on-street bike lanes, on-street bike paths (cycle tracks), and off-street bike paths. Paths have an emphasis on recreational cycling and are typically useful for leisure purposes, while on-street bike lanes more serve utilitarian cycling, offer a more direct route, and are more useful for reaching destinations. *Cycle tracks*, typical in Northern Europe, are on-street bike paths separated by physical barriers from motor vehicle traffic, essential in cities with high motor vehicle traffic with vehicles blocking bike lanes. The most separation from motor vehicle traffic is reached on off-street paths; most cyclists report the highest level of comfort and perceived safety using these facilities (Fig. 5.11).

Metrics to measure walking/cycling vary among studies. Some studies estimate average walking and sum up walking and biking trips per week (averaged) in high- versus low-walkable neighborhoods (Cervero and Gorham 1995; Cervero and Radisch 1996), while other studies assess the percentage of all trips made by walking/cycling relative to other transport modes (e.g., transit, automobile) based on the assumption that using different travel modes (e.g., car, transit, walk/bike) people make around 30 trips weekly (Ross and Dunning 1997) and arrive at absolute weekly walk/bike trip estimates (Handy and Clifton 2001; McNally and Kulkarni A 1997; Kitamura et al. 1997), and finally, observational studies observe or survey pedestrians and cyclists regarding influences of high- and low-walkable communities and other environmental factors on pedestrian or cycling behavior within targeted neighborhoods (Cervero 1991; Shriver

Fig. 5.11 Off-street separated red-colored bike path in Rotterdam, Netherlands. (Author 2010)

1997; Zook et al. 2011; Clayton and Musselwhite 2013; Christiansen et al. 2014). For example, during family interviews many elements of everyday travel behavior can be discussed including the way family members interact with the environment during travel, what they enjoy (or don't) during family rides, how they see the ideal cycling environment, and whether they know the cycle paths and routes in the area (Clayton and Musselwhite 2013). Even when sociodemographic determinants of walking/cycling for transport have been controlled for, environmental variables remain significant predictors.

Observations suggest that residents have higher rates of walking and cycling in communities with higher density, greater connectivity, and more land-use mix compared with neighborhoods that have low density, poor connection, and single land use (Saelens et al. 2003). Additionally, survey and interview results indicate that a *route design* such as signposting is important to active travel including cycling. Providing official information in the form of signpost displays about the local area and attractions, shops, and services in the cycle infrastructure seems desirable and beneficial.

Additionally, signposting with extra signs along cycle paths to wayfind indicating location within a designated cycle path reassures less-experienced cyclists and helps them orient themselves while not feeling lost. At the same time, signposting and other changes should be only so much, without putting too much pressure on the local environment, so that the experience of the natural areas, and therefore, enjoyment of a route, will not spoiled (Clayton and Musselwhite 2013).

5.3.1 Active Travel

Regardless of neighborhood built environment, compared with other travel modes (e.g., automobile), the frequency of walking trips per week is relatively low in the United States (Ross and Dunning 1997). Nevertheless, walking trips per week done by residents of high-walkable neighborhoods are about two times those reported by low-walkable neighborhood residents (3.1 and 1.4 trips, respectively) and with the magnitude dependent in part on the walk trip purpose (Saelens et al. 2003). While walking for exercise does not seem to differ between high- and low-walkable neighborhoods (Handy and Clifton 2001), utilitarian walking with purposes including getting to work and walking for errands seem more associated with high-walkable neighborhoods (Cervero and Gorham 1995; Cervero and Radisch 1996). Clayton and Musselwhite (2013) explored whether people like cycling just for cycling's sake. Often, cyclists report using bikes as a means of transport for commuting and other functional trips (e.g., going somewhere in particular, having a solid purpose), so they have a more utilitarian purpose rather than cycling just for cycling's sake. When there is a reason for the cycle (e.g., going to a certain destination), the family members seem to enjoy it more since the family cycle is purposeful and is validated (Fig. 5.12).

5.4 Limitations of Transportation Studies

However, partly because these transportation studies do not apply correlational or regression techniques, as well as other reasons, they cannot quantify the relation between neighborhood characteristics and non-motorized transport, so it is hard to determine exactly how much the factors of neighborhood environment influence the magnitude of walking and cycling for transport, thus, it is difficult to develop objective and reliable walkability index that could be assigned to high- and low-walkability. Other reasons

Fig. 5.12 Walking in Boston past a popular restaurant. (Author 2013)

include inconsistency in methodology and study design across transportation studies examining rates of physical activity as well as inconsistency in evaluation of potential confounding variables. Well-specified guidance for determining walkability and neighborhood selection is still lacking, rending difficult an objective comparability across neighborhoods (Matley et al. 2000). To illustrate inconsistency in assessing physical activity, as mentioned earlier, while some transportation studies measure the percentage of all trips taken by walking/cycling, other researchers measure the absolute number of walking/cycling trips precluding researchers from comparing high- and low-walkable neighborhoods and rates of physical activity across studies in an objective way.

Another example of inconsistency in evaluation of potential confounding variables includes the lack of assessment and statistical control for individual-level factors that may influence walking/cycling rates for transport in high- and low-walkable communities (e.g., residents' median age and ethnicity) (Handy 1992) known to be associated with walking and cycling although differences in neighborhood-level socioeconomic status are frequently controlled for by transport studies examining walking/cycling in high- versus low-walkable neighborhoods (Cervero and Gorham 1995; Cervero and Radisch 1996; McNally and Kulkarni 1997). Transportation and planning studies that have taken a correlational and regression approach make it possible to measure the relation between neighborhood characteristics and walking/cycling rates as well as examine consistently the influence of potential individual-level (age, income, automobile ownership) confounding variables that may impact rates of physical activity in high- and low-walkable neighborhoods. Neighborhood-level characteristics related to non-motorized transport and most consistently and positively correlated with walking/cycling trips include population density as well as land-use mix and the close proximity of nonresidential land use to housing, followed by the walking and cycling infrastructure (e.g., existence of bike paths, pedestrian-friendly facilities such as continuous sidewalks, street lighting).

5.5 Physical Activity Programs

The current overweight/obesity epidemic in the United States is attributed partly to low levels of physical activity. Adults and children in the United States have decreased walking and biking activities by 30% since the late 1970s (Hu and Reuscher 2004). According to the *Passenger Travel Facts and Figures* 2016, a report by the US Department of Transportation, walking and biking accounts for a small portion of passenger travel, comprising just 11.5% of local trips and 1% of miles in 2009. Walking trips comprised 10.4% of trips in 2009. Walking predominantly is done for family/personal errands (the category of family/personal errands includes personal business, shopping, and medical/dental appointments) and social/recreational purposes accounting for 83% of walking trips (*Passenger Travel Facts and Figures* 2016). However, active transportation such as walking/cycling has been believed to be one approach to address the American obesity epidemic (Green and Klein 2011). To achieve moderate intensity activity while accruing the largest health benefits is possible for adults at any age by walking briskly, which is the most common moderate intensity activity (Sallis et al. 2004). Since walking can be done for transportation and health purposes, transportation and health professionals are trying to find solutions that can increase physical activity, reduce obesity and associated health risks, and improve public health.

Many countries introduce policies to reduce dependence on private vehicle. The development of compact walkable communities is considered one of the most effective strategies. Before such policies are introduced, it is important to establish whether the policy-based assumptions are supported by the robust research findings. Rigorous research makes valuable contributions by helping formulate a useful policy tool. Research results have important policy implications. For example, if closeness of different sites such as schools, stores, parks, and workplaces to residential neighborhoods is found critical to forming favorable attitudes towards non-motorized travel (also known as active travel), such as walking and bicycling, local governments and school districts should consider policies aiming at how best to arrange activity sites relative to surrounding residential neighborhoods (Seraj et al. 2012). Similarly, if residents of higher-density areas shape their travel behavior so that the total use of private vehicles is reduced while increasing modal shares of other transportation compared to lower-density neighborhoods, then density-increasing measures might be efficient strategies to reduce car dependence. However, so far there is no clear understanding of the relationship between built-environment factors and personal attributes and travel behavior. Thus, it is hard to develop efficient urban policies targeted at changing travel behavior (Dieleman et al. 2002). For example, both travel mode and distances traveled may be impacted by urban form and design, but also individual attributes (that is, socioeconomic, demographic) are linked to travel behavior.

To motivate residents, programs should be initiated to create safe and accessible routes for people of all ages as well as for children to walk and bike to school; the public should be made aware of the walking and bicycling environment and provided with safety training through discussions and presentations. An example is a special community-based program, "The Safe Routes to School Program," created in Marin County, CA in the early 2000s. The program provides a successful model for promoting walking and biking to school. The program encourages involvement of the entire community (e.g., more than 7600 students in 21 participating public and private schools have been served by the third year of the program) (Staunton et al. 2003). The Marin County Safe Routes to School Program is funded by grants and donations from the Marin Community Foundation, National Highway Traffic Safety Administration, local businesses, Marin County, the City of San Rafael, and the California Departments of State Services and Transportation. Most funding is dedicated to infrastructure-related activities to increase safety for student walking and biking. The activities that decrease the traffic danger faced by students include mapping safe routes to schools consisting of identifying, creating, and walking the safe routes, designing solutions such as sidewalks, improving pedestrian signage and crossings, extending existing bike trail, bike lanes, and so on, and fund raising for needed traffic infrastructure changes. The success was compelling with an increase in active school trips while the use of private vehicles for transporting students to school decreased substantially. The student transportation surveys asked school students the transportation mode they used in traveling to school with the following results: walking (64%), biking (114%), and carpooling (91%); only 39% of students came to school by private vehicles. Additional benefits of this and similar programs include health improvement, reduction of traffic congestion, and a greater sense of community (Staunton et al. 2003) (Fig. 5.13).

Fig. 5.13 Many people are walking along the Esplanade in NYC. (Author 2013)

Another example of the promotion of active transportation and how public health can be linked to urban planning is the Community Healthy Places (CHP) program created in 2006 by the Ohio Health Department in Columbus, OH to set up development policies and practices to mitigate adverse health impacts and to create places that promote physical activity as part of daily life. Local communities called for the changes in the built environment and in 2004–2006 submitted neighborhood plans to the Columbus with the objective to improve walking and biking infrastructure (Green and Klein 2011). The CHP established relationships with Columbus government agencies including the Building Services Division, which is responsible for building permits and new development, so the CHP was granted permission to make recommendations to any rezoning application during monthly rezoning reviews (needed when the current land use does not permit the proposed development) enabling a discussion and implementation of the ideas of improving walkability of the built environment. Next, the Development Commission can ask the developer to include active transportation features based on the request by CHP and once the developer has the chance to address the recommendations to include active transportation features, the Commission makes recommendations on whether to approve or disapprove the rezoning application to the Columbus City Council who in turn must pass legislation to enact any rezoning application and change the zoning of a piece of land. Some elements of walking and biking infrastructure changes, as recommended by the CHP, are low-cost, for instance, a bike rack can be added for $200, while making other changes can be substantially more costly for a developer, including making wider or longer sidewalks. To demonstrate a success of this strategy, after the CHP began reviewing city development rezoning applications in Columbus, 64% of development applications adopted active transportation components not required by the zoning code, such as including bike racks, widening or adding sidewalks, and providing sidewalk connectivity (versus 7% of development applications, which included active transportation components prior to CHP's review and recommendations) (Green and Klein 2011).

5.6 Built Environment and Cycling

Cycling has increased substantially on the North American continent including both in the United States and Canada over the past few decades. Since 1990, large American and Canadian cities have introduced and

Fig. 5.14 New York is among the US cities where cycling rates have increased. (Author 2013)

realized numerous infrastructure and innovative programs to encourage cycling and enhance cycling safety, experiencing as a result a doubling and tripling in cycling rates, including in Chicago, Minneapolis, New York, Portland, San Francisco, and Washington in the United States and Montréal, Toronto, Vancouver in Canada (Fig. 5.14).

Despite the NHTS results suggesting no link between city size and cycling levels, examining cycling in the US large cities, Pucher et al. (2011) observed that larger cities seem to have lower cycling activity because of larger land area, longer trip distances, and more widespread public transport systems. Cycling levels are much higher in the central cities than the overall metro areas, while within the cities there is spatial variation where established gentrifying neighborhoods close to downtown tend to have higher cycling rates than in the outer neighborhoods and in suburbs due to location within convenient cycling distance of employment at the city center and university campuses and with diverse land uses. The expectations are that higher densities of larger cities are favorable to cycling due to the concentration of many origins and destinations, however, high road traffic and restricted space for bikeway facilities associated with higher density might also discourage cycling. Other factors impacting cycling include percentage of college students (Minneapolis' share of college students is 12%) and car-free households (for example, the percentage of car-free households is 54% in NYC) who are more likely to cycle (Pucher et al. 2011). These large cities applied innovative policies and succeeded at increasing cycling far more than the United States and Canada did as a whole providing valuable lessons on how to encourage cycling for other communities. A range of infrastructure improvements includes implementation of the following: expansion and improvement of bike lanes and paths, traffic calming, parking, bike-transit integration, bike sharing, training programs, and promotional events.

Portland, OR implemented a comprehensive package in terms of almost everything including cycling infrastructure, programs, and policies promoting cycling resulting in a sixfold increase in cycling rate since 1990; from 1.1% to 5.8% in 2009, the highest cycling level of any large North American city.

Portland provides a model of integrated cycling policies for other North American cities that can build upon Portland's experience (Pucher et al. 2011). The central feature of cycling policy in Portland has been the provision of more and better bikeway facilities. Portland has one of the largest supply per capita with 73 km of bike lanes and paths per 100,000 population (Pucher et al. 2011). A brand new six-mile Sunday Parkways route was added in August 2017 in the Outer Northeast Portland neighborhoods (The City of Portland 2017b). During 2017–2020, the Portland plans to construct close to $20 million worth of cycling and walking infrastructure to enable neighborhood residents to reach their jobs, schools, parks, and local businesses more easily.

Portland also advocates for cycling via numerous marketing programs, educational campaigns, and community cycling events such as Sunday Parkways and others. Sunday Parkways is a series of free community events organized since 2008. The first successful Sunday Parkways in the nation was held in June 2008 with 15,000 participants. During the events, "soft" street closures are implemented (i.e., residents living along the routes can access their driveways, but others cannot drive on the street, thus creating a relatively traffic-free space) and people enjoy the city's traffic-free streets as they discover active transportation by strolling, biking, and the like. The main goals for the Sunday Parkways project include an increase of the health and physical activity of all Portland residents; reduction of car dependence for all trips; participation of especially low-income and communities of color who have greater risk for diseases associated with lack of physical activity and obesity including diabetes, heart disease, and other diseases; and increase of acceptability of alternative modes of travel including bicycling and walking, among others (The City of Portland 2013). Programs also include cycling training courses for both children and adults. Safety education is equally important in Portland, especially to new immigrants, refugees, and English-language learners: since 2016, a new program has been launched providing driver safety education training to these categories of population over the age of 18 years with an additional impetus of lower auto insurance rates upon completion of the course (The City of Portland 2016).

Portland has a reputation as a walkable, bikeable, livable city; however, there are more annual traffic deaths than homicides. The city made a commitment to end traffic violence by adopting its *Vision Zero* strategy with a goal to eliminate all traffic deaths and serious injuries on Portland streets by 2025 by fixing unsafe streets (The City of Portland 2017a). Many deadly crashes occur due to dangerous behaviors including driving impairment due to drugs or alcohol, distracted and aggressive driving, speeding, and violation of traffic laws. A disproportionate number of traffic deaths happen on certain types of streets such as wide arterials with fast-moving traffic, multiple lanes in each direction, unprotected pedestrian crossings and bike lanes, and longer distances between signals. So designing streets strategically may discourage dangerous driving where all people are safer even if people make mistakes (The City of Portland 2017a).

To avoid conflicts between cars and bikes and make bike lanes more visible to motorists and increase cycling safety, bike lanes have been painted bright green, blue, or red on some problematic streets in Portland, while on some streets buffered bike lanes have been installed where a diagonally striped lane separates the bike and car lanes. Portland also uses bike boxes, which is an intersection safety design preventing collisions between cyclists and cars. The bike box is a green box painted on the road with a white bicycle symbol inside serving to make cyclists more visible and motorists more aware of bicyclists. Regarding bike parking for cyclists, to avoid theft and protect bicycles from inclement weather, provide safe sheltered parking, and encourage future cyclists, Portland used innovative on-street bike parking called "bike corrals" converted from one- or two-car parking spaces. As of 2017, Portland had 142 such bike corrals. Bicycle corrals consist of several groups of bicycle racks installed in the parking strip in areas with high bicycle parking demand accommodating as many as 12 to 24 bicycles in one or two auto parking spaces (The City of Portland 2017a).

Bike sharing is a program providing impetus to cycle more. Among other benefits of bike sharing programs identified in Europe was better coordination of cycling with public transport. Expanding bike sharing in the United States may further contribute to the growth of cycling. In July 2016, Portland launched a bike share program called BIKETOWN with 1000 bikes at 100 stations across downtown and several neighborhoods. The program uses Social Bicycles (SoBi) designed for easy riding in an urban environment and modeled after the step-through Dutch frame made comfortable for a wide range of riders (BIKETOWN 2017).

5.7 Active School Transport (AST)

Characteristics of built environment and accessibility affect the choice of transportation, especially the decision whether to use non-motorized modes of transportation. What changes in urban form (e.g., reducing traffic speed, adding more paths to the street network, using land-use mixes, and increasing density) may help promote active modes of travel (cycling). The recent research examined an association between urban form/land use measured by the objective walkability index and *active school transport* (AST) as opposed to *passive travel* (that is, riding in a car). It has been shown that AST may importantly contribute to physical activity and health. AST varies by country with European countries reporting the highest prevalence worldwide. Determining if the urban form contributes to variations between schools could assist in promoting AST. Specifically, Christiansen et al. (2014) were interested in finding land-use (LU) effects on AST including the barriers to AST and whether improving infrastructure results in an increase in cycling and what is needed to increase it. The authors also wanted to determine whether people like cycling just for cycling's sake as riders report experiencing pleasure from motion they feel when cycling, thus setting bikes apart from other travel modes, particularly automotive vehicles (Clayton and Musselwhite 2013).

The study was conducted in Denmark where the proportion of non-motorized travel modes is much higher than in other countries. Children of school age use active types of travel to journey to school. Among different modes of travel used to get to school in Denmark, cycling's share is the highest with 50–70%, and walking with another 10–20%, while passive travel accounts for just 6% (Christiansen et al. 2014). In the study, 50% of the participants reported parents' weekly cycling and/or encouragement of cycling, while 90% reported that their friends cycle to school every day. Land-use effect on AST was studied by constructing a ***walkability index***, which measures how walkable an area is; it is a composite index based on road connectivity, traffic exposure, and residential density. Higher walkability around schools as well as the existence of many paths and safe crossings results in a greater likelihood for AST. To compare, in the United States, 50% of children travel to and from school by private car while about 30% of children take a school bus (Our Nation's Travel 1997). Further, the US national average of children who either bicycle or walk to school is just 13% (Safe Routes to School 2010). Only about one US child in nine starts the day by walking or biking to school (Staunton et al. 2003). However, in some areas such as Southern California (the Los Angeles–Riverside–Orange County consolidated metropolitan statistical area, or CMSA), the percentage of children bicycling or walking to school is much higher at about 21%.

The California add-on sample of the 2009 National Household Travel Survey (NHTS) of the United States has been used to study the link between the attributes of built environment (e.g., exposure to traffic such as volume of traffic and speed, street connectivity, and accessibility measure by distance to school) as well as crime statistics, weather, and parental attitudes towards bicycling and walking (Seraj et al. 2012). The authors make a recommendation that transportation engineers incorporate features of urban forms that contribute to safer walking or bicycling to school. Using grid street patterns provides direct connectivity between home and school, so children are less exposed to traffic along busy arterials, and decreases travel distances. This urban design strategy may reduce parental worries about school distance, traffic conditions (volume and speed), and crime-related concerns. Barriers to walking and biking to school, such as physical distance, traffic danger (volume and speed), crime, and so on, vary by community (Centers for Disease Control and Prevention 2002). Among barriers to AST, exposure to heavy traffic and high traffic speed have both been found to decrease trips by active modes (Christiansen et al. 2014).

While some studies examine what factors of the built environment influence AST, other studies investigate the opposite: what factors may deter children from walking or bicycling to school? The distance to school appears to be one of the key variables impacting children's choice of transportation to school (Timperio et al. 2006; McDonald 2008). The separation between activity sites (e.g., between homes and schools) influences AST: active travel is prevalent up to 4 km (about 2.5 miles) while as distances become longer than 2 km (about 1.2 miles), the odds for AST get smaller; however, acceptable distances differ across countries due to differences in culture and physical environment (Christiansen et al. 2014). When school is located within or in close proximity to residential neighborhoods, it is less likely to encounter high volume or high speed traffic to which the child would be exposed. The opposite is true: the likelihood of encountering bad traffic conditions increases with distance. Thus, the longer the school-to-home distance, the less likely a child is to engage in AST. Greater traffic volume and higher traffic speed, which are often associated with urban locations, are linked to parental concern about the personal safety of their children when they walk or cycle. So implementation of speed reducing measures such as traffic-calming devices would lessen the speed and the volume of traffic making streets safer to navigate. These interventions may also decrease a parent's anxiety about the traffic conditions for children's AST. Personal factors, environmental factors, and trip characteristics (such as the purpose and length of the trip) interact in complex ways about the decision whether to engage in walking or biking, thus making it desirable to incorporate these determinants in active transport research for accuracy purposes (Moudon and Lee 2003). In a similar way, if the distance between home and school is shortened, not only are parental concerns about distance reduced, but also worries about volume and speed of traffic are decreased.

5.7.1 AST and Safety Perceptions

With 30% of US school students being taken to school by school bus, school transportation has to be studied. It is helpful to consider factors of physical environment at three different levels: neighborhood, route, and destination (here, schools) (Panter et al. 2008). Studies are lacking on the link between the urban form around schools and AST. Therefore, studying walking/cycling to get to school should include the examination of urban form around school locations. Even less attention was given to the importance of perceptions of physical environment and social environment (Christiansen et al. 2014). The subjective indicators can be used to analyze the link between urban form and AST. The subjective indicators are measured by perceptions of physical and social environment. *Safety perception* can be measured through perceived physical environment as safety en route to school (e.g., as safe and very safe). Perceived social environment can be measured as perceived parental encouragement of cycling (that is, *social support*) and cycling by parents and friends (that is, *social norm* of cycling). These standards of social support reflected by parental encouragement and social norms of cycling reflected by parents and friends' frequent cycling lead to higher AST with the latter even more important than the former (Christiansen et al. 2014).

Previous research suggests that improving physical structure and street design features may lead to increased active modes of travel including walking and cycling. Improving safety perceptions including perceptions of the physical environment may also help overcome concerns over safety of active travel. If route is perceived to be 90% safe, it means more trips by active modes (Christiansen et al. 2014). Long-term improvements to the physical structure and simple modifications of street design for walking and cycling increase the potential for active transport (Lee and Moudon 2008). Clayton and Musselwhite (2013) explored whether changes to cycle infrastructure improve the experience of cycling for families. Improving cycling environment, for example, by adding excitement in a cycle network infrastructure (such as introducing enjoyable activities along the route, or new aspects to a routine journey), a cycle network may be used more often and motivation for cycling is at least maintained and therefore the incidences of active travel increase. Among other benefits, those who are less experienced in cycling may reduce their fears of cycling with infrastructure improvement.

Street design features should be distinguished from the physical structure (urban form) (Cervero and Kockelman 1997). Often, attributes of both urban form and street design covary. The spatial covariation makes conventional statistical analyses problematic due to spatial multicollinearity (Saelens et al. 2003). The urban form context can generate "opposing and synergetic effects" on active transport. For example, while higher density generally is more conducive to non-motorized travel such as walking or cycling, areas with greater density and street access are perceived to have higher crime rates, which may increase parent's anxiety about personal safety for their children (Harries 2006; Foster and Giles-Corti 2008).

5.8 Use of Public Transportation

The post–World War II suburbanization of cities into surrounding areas has created land-use patterns and trips that public transport systems could not easily serve resulting in a steep decline in demand for public transport during the decades right after the war; first in North America because wartime fuel and tire rationing ended and car production interrupted during the war resumed, as well as rapidly rising per capita income and car ownership in the 1950s and 1960s resulting in widespread auto-dependent suburban sprawl. Similar reasons undermined public transport demand later in Western Europe. For example, by the late 1960s, public transport services had been reduced or cut, buses replaced trolley services, fares were raised due to increasing competition from the automobile, and trips per capita dropped by 21% from 136 to 107 between 1956 and 1968 in West Germany. Between the late 1960s and early 1980s, however, partly due to the two oil price shocks of the 1970s and because of the subsidization by the federal government, public transport services grew in quantity and quality leading to an increase in public transport demand in West Germany (Fig. 5.15).

Fig. 5.15 Light rail ("Ubahn") on a street by residential buildings in Essen, Germany. (Author 2010)

Fig. 5.16 Waiting for a commuter train to New York City. (Author 2013)

Between 1945 and 1973, the US public transportation suffered from insufficient public financing leading to an increase in fares, deterioration of service, and widespread bankruptcies of public transport firms nationwide. For example, streetcar services became almost non-existent and partially replaced by bus services. By 1970, most services in public transport in the United States suffered from poor maintenance, unreliability, and lack of coordination. However, since the 1960s and 1970s, the market share of public transport, expressed as the number of annual public transport passengers, has been increasing in North America and Western Europe despite growing incomes and car ownership rates and proliferating car-oriented suburban sprawl (Buehler and Pucher 2012). In the 1970s, the federal government supported public transport by subsidizing its capital and operating costs while financial support by state and local governments also rose. Public transport, which had been mostly privately owned and operated until 1960s, by the end of the 1970s almost entirely became publicly owned, managed, and financed in the United States. Government funding at all levels enabled substantial expansion and improvement of public transport services resulting in an increase in public transport demand since 1973 (Buehler and Pucher 2012).

Generally, public transport demand is linked to economic cycles with short-term declines in demand occurring during economic recessions and short-term increases in passengers taking place during booms in the economy or sharp fuel price increases. For example, the observed decline in work trips (measured as the percentage of public transport trips for work purpose) from 40.5% to 35.3% between 2001 and 2009 is partly explained by the 2008–2009 recession in the United States (Figs. 5.16, 5.17 and 5.18).

In both countries, subsidies for public transport by governments decreased during the 1980s. After Germany's reunification there was a steady increase in public transport demand observed in rising passengers between 1991 and 2010, partly due to an increase in the gasoline tax, which doubled from \$0.41 per liter to \$0.88 per liter over 1990–2010 (Buehler and Pucher 2012), and because of service improvement reflected in better comfort, convenience, and reliability (see below for strategies to increase public transport demand).

Fig. 5.17 Public transportation in Boston. (Author 2013)

Fig. 5.18 Walking to a public transportation stop in Boston. (Author 2013)

Measures to increase public transport use include higher productivity and lower costs so that more services could be provided at lower fares, thus attracting more riders. All strategies that have been developed to increase demand can be grouped into three categories:

(1) Expansion and improvement of services across all transport systems. Service has been greatly expanded in Germany, offering 59 vehicle kilometers of combined service by regional rail and metro, and bus and light rail per year per inhabitant in 2009 compared to just 20 in the United States. Services have been greatly improved since the mid-1990s in Germany: stations and vehicles have been modernized, vehicles offer low-floor boarding and comfortable seating, and real-time information is provided both at stations and on-board many vehicles. Schedules and route information can be found at all bus stops.

(2) Attractiveness of fares and convenience of ticketing. In Germany, there is a choice of tickets including monthly (offering deep discounts for all groups), semester, and annual tickets with discounts for children, university students, and seniors. In addition, when there are large events, entrance tickets include free public transport.
(3) Full coordination of services, ticketing and timetables, and fares among regions and different modes. Multimodal service coordination includes convenient transfers between bus and rail, extensive bicycle parking offered at rail stops, and availability of park and ride facilities for cars at suburban rail stations (Fig. 5.19).

To complement the aforementioned strategies, public policies can promote use of public transport by limiting car ownership, use, and parking, and increasing pricing. For example, purchasing a new car entails 19% sales tax in all German states. Driver training and licensing is expensive with more than $2000 for a license. Land-use policies can either encourage or restrict public transport demand. Car use is restricted through application of traffic calming on most residential streets (with 30 km/hour speed, or less than 19 mph) and general speed limits of 50 km/h (about 31 mph) in German cities. Central parts of most city centers have integrated large car-free zones. Parking supply is limited and expensive as most municipalities have decreased car parking in downtowns and raised fees for parking since the 1960s. These successful policies explain the much higher and faster growing levels of public transport use in Germany.

Regarding recent trends in public transport demand, other high-income countries have been much more successful than the United States at increasing public transport use, measured both per person and as a market share of total urban travel (Buehler and Pucher 2012). Remarkably, public transport is used more than twice in Canada than in the United States, expressed as a number of annual public transport trips per capita during 2005–2010, with 55 versus 24 trips per year in the United States (Buehler and Pucher 2012).

The greater public transport demand in Europe is attributable to higher-density cities, much higher taxes on fuel and car purchases, limited and more expensive car parking, and restricted car use in cities. Public transport services in European cities are also of better quality than in the United States. The mode share of public transport is measured as a percentage of trips by all means of travel for all trip purposes and serves as an

Fig. 5.19 Real-time light-rail schedule and train information in Essen, Germany. (Author 2010)

alternative indicator of public transport demand (Buehler and Pucher 2012). By using aggregate national statistics, public transport demand trends can be examined. So, the mode share of public transport in the United States is slightly less in 2007–2010 compared with 1980–1983, but higher than for any intervening period between 1983 and 2000 (less than 2%), which is 3–6 times less than in European countries on average (5–11%), and ten times less than in Switzerland, which has the highest

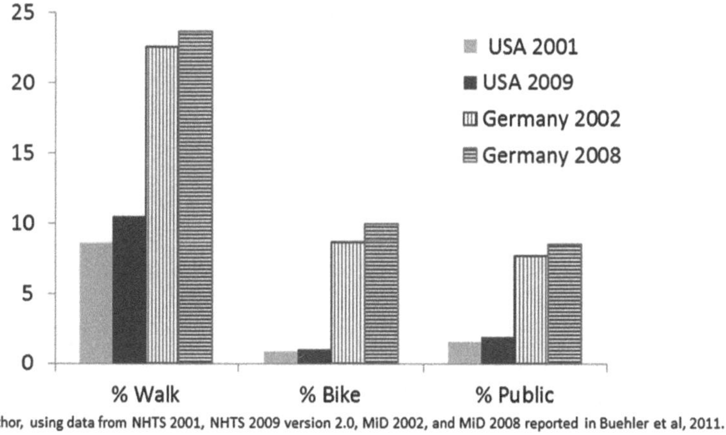

Fig. 5.20 Active travel percentage of all trips (%) for USA (2001, 2009) and Germany (2002, 2008)

share of trips by public transport at 20% for all trip purposes combined. An analysis of 2009 NHTS data for the United States and the similar 2008 Mobilität in Deutschland (MiD) survey for Germany revealed that 8.5% of all trips taken in Germany were by public transportation, while the share of public transportation-based trips in the United States was 1.9% in 2009 (Buehler et al. 2011) (Fig. 5.20).

Mode shares of workers who commute using public transportation grew from roughly 13% to 16% in Germany, the largest increase in public transport mode share compared with Ireland, Canada, Australia and the United States between 1993 and 2008. During this time, the percent of American workers commuting by public transport had been stable at about 5%.

5.9 Built Environment and Travel Behavior

Built-environment factors affect travel behavior represented by mode choice, trip generation, vehicle hours traveled (VHT), and vehicle miles traveled (VMT). The study of the relationship between land use and travel behavior is of great importance due to environmental and social concerns linking transportation and climate change. The United States Environmental Protection Agency (EPA) estimates that transportation is a single large producer of the total carbon dioxide (about 30%) in the United States, which is the main contributor to the greenhouse effect (EPA 2006). The estimate makes it critical to identify ways to decrease the negative impacts of transportation on the environment. Compared with other policy measures aimed at the mitigation of the environmental impact of transportation such as fuel tax and congestion pricing, land-use planning is considered an effective and a long-range approach. The approach involves promoting higher density, mixed use, and compact urban development with the potential for enhancing the efficiency of public transportation, decreasing automobile dependence, and thus changing travel behavior through reduction of VMT and, accordingly, pollution emissions (Fig. 5.21).

Studies over the past decades produced mixed empirical results indicating only minor impact of land use on travel. Much of the research indicates that a compact city with well-mixed land use tends to produce VMT (Hong et al. 2014). There exist four methodological issues including

Fig. 5.21 Compact development consisting of multistory residential buildings stores with commercial uses occupying the first floor and limited car parking in Essen, Germany. (Author 2010)

(1) self-selection, (2) spatial autocorrelation, (3) inter-trip dependency, and (4) geographic scale, which may have affected the empirical results resulting in conflicting findings. Using various geographic scales, for example, measuring a phenomenon (e.g., land-use mix at the census block level in one study, but measuring land-use mix at the census tract level in another study) makes comparisons of the impacts of built-environment factors on travel behavior across different studies less meaningful due to the modifiable areal unit problem (MAUP). *Inter-trip dependency* refers to the situation when trips are examined separately, isolating each trip from a complete tour and thus ignoring the interdependency between trips, which shapes an individual's travel decision. *Self-selection* occurs when people select themselves into preferred choices instead of being randomly distributed. This introduces bias into a regression model, which establishes behavioral relationships. To illustrate this challenging methodological issue in the land-use travel analysis, people with a preference for public transportation over driving seem to select transit-served residential neighborhoods. It follows then that people's attitudes and perceptions about transport modes make them select their home locations within neighborhoods associated with more transit usage and less reliance on car making the extent to which the built-environment characteristics contribute to less driving less clear. Accordingly, ignoring the self-selection problem may estimate impacts of land use on travel incorrectly. Some researchers believe that utilizing a rich set of socioeconomic factors may relieve the problem of self-selection. However, accounting for the effect of self-selection may make some built-environment variables insignificant in influencing travel behavior. Similarly, *spatial autocorrelation*, common in geographic studies, is a challenge in statistical analysis because it makes the assumption of independence of observations no longer valid, so applying linear regression analysis to a spatially autocorrelated data can produce biased estimates resulting in the incorrect inference of land-use significance.

Researchers apply multilevel/hierarchical modeling to resolve spatial autocorrelation in spatial analysis. The multilevel modeling is commonly used when the data are nested in groups. Antipova et al. (2011) used a multi-level modeling perspective to investigate the combined effects of land-use types and sociodemographics (including both individual and neighborhood attributes) on commuting, measured by both travel distance and time, to capture the effect of the spatial separation of residential location and workplace as well as road conditions. The authors conclude that land-use types explain commuting time in a significant way, and sociodemographic variables of both individuals and neighborhoods importantly impact individuals' commuting behavior. In agreement, Hong et al. (2014) find that one of the most significant factors influencing an individual's VMT for work and non-work types of tours is a number of vehicles in the household indicating that automobile ownership encourages people to drive more. However, more caution is needed in interpretation because the built environment can impact auto ownership. Being less sensitive to travel cost, people with a higher household income tend to travel more. Similarly, graduate education level is associated with longer commuting VMT compared to people who graduated only high school. This may be explained by the fact that jobs that match their education level tend to be less ubiquitous, and due to their economic and social activities being spatially more extensive.

Demographic features play a substantial role: people from larger-sized households tend to reduce VMT in work tours because in a large household, the number of trips per person and VMT per person can be reduced as household members coordinate activities and share rides. As the number of workers in a household increases, VMT per person gets lower in non-work tours as workers can share their responsibility. Also workers have less time to take part in non-work related activities. The key consideration when people choose a travel mode are attitudinal measures (i.e., comfort, predictable travel, flexibility, quickest travel time, and privacy) (Hong et al. 2014).

Using aggregate data across the Seattle metropolitan region, Hong et al. (2014) studied the effects of the built environment on transportation, expressed as the work tour and non-work tour records and calculating the total VMT per person. Using data on land-use types, area, and total building floor area, different measures of built environment have been calculated at different geographic scales including residential and nonresidential land uses densities, an entropy measure of land-use mix and distance from Central Business District (CBD). Use of the two geographic scales—1-km buffer and the Traffic Analysis Zone (TAZ) based on respondent's home location—allowed comparisons of the influences of built-environment factors on individuals' VMT between two different geographic scales. Public transit accessibility represented by distance to the nearest bus stop was positively associated with VMT for non-work tours, with an increase in the distance to the nearest bus stop associated with an increase in non-work VMT. Residential density, entropy, and intersection density all impact non-work related VMT in a significant way:

in communities that are denser, with good land-use mix, and that are pedestrian friendly, people tend to drive less.

Nonresidential density and distance from the CBD influence work-related VMT: density that is positively associated with jobs opportunities has a negative impact on work-related VMT (higher nonresidential density is linked to reduced VMT), while distance from CBD has a positive association with work-related VMT (the farther from the CBD a house is located, the greater the work-related VMT). The authors come to the conclusion that even accounting for travel attitude and spatial autocorrelation, land-use factors impact VMT in a highly significant way.

5.10 What Is TOD?

Due to substantial increases in road traffic congestion since the mid-1970s, many large US cities are seeking to ameliorate rising congestion by encouraging people to drive their cars less and use other transportation options, and improve accessibility (Ratner and Goetz 2013). Various urban development strategies include transit-oriented development (TOD), which is an alternative to traditional, decentralized, car-oriented suburban growth through land-use development that is high density and pedestrian friendly with mixed land uses. TODs lessen a complete reliance on automobiles and provide residents with a strong sense of community (DRCOG 2014). There are different definitions of TOD, however, all have similar components with a central focus on higher density, land -use mixes including residential, employment and shopping opportunities, and livability around the stations, but also aiming at the reduction of regional congestion and an improved regional environment (Ratner and Goetz 2013) (Fig. 5.22).

With TODs, growth and new construction projects are located proximate to transportation nodes, that is, within a walkable distance to a transit station, often in a quarter- to half-mile area around transit stations. TODs provide various transportation choices including rail, bus, walking, and bicycling, and access to regional amenities including employment, civic uses (e.g., government centers), healthcare, entertainment, shopping, and cultural and recreational facilities through the regional transit network. By deliberately placing homes, jobs, and other mentioned amenities close to transit stations, active travel and transit use are being promoted, thus achieving local efficiency (Ratner and Goetz 2013) (Fig. 5.23).

Fig. 5.22 A residential area is closely located to light rail and bus stops in Essen, Germany. (Author 2010)

In Denver, CO the re-introduction of rail service into the urban region in the early 1990s after it was stopped in the 1950s, and a subsequent 2004 long-range regional land use and rail and bus rapid transit development program entitled "FasTracks," led to the creation of the different TOD station types developed by the city and county of Denver. TOD station types were only assigned for stations within the boundaries of Denver including downtown, urban neighborhood, urban center, major urban center, and commuter town stations. Each station area is classified into one of the seven following types based on characteristics commonly found in rail transit-served places (Ratner and Goetz 2013).

1. *Downtown transit station* area is characterized by diverse mixed use (office, residential, retail, entertainment, and civic uses, where residential consists of multi-family and loft housing), highest density, tallest buildings (five stories and higher), high pedestrian activity, transit hub, and historic areas. In terms of employment type, downtown station areas represent a prime office and shopping location.

Fig. 5.23 Access to underground light rail is highly visible in Essen, Germany. (Author 2010)

The area is an intermodal facility/transit hub: it is a major regional destination with high-quality feeder bus and streetcar connections.

2. *Major Urban Center* is characterized by high density, tall buildings, and mixed use (office, retail, entertainment, residential consisting of mainly multi-family and townhomes), with a focus on employment and more than 250,000 square feet of office space and 50,000 square feet of retail space. Major Urban Center station areas are linked with district circulator transit and express feeder bus and represent sub-regional destination. A large amount of TOD development occurs at the Major Urban Center stations. Compared with the Main Street stations, there is a more even land-use mix at the Major Urban Center transit station area.
3. *Urban Center* is characterized by mixed use (retail and office, and residential consisting of mainly multi-family and townhomes), high density, grid and alley block pattern, high pedestrian activity, and multi-modality. Urban Center station areas are linked with district circulator transit and express feeder bus and represent sub-regional

destinations. A moderate number of non-residential development including retail, office, and government projects takes place at the Urban Center stations. Population density at this station type is very close to the metro neighborhood average, and there are only a few mixed-use projects found here.

4. *Urban Neighborhood station* areas are characterized by multi-family townhomes and small-lot single-family residences and two- to seven-story-high buildings. Employment is mostly local, serving retail (no more than 50,000 square feet). The areas function as neighborhood walk-up stations with local and express bus connections. A moderate amount of residential development of all types occurs at the Urban Neighborhood stations. Similarly to the Urban Center areas, population density at the Urban Neighborhood stations is very close to the metro neighborhood average and there are only a few mixed-use projects found here.
5. *Commuter Town Center station* areas have mixed use (office, retail, residential consisting of multi-family, townhome, small-lot single-family residences), and employment is local and commuter-serving (no more than 25,000 square feet). These areas function as a capture station for in-bound commuters with large park-n-ride facilities.
6. *Main Street station* areas have mixed use including residential (desired housing type is multi-family, not single family) as well as neighborhood retail. Main Street retail infill employment, the areas are bus or streetcar corridors with district circulator or feeder transit service, walk-up stops, they have no transit parking.
7. *Campus/Special Events station* areas' land-use mix consists of university campus and sports facilities; the areas have only limited multi-family residential housing and limited office/retail employment, and they are large commuter destinations with large park-n-ride facilities.

5.10.1 Impact of TOD on Land Use

TODs are impacting land use and urban form throughout the metropolitan region. The closer to downtown the greater the impact of TODs. The areas surrounding the transit station experience tremendous changes in land use. For example, the area around Union Station in downtown Denver was transformed most dramatically during 2000–2010. Most of the development activity (especially residential, of which a majority has been developed at the downtown and Main Street stations) has taken place in station areas close to the downtown (Ratner and Goetz 2013).

Different forms of land use are attracted to the specific location-based individual transit corridors with office development becoming more transit-oriented regionally. Ratner and Goetz (2013) examined new land-use development by TOD station type, as defined by the city and county of Denver. Regarding office development, the largest percentage of new offices has been developed at downtown station areas (62%) and major urban centers (28%). Among other types of non-residential land use, a majority of the non-residential development is occurring at downtown stations including government, cultural, and convention/sports development. The second largest non-residential development, which mostly consisted of retail and office space, occurred at major urban center stations. New non-residential development including some retail, office, and government development has occurred at urban center stations, while 100% of the education-oriented development has taken place at the campus stations.

New residential development is attracted to transit-oriented communities where apartments and condos are the type of housing that has been developed across all TOD types, and the prevalence is especially high in major urban centers where apartments and condos together account for almost 50% of all new residential development. Although all station types have experienced some apartment development, urban centers are less residentially oriented and experience significantly less residential development than the other station areas. The most condo development took place at the downtown and Main Street stations, the areas that have been recently gentrified. Housing for seniors has been developed predominantly in downtown station areas (67%), followed by urban neighborhood stations (33%). Student housing took place at urban neighborhood stations (48%), campus station areas (37%), and downtown stations (15%). Almost all new hotel development (90%) has occurred at the downtown stations due to the tourism promotion in the downtown area along with the opening of the Convention Center. The largest share of affordable housing has been developed at Main Street transit station areas (76%), followed by urban neighborhood stations (15%), and at major urban centers and downtown stations (each accounting for 5%) (Ratner and Goetz 2013).

5.10.2 Impact of TOD on Urban Mobility

Noted previously, traffic congestion has increased substantially since the mid-1970s by every measure of mobility performance. The 2015 *Urban Mobility Scorecard* report jointly produced by Texas A&M Transportation Institute and INRIX, Inc. analyzes traffic conditions in 471 urban areas across the United States. According to the report, the total congestion problem exceeds that of the pre-recession period (Schrank et al. 2015). Due to the problem of congestion, a massive amount of time, fuel, and money is wasted. For example, urban Americans had to travel an extra 6.9 billion hours and purchase an extra 3.1 billion gallons of fuel due to congestion while a congestion cost was $160 billion in 2014. During 2013–2014, traffic congestion increased in 95 of America's 100 largest metro areas. Compared with 1982, more of the trips have to be planned by travelers around traffic jams, involving more hours of the day and more parts of town, sometimes including weekends and rural areas (Schrank and Lomax 2009). Regarding travel delay, travelers had to plan 48 minutes to make a freeway trip that takes 20 minutes in light traffic in order to arrive on time.

5.10.3 Solving Urban Congestion

Although the problem of urban congestion is very large, it is solvable. For example, slower growth or modest declines in congestion may occur in response to higher fuel prices and economic slowdown. During economic recession in a region, congestion may decrease, however, once the recession is over and the economy rebounds, the congestion problem does the same. Examples include the economic slowdown in the states of the Northeast in the early-to-mid-1980s, Texas in the mid-1980s, then California in the early-to-mid-1990s during which there were declines in congestion that were reversed as soon as the economy rebounded. During the most recent economic recession of 2008–2011, there was a small decline in traffic volume in 2008. However, with different areas experiencing declines of varying degrees, not surprisingly, the places with the most congestion, urban areas on weekdays, had the smallest declines (Schrank and Lomax 2009). Consistent with prior economic recessions, in areas with employment losses, traffic volume and congestion either grew slowly or had zero growth during the second half of 2008. And as soon as fuel prices declined in the fall of 2008, traffic volume started to increase.

To reduce congestion, a diversified approach is needed. The best mobility improvement programs consist of a mix of various strategies including increased capacity, applying travel and work schedule options, and decreasing physical distance between activity sites such as between homes and jobs (Schrank et al. 2015). Mitigating the problem and improving urban mobility requires involvement of all urban actors from agencies to businesses to manufacturers to commuters and travelers. Since congestion affects areas of every size, not just big cities, all communities including metro regions, cities, and neighborhoods need to use a different mix of the potential congestion-reducing solutions (Schrank and Lomax 2009). Each urban region should apply a combination of different strategies consisting of transportation and land-development solutions. These strategies should correspond to the region's goals and vision and may include construction solutions, more travel options (e.g., different routes, introduction of high-occupancy vehicle lanes and toll highways/lanes, capacity expansions in important corridors through new roads to serve new developments and public transportation improvements, changing the usage patterns, e.g., flexible work hours, etc.), low-cost productivity improvements through efficient and reliable operation of roads and public transportation (e.g., incident management, that is, rapid removal of crashed vehicles, coordinated timing of the traffic signals, improving road and intersection designs), or through diversified land-use patterns (e.g., denser developments with a mix of jobs, shops, and homes), or redevelopment solutions.

An important congestion-combatting strategy includes TOD. Dense, mixed-use, and walkable developments are founding principles of TOD, allowing more people to use alternative modes of transportation including walking, biking, or taking transit to more destinations that are closer to each other. Land-use mix encourages interaction between origins and destinations and decreases the distance between them, which together with pedestrian-friendly urban design increases the potential for walking among the various land uses and to and from the transit station as well as cycling for short- to medium-distance trips (Pucher et al. 2011). Higher population and employment densities generate a larger market for inbound and outbound transit ridership, which can potentially improve farebox returns and better balance transit network flows (Higgins and Kanaroglou 2016).

5.10.4 TOD and Urban Mobility

TODs have the great potential to impact urban mobility. Urban mobility can be measured by peak traveler delay, cost per traveler, regional delay, and a regional time travel index as suggested by the Texas Transportation Institute (Schrank and Lomax 2009). *Delay per Peak Traveler* is the ratio of the extra time that a traveler spends at congested speeds rather than free-flow speeds (60 mph on freeways and 35 mph on principal arterials) divided by the number of persons making a trip during the peak period (6 to 10 a.m. and 3 to 7 p.m.). It is measured in person-hours. *Travel Time Index* measures congestion on each trip and each mile of travel. Travel Time Index is calculated as the ratio of travel time in the peak period to travel time in free-flow. For example, a value of 1.30 means that a 40-minute free-flow trip takes 52 minutes in the peak (52/40 = 1.3). *Total Delay,* measured in person-hours, indicates the overall magnitude of the congestion problem. Total Delay is measured by the total travel time above that is required to conduct a trip at free-flow conditions. The distribution of ranked total delay usually follows the population ranking since larger regions usually have more delay. *Congestion Cost* is defined as a value of travel delay for 2014 (estimated at $17.67 per hour of person travel and $94.04 per hour of truck time) and excess fuel consumption estimated using state average cost per gallon of fuel.

The opening of a transit line can contribute towards mitigation of traffic congestion. For example, in Denver, CO the re-introduction of rail service into the urban region has led to an improvement of mobility performance. According to the 2009 Urban Mobility Report, from 1991 to 1995, Denver's mobility grew consistently worse with an increase in many congestion measures. To illustrate, its peak traveler delay increased from 22 to 37 person-hours (68%), cost per traveler increased by 69%, and regional delay grew from 17,500,000 to 32,400,000 person-hours (84%). During this period, Denver advanced in the ranking of a regional Time Travel Index, falling from 27th to 15th across 90 US cities. This trend continued through 2000, however, in 2001 there were rapid changes in the situation. The worsening of mobility performance measures decreased significantly in the Denver region from 2000 to 2005. In fact, mobility improved during this time period. To illustrate, total regional delay underwent a much slower growth compared to the period 1986–2000, while peak travel time (expressed as a number of rush hours) and system delay (expressed as a percentage of congested lane-miles) experienced only minimal growth. Further mobility improvement is reflected in Denver's moving from 9th to the 13th on the Time Travel Index of the worst metro areas. Temporally, this mobility improvement occurred as the southwest corridor light rail line opened in the Denver region.

5.10.5 Corridor Planning

To maximize the benefits from the region's transit investment, efforts are made to enhance access to employment and education, diminish housing and transportation costs (especially for low-income groups), promote active living and healthy aging, reduce air, water, and soil pollution, and efficiently use natural resources by investing in TOD. Detailed planning efforts are made on the development of mixed-use, pedestrian- and bicycle-friendly "urban centers" located along transit lines, along with the corridor planning where a corridor connects the diverse neighborhoods with the communities of the metro area and the world.

The corridor is defined as the area within one mile of transit lines. The corridor should serve both the residents, employees, and neighborhoods along the corridor and also commuters residing outside the corridor. For example, the East Rail Line Corridor, one of the best airport-to-downtown transit connections in the United States, links downtown Denver to Denver International Airport as the commuter rail line connects workers to employment centers and employment centers to each other and diverse destinations, and provides access to goods and services.

It is widely documented that increases in housing costs due to area development may cause many low-income households and non-White families to move out of these areas despite their long presence in neighborhoods. Thus, the corridor should be designed to preserve existing neighborhoods while preventing potential adverse impacts of new development such as rising rents and resident displacement. Corridor-level planning goals include promotion of economic development around the stations; enhancement of connections to the stations so that the established neighborhoods in proximity to the stations have easy, safe, and reliable access by multiple modes including walking, bicycling, transit, and driving; provision and increase of multiple quality housing throughout the corridor that is affordable for people who work at large employment centers along and near the corridor; and promotion of green building along the corridor in order to support sustainable transit-oriented communities.

5.10.6 Recommendations on TOD Implementation

TOD implementation involves multiple steps including development of visions and goals, TOD site selection among the potential sites by examining all station area plans along the corridor and economic, demographic, and real estate conditions at each station; development of a detailed scope of work for specific activities at that site; conduct of an in-depth study to identify needs and gaps and the development potential including ideal business and industry types for each station; and stakeholder engagement including regional and jurisdictional economic development agencies focused on the creation of middle-income jobs and assistance with expansion of existing businesses and attraction of new businesses in the corridor (East Corridor Blueprint 2014). Other key stakeholders may include planning, public works, public health agencies, and housing authorities.

Corridor working-group activities include development of corridor-wide vision and goals to achieve it; examination of demographics and economic characteristics along the corridor and a review of station area plans and maps; assessment of corridor-wide needs, opportunities, and challenges to TOD implementation incorporating housing opportunity, economic development, public health and active living, and transit accessibility; recommendations for implementation including an initial activities list; and development of detailed pre-development planning at a specific site accompanied by a detailed scope of work for specific activities at that site. To implement successful TOD, it is critical to engage broader stakeholders (that is, people who have an interest in the process and who may be affected by the relevant outcome such as affordable housing, displacement, greater access to jobs and job creation, healthy living, personal safety at stations, and transit access, etc.) in a corridor planning process.

To provide feedback on the vision, goals, and recommendations from residents, business owners, and other members of the communities in the corridor, it is important to ensure representation across all ages, incomes, racial/ethnic membership, and various interest areas such as small-business ownership, public health, affordable housing, and traditionally underrepresented groups such as low-income communities, communities of color, and so on, and include 10–20 community members in a broader stakeholder committee and provide them with the needed tools and knowledge to ensure effective and meaningful participation in the planning process. It is also important to use the experience of existing city and regional transportation programs and build upon previous work to improve infrastructure incorporating many travel modes. This may include cataloging infrastructure elements such as sidewalk conditions, access walks, and proximity to safe crossing locations, and amenity needs to improve neighborhood connection to bus and rail-transit stations. For example, gaps in resources such as bus stop amenities may be identified for areas surrounding transit stations including benches, shelters, and so on. These identified gaps should be included into a dynamic database so that the highest need areas are prioritized in terms of investment and implementation. Conduct a parking structure feasibility study in station areas for the potential of converting surface parking to structured parking, so that land could be used more productively. Develop a strategy emphasizing permanent preservation and creation of affordable housing for older people, families with children, and students, and encourage diverse, mixed-income housing choices in transit-oriented communities along the corridor. To implement identified recommendations to accomplish the corridor vision, continue collaborative efforts among TOD working-group participants through regular meetings and communication to monitor progress on activities, determine new planning and development opportunities, and identify financial support opportunities for initiatives along the corridor (East Corridor Blueprint 2014).

5.10.7 TOD Impacts on Residential and Business Location and Travel Decisions

To understand businesses, employees, and residents' experiences and perceptions in areas around existing and planned light rail and bus rapid-transit station areas across the metropolitan region, the Denver Regional Council of Governments (DRCOG) conducted the Who is TOD Survey in 2009, 2010, and 2016 (DRCOG 2014). People have various motivations to live or work in TODs, but particularly they are attracted by housing type and cost, neighborhood quality, and convenience to various amenities (Boschmann and Brady 2013). The results of the Who is TOD Survey reveal that living and working benefits of transit-oriented, walkable, mixed-use communities attract large numbers of people and influence their location and travel decisions (Who Is TOD in Metro Denver? 2010). To illustrate, in dense, mixed-use communities with limited or expensive parking and served with high-quality transit in the Denver region, people are more likely to use transit. The *Denver Post* wrote on June 14, 2010, "People are willing to pay about 4 percent more a month to rent an apartment within a quarter-mile of a light-rail stop." Workers indicate a strong influence of the advantages of an easy commute by rail or bus transit, access by walk/bike, and access to amenities (land-use mix) on job choice, with downtown Denver employees rating these benefits as a strong influence on job choice compared with employees outside downtown (25% versus 10%), while the importance of other choice factors including salary/wage, healthcare and retirement benefits, company reputation, and interesting work are comparable between the downtown and outside downtown locations. Similarly, business location decisions are strongly influenced by transit amenities with a quarter of downtown Denver businesses rating transit access as a strong influence in their business location decision compared with just 7% of businesses located outside of downtown. However, businesses located in station areas outside of downtown value access and car parking much more strongly in their location decisions compared with downtown businesses (64% versus only 32%) (Who Is TOD in Metro Denver? 2010).

Proximity to transit itself is rarely a main motivating factor, and to make non-work trips TOD residents use cars more frequently than other modes (Boschmann and Brady 2013). To illustrate, in Germany, where the percentage share of trips by public transport was 8.5% in 2008 (versus 1.9% in 2009 in the United States), 88% of people reside within 1 km (0.62 mile) of a public transport stop, while only 43% of Americans live within 1 km of a public transport stop (Buehler and Pucher 2012). However, the prevalence of using alternative transportation modes is higher among travelers residing in transit-served communities compared with those outside of these areas, both for commute and non-work purposes (Fig. 5.24).

According to the Who is TOD Survey, in downtown Denver, residents and workers use alternative modes for commute and non-commute trips, with residents using transit, carpooling, walking, and biking for 44% of commute trips compared to 32% of commute trips that people living close to station areas outside downtown make using alternative modes (Who Is TOD in Metro Denver? 2010). Employees who work in downtown Denver use alternative modes for commute trips 50% of the time; this compares with only 12% of commute trips by employees working in station areas outside downtown. Further, 33% of businesses in downtown Denver offer their employees incentives to drive less in the form of free or subsidized transit passes. In contrast, just 4% of businesses located outside downtown provide their employees with this benefit. Respectively, employees who work in downtown Denver park for free next to their workplace much less often (13% reported free parking) than their peers in station

Fig. 5.24 Residential suburbs are served by efficient light rail in Essen, Germany. (Author 2010)

areas outside of downtown, 86% of whom reported free parking right next to their place of employment (Who Is TOD in Metro Denver? 2010). Similar findings are for non-commute trips with residents of downtown Denver using alternative modes almost twice as often as residents who live close to transit station areas outside of downtown to buy food, do shopping, visit restaurants, visit open space and parks, and access entertainment. Similarly, downtown Denver employees are ten times more likely to walk to lunch or to conduct personal errands (60% versus 6%) compared with employees in station areas outside of downtown Denver (Who Is TOD in Metro Denver? 2010).

5.11 Changing Urban Demographics and Travel Behavior

The urban context within which urban travel takes place is changing, including population demographics and residential patterns. Many factors lead to an increase of the older population group, including higher longevity, smaller families, and lower fertility among younger groups (Boschmann and Brady 2013). Globally, the percentage of person 65 years and older making up total population was 8% in 2009 and will be 16% in 2050. An even greater increase is expected in the United States due to the aging of the Baby Boomers: from 13% in 2010 to 20.2% by 2050, with especially rapid growth in the Sunbelt cities. With older adults becoming a larger share of the US population, urban planners need to be aware of the changing travel behaviors and needs of this population group in cities. It is important to study older people's travel behavior as both absolute and proportional increase of older adults necessitates urban changes to make cities aging-friendly and more livable. Additionally, there is great concern about the total mileage of non-work-related travel among Baby Boomers in the near future due to their potential impact upon the environment, while the increased travel of older adults may add to existing traffic congestion (Boschmann and Brady 2013).

Travel behaviors among older people is often quantified by total trips, trip frequency, mean trip distance, mode choice, trip purpose, activity patterns, and time-of-day travel. Various issues may arise due to a higher mobility of older adults, which include safety in driving, driving cessations, mobility and disability, resource accessibility, and overall quality of life. In the developed regions, due to suburbanization of the city and increased rates of older adults, licensed older people tend to use more private vehicles compared with other travel modes: in low-density and car-dependent suburbs nearly 90% of older people use cars, while only 2% use public transport (Boschmann and Brady 2013). Many older people use cars since that is what they have done for most of their lives. Cars will remain their main transport choice; older people do not use public transit as it is better suited for younger working-age adults.

5.11.1 TOD Impacts on the Elderly's Travel Behavior

Despite corridor-wide and regional benefits that result from investment into rail transit, it has highly variable and station-specific effects on land use, real estate development, and economic opportunity on the station area scale. Strategic neighborhood design can contribute to more active aging for older adults through increased walking and thus less driving. It is hypothesized that people who have access to transit, sidewalks, and safe communities that facilitate the regular use of transit have healthier outcomes through leading more active lifestyles. Boschmann and Brady (2013) explore how TOD impacts travel behaviors (trip generation, distance, and trip purpose) of older adults in the Denver metropolitan area, a popular migration destination, including for Baby Boomers, during the twentieth century. It is estimated that the share of the population who is 65 years and older in the Denver metro area will grow from 11.9% in 2015 to 17.7% by 2030 and to 21.7% by 2050 (CSDO 2017). Besides being a place of rapid demographic and aging growth, Denver has experienced a large amount of TOD (Ratner and Goetz 2013). Overall, the average number of trips and trip distances decrease as age increases, for both males and females. Regarding the number of trips, these decrease steeply after age 75 (especially at age 78), and a similar pattern is observed for average trip distances. Across all age groups, females take significantly fewer and shorter trips compared to males; they also have higher percentages as passengers. However, lifestyles may impact mobility more than age alone, making it important to examine travel behaviors based on lifestyle clusters.

It is possible to implement changes in travel behavior of older adults via strategic urban development including high-density, walkable, mixed land uses, and proximity to transit centers. Does residing in or near a TOD reduce trips, travel distances, and car use among older adults? Travel behavior in older adults living in TODs is different from those who do not live in TODs. In fact, living in a TOD decreases car dependency and increases use of non-car modes (by 61% in the Denver sample). Older adults living in TODs exhibit shorter travel distances compared with non-TOD residents. Mean trip distance is lower for TOD residents across all age groups; sometimes the difference is almost one mile less for TOD residents. Regarding trip generation (total number of trips), residing in a TOD increases total trips by 26.2%: TOD residents make more daily trips (4.08) on average than non-TOD residents (3.61) across all age groups except for the 85 and older group (Boschmann and Brady 2013). These findings on travel behavior among older adults in the Denver metro area support key arguments of the advantages of mixed land uses and activities provided by high-density communities stimulating more, but shorter, daily trips, and increased non-car trips.

Bibliography

Antipova, A., F. Wang, and C. Wilmot. 2011. Urban Land Uses, Socio-demographic Attributes and Commuting: A Multilevel Modeling Approach. *Applied Geography* 31 (3): 1010–1018.

Baron, P. 1995. Transportation in Germany: A Historical Overview. *Transportation Research Part A: Policy and Practice* 29 (1): 9–20.

BIKETOWN. 2017. At: https://www.biketownpdx.com/how-it-works/meet-the-bike. Accessed 2 Sept 2017.

Boschmann, E.E., and S.A. Brady. 2013. Travel Behaviors, Sustainable Mobility, and Transit-Oriented Developments: A Travel Counts Analysis of Older Adults in the Denver, Colorado Metropolitan Area. *Journal of Transport Geography* 33: 1–11.

Buehler, R., and J. Pucher. 2011. Sustainable Transport in Freiburg: Lessons from Germany's Environmental Capital. *International Journal of Sustainable Transportation* 5 (1): 43–70.

———. 2012. Demand for Public Transport in Germany and the USA: An Analysis of Rider Characteristics. *Transport Reviews* 32 (5): 541–567.

———. 2017. Trends in Walking and Cycling Safety: Recent Evidence From High-Income Countries, with a Focus on the United States and Germany. *American Journal of Public Health* 107 (2): 281–287. At: http://ajph.aphapublications.org/doi/full/10.2105/AJPH.2016.303546. Accessed 2 Sept 2017.

Buehler, R., J. Pucher, D. Merom, and A. Bauman. 2011. Active Travel in Germany and the U.S. Contributions of Daily Walking and Cycling to Physical Activity. *American Journal of Preventive Medicine* 41 (3): 241–250.

Centers for Disease Control and Prevention (CDCP). 2002. Barriers to Walking and Biking to School –United States, 1999. *MMWR Morbidity and Mortality Weekly Report* 51: 701–704.

Cervero, R. 1991. Land Uses and Travel at Suburban Activity Centers. *Transportation Quarterly* 45: 479–491.

Cervero, R., and R. Gorham. 1995. Commuting in Transit Versus Automobile Neighborhoods. *Journal of the American Planning Association* 61: 210–225.

Cervero, R., and C. Radisch. 1996. Travel Choices in Pedestrian Versus Automobile Oriented Neighborhoods. *Transport Policy* 3: 127–141.

Cervero, R., and K. Kockelman. 1997. Travel Demand and the 3DS: Density, Diversity, and Design. *Transportation Research Part D: Transport and Environment* 2(3): 199–219.

Christiansen, L.B., M. Toftager, J. Schipperijn, A.K. Ersbøll, B. Giles-Corti, and J. Troelsen. 2014. School Site Walkability and Active School Transport – Association, Mediation and Moderation. *Journal of Transport Geography* 34: 7–15.

Clayton, W., and C. Musselwhite. 2013. Exploring Changes to Cycle Infrastructure to Improve the Experience of Cycling for Families. *Journal of Transport Geography* 33: 54–61.

Cradock, A., P. Troped, B. Fields, S. Melly, S. Simms, F. Gimmler, and M. Fowler. 2009. Factors Associated with Federal Transportation Funding for Local Pedestrian and Bicycle Programming and Facilities. *Journal of Public Health Policy* 30: S38–S72.

CSDO (Colorado State Demography Office). 2012. *Population Data*. At: https://demography.dola.colorado.gov/population/data/sya-regions/. Accessed 27 Aug 2017.

Dieleman, F.M., M. Dijst, and G. Burghouwt. 2002. Urban Form and Travel Behaviour: Micro-level Household Attributes and Residential Context. *Urban Studies* 39 (3): 507–527.

East Corridor Blueprint. 2014. *Denver Region Sustainable Communities Initiative*. Denver Regional Council of Governments (DRCOG). At: https://drcog.org/sites/drcog/files/resources/East%20Corridor%20Blueprint_0.pdf. Accessed 23 Aug 2017.

EPA. 2006. *Greenhouse Gas Emissions from the U.S. Transportation Sector, 1990–2003*. Washington, DC: Environmental Protection Agency.

Foster, S., and B. Giles-Corti. 2008. The Built Environment, Neighborhood Crime and Constrained Physical Activity: An Exploration of Inconsistent Findings. *Preventive Medicine* 47 (3): 241–251.

Frumkin, H., L. Frank, and R.H. Jackson. 2004. *Urban Sprawl and Public Health: Designing, Planning, and Building for Healthy Communities*. Washington, DC: Island Press.

Green, C.G., and E.G. Klein. 2011. Promoting Active Transportation as a Partnership Between Urban Planning and Public Health: The Columbus Healthy Places Program. *Public Health Reports* 126 (1): 41–49.

Handy, S.L. 1992. Regional Versus Local Accessibility: Neo-traditional Development and Its Implications for Non-work Travel. *Built Environment* 18: 253–267.

Handy, S.L., and K.J. Clifton. 2001. Local Shopping as a Strategy for Reducing Automobile Travel. *Transportation* 28: 317–346.

Harries, K. 2006. Property Crimes and Violence in United States: An Analysis of the Influence of Population Density. *International Journal of Criminal Justice Sciences* 1 (2): 24–34.

Higgins, C.D., and P.S. Kanaroglou. 2016. A Latent Class Method for Classifying and Evaluating the Performance of Station Area Transit-Oriented Development in the Toronto Region. *Journal of Transport Geography* 52: 61–72.

Hong, J., Q. Shen, and L. Zhang. 2014. How Do Built-Environment Factors Affect Travel Behavior? A Spatial Analysis at Different Geographic Scales. *Transportation* 41: 419–440.

Hu, P., and T. Reuscher. 2004. *Summary of Travel Trends: 2001 National Household Travel Survey*. Washington, DC: Department of Transportation, Federal Highway Administration (US). At: http://nhts.ornl.gov/2001/pub/STT.pdf [cited 15 Sept 2010]. Accessed 15 Aug 2017.

Jacobsen, P.L., D.R. Ragland, and C. Komanoff. 2015. Safety in Numbers for Walkers and Bicyclists: Exploring the Mechanisms. *Injury Prevention* 21 (4): 217–220.

Kitamura, R., P.L. Mokhtarian, and L. Laidet. 1997. A Micro-Analysis of Land Use and Travel in Five Neighborhoods in the San Francisco Bay Area. *Transportation* 24: 125–158.

Larson, J., and A. El-Geneidy. 2010. A Travel Behavior Analysis of Urban Cycling Facilities in Montréal, Canada. *Transportation Research D* 16 (2): 172–177.

Lawlor, M. 2015. Case Study Analysis of the Attributes of Walkability. *The Journal of Sustainable Mobility* 2 (2): 54–66.

Lee, C., and A.V. Moudon. 2008. Neighbourhood Design and Physical Activity. *Building Research and Information* 36 (5): 395–411.

Matley, T.M., L.M. Goldman, and B.J. Fineman. 2000. Pedestrian Travel Potential in Northern New Jersey: A Metropolitan Planning Organization's Approach to Identifying Investment Priorities. *Transportation Research Record* 1705: 1–8.

McDonald, N.C. 2008. Children's Mode Choice for the School Trip: The Role of Distance and School Location in Walking to School. *Transportation* 35 (1): 23–35.

McNally, M.G., and A. Kulkarni. 1997. Assessment of Influence of Land Use–Transportation System on Travel Behavior. *Transportation Research Record* 1607: 105–115.

Menghini, G., N. Carrasco, N. Schüssler, and K. Axhausen. 2010. Route Choice of Cyclists in Zurich. *Transportation Research A*(9): 754–765.

Moudon, A.V., and C. Lee. 2003. Walking and Bicycling: An Evaluation of Environmental Audit Instruments. *American Journal of Health Promotion* 18 (1): 21–37.

Our Nation's Travel. 1997. *1995 NPTS Early Result Report*, Publication FHWA-PL-97-028. Washington, DC: US Department of Transportation.

Panter, J.R., A.P. Jones, and E.M. van Sluijs. 2008. Environmental Determinants of Active Travel in Youth: A Review and Framework for Future Research. *International Journal of Behavioral Nutrition and Physical Activity* 5(1): 34.

Passenger Travel Facts and Figures. 2016. U.S Department of Transportation. Bureau of Transportation Statistics. At: https://www.rita.dot.gov/bts/sites/rita.dot.gov.bts/files/PTFF%202016_full.pdf. Accessed 15 Aug 2017.

Pucher, J., R. Buehler, and M. Seinen. 2011. Bicycling Renaissance in North America? An Update and Re-appraisal of Cycling Trends and Policies. *Transportation Research Part A* 45 (6): 451–475.

Randall, T.A., and B.W. Baetz. 2001. Evaluating Pedestrian Connectivity for Suburban Sustainability. *Journal of Urban Planning and Development* 127: 1–15.

Ratner, K.A., and A.R. Goetz. 2013. The Reshaping of Land Use and Urban form in Denver Through Transit Oriented Development. *Cities* 30 (1): 31–46.

Ross, C.L., and A.E. Dunning. 1997. *Land Use Transportation Interaction: An Examination of the 1995 NPTS Data*. Atlanta: U.S. Department of Transportation: Federal Highway Administration.

Saelens, B.E., J.F. Sallis, and L.D. Frank. 2003. Environmental Correlates of Walking and Cycling: Findings from the Transportation, Urban Design, and Planning Literatures. *Annals of Behavioral Medicine* 25 (2): 80–91.

Safe Routes to School. 2010. *U.S. Travel Data Show Decline in Walking and Bicycling to School Has Stabilized*. Press Release, 8 Apr 2010.

Sallis, J.F., L.D. Frank, B.E. Saelens, and M.K. Kraft. 2004. Active Transportation and Physical Activity: Opportunities for Collaboration on Transportation and Public Health Research. *Transportation Research Part A* 38 (4): 249–268.

Savitch, H.V. 2011. A Strategy for Neighborhood Decline and Regrowth: Forging the French Connection. *Urban Affairs Review* 47 (6): 800–837.

Schrank, D., and T. Lomax. 2009. *2009 Urban Mobility Report*. Texas Transportation Institute. https://static.tti.tamu.edu/tti.tamu.edu/documents/ums/archive/mobility-report-2009.pdf. Accessed 26 Aug 2017.

Schrank, D., B. Eisele, T. Lomax, and J. Bak. 2015. *2015 Urban Mobility Scorecard*. August. College Station: Texas A&M Transportation Institute and INRIX, Inc.

Seraj, Sidharthan, Pendyala Bhat, and K.G. Goulias. 2012. Parental Attitudes Toward Children Walking and Bicycling to School. Multivariate Ordered Response Analysis. *Transportation Research Record Journal of the Transportation Research Board* 2323 (1): 46–55.

Shriver, K. 1997. Influence of Environmental Design on Pedestrian Travel Behavior in Four Austin Neighborhoods. *Transportation Research Record* 1997 (1578): 64–75.

Southworth, M. 2005. Designing the Walkable City. *Journal of Urban Planning and Development* 131 (4): 246–257.

Staunton, C.E., D. Hubsmith, and W. Kallins. 2003. Promoting Safe Walking and Biking to School: The Marin County Success Story. *American Journal of Public Health* 93 (9): 1431–1434.

The City of Portland. 2013. *2013 Portland Sunday Parkways Report*. At: https://www.portlandoregon.gov/transportation/article/470728. Accessed 2 Sept 2017.

———. 2016. *City and Partners Offer Driver Safety Class for New Portlander Community*. At: https://www.portlandoregon.gov/transportation/article/568960. Accessed 2 Sept 2017.

———. 2017a. *Vision Zero*. At: https://www.portlandoregon.gov/transportation/71733. Accessed 2 Sept 2017.

———. 2017b. *PBOT News Release: Sunday Parkways Makes a 'Big Jump' in East Portland This Sunday*. At: https://www.portlandoregon.gov/transportation/article/651623. Accessed 2 Sept 2017.

The Denver Regional Council of Governments (DRCOG). 2014. At: https://drcog.org/planning-great-region/implementing-metro-vision/transit-oriented-development. Accessed 22 Aug 2017.

Timperio, A., K. Ball, J. Salmon, R. Roberts, B. Giles-Corti, D. Simmons, L.A. Baur, and D. Crawford. 2006. Personal, Family, Social, and Environmental Correlates of Active Commuting to School. *American Journal of Preventive Medicine* 30 (1): 45–51.

Vandenbulcke, G., C. Dujardin, I. Thomas, B. de Geus, B. Degraeuwe, R. Meeusen, and L. Int Panis. 2011. Cycle Commuting in Belgium: Spatial Determinants and 'Re-cycling' Strategies. *Transportation Research Part A* 45 (2): 118–137.

Who Is TOD in Metro Denver? Benchmarking the Evolution of TOD in Metro Denver. A Summary of Key Findings from the Business, Employee and Resident Surveys. Denver Regional Council of Governments (DRCOG). June 2010. At: https://drcog.org/sites/drcog/files/resources/Who%20Is%20TOD%20Summary%20Report.pdf. Accessed 23 Aug 2017.

Zook, J.B., Y. Lu, K. Glanz, and C. Zimring. 2011. Design and Pedestrianism in a Smart Growth Development. *Environment and Behavior* 44 (216): 216–234.

CHAPTER 6

The Relationship Between Urban Environment and Health

6.1 Introduction

Cities comprise fewer than 3% of the land surface of the Earth, yet they account for 78% of anthropogenic carbon emissions and considerable amounts of airborne toxins and pollutants negatively affecting local and regional air quality (Bereitschaft and Debbage 2013). How urban areas grow and develop spatially represents a crucial part in the relationship between cities and air quality with lower- density development associated with a greater magnitude of vehicle emissions (Stone 2008), which have potential implications for urban public health. The following paragraphs provide a brief overview of air pollution including the definition and types of air pollutants most relevant to human health to establish a basis for understanding the relationship between urban environment and health.

Urban form is understood as the spatial organization and arrangement of city components including population, buildings, and infrastructure. There is a difference between the terms "air quality" and "air pollution." *Air quality* is defined as the outdoor (ambient) level of air pollution with concentrations measured using air-quality monitors, and *air pollution* is understood as both the man-made emission of air pollutants and ambient concentrations of air pollutants (Bereitschaft and Debbage 2013). One of the sources of air pollution data is the United States Environmental Protection Agency's (EPA) Air Data database (www.epa.gov/outdoor-air-quality-data), which offers access to ambient air quality data collected from

A. Antipova, *Urban Environment, Travel Behavior, Health, and Resident Satisfaction*, https://doi.org/10.1007/978-3-319-74198-7_6

the US state, local, and tribal monitoring agencies. ***Air quality index*** is based on the measurement of six ***criteria air pollutants*** (sulfur dioxide (SO2), nitrogen dioxide (NO2), coarse particulate matter (PM10), fine particulate matter (PM2.5), carbon monoxide (CO), and ozone (O3)). Several studies have empirically observed the direct links between spatial structure (urban form) and air pollution emission, for example, ground-level ozone formation (Stone 2008). In the United States, large sprawled cities that are less dense and with less street connectivity exhibited considerably higher O3 *precursor emissions* (i.e., NOx and VOC produced from the combustion of fossil fuels; in the study, their total average annual tonnages at the metropolitan level have been summed to produce a single measure of regional ozone precursor emissions), higher O3 concentrations as well as considerably more O3 *exceedances* (i.e., the average number of high ozone days per year, with O3 $>$ 0.075 ppm) (Stone 2008), while global cities with less contiguous built-up areas had higher average concentrations of NO2 (Bechle et al. 2011). O3 *exceedance* is most relevant to urban public health as it captures the annual frequency of high ozone days reflecting overall air quality. Areas with urban sprawl characterized with lower urban continuity and higher shape complexity generally have higher PM2.5 emissions from nonpoint sources (Bereitschaft and Debbage 2013).

In developed countries, there are several air pollutants with adverse health effects at current concentration levels including particulate matter (PM), nitrogen dioxide (NO2), and ozone (O3) (Hoek et al. 2008); all three are criteria pollutants regulated in the US Clean Air Act. NO2 is associated with numerous negative health effects including lung cancer, cardiopulmonary mortality, and type 2 diabetes (Bechle et al. 2011). *Particulate matter* contains solid particles and liquid droplets of varying size, physical and chemical composition, and coming from different sources. The US national health standards for the ambient (outdoor) air quality use the mass concentration of inhalable particles, that is, particles with an aerodynamic diameter of 10 μm ("μm" represents a micrometer, commonly known as a micron) or less (PM10) originating from soil and other material and *fine particles* (generally referring to particles with an aerodynamic diameter of 2.5 μm or less in diameter (PM2.5), consisting of soot, acid condensates, sulfate and nitrate particles, and originating from combustion of fossil fuels in transportation and industry (Dockery et al. 1993). PM2.5 is a major urban air pollutant that reduces visibility and negatively impacts public health including morbidity and mortality (Pope et al. 2009).

The EPA uses *primary standards* (i.e., health-based) for fine particles, and the *secondary standards* (welfare-based) PM. Regarding primary standards, the annual PM2.5 standard was lowered to 12.0 micrograms per cubic meter (mg/m^3) to protect against health effects due to long- and short-term exposures (such as premature mortality, increased hospital admissions and emergency department visits, and growth of chronic respiratory disease). The secondary PM standard is 35 mg/m^3 (EPA 2013).

In regions with high anthropogenic pressure including eastern North America, Western Europe, and East Asia, there is a clear relationship between population density and topography (i.e., surface elevation) and PM2.5: densely populated, low-lying areas of East China and the valley of the Po River of northern Italy have higher PM2.5 concentrations compared with neighboring higher-altitude areas, for example, the Appalachian Mountains in the eastern United States is a relatively clean area with lower levels of PM2.5 (Van Donkelaar et al. 2010).

Although long-run (i.e., months, years) exposure is more harmful compared with harm from a short-term (hours, days) increase in exposure (Greenstone et al. 2015), both long-term and short-term exposure to air pollution is related to lung-related diseases (Deepa et al. 2017). Importantly, short-term (day-to-day) fluctuations in ambient particle concentrations are related to changes in hospital admissions and mortality (measured as total daily number of deaths). The significant associations were found between particles and respiratory deaths, and between NO2 and O3 and cardiovascular deaths in 1992–1994 London (Bremner et al. 1999). A study on 29 European cities demonstrated that for a 10 μg/m^3 increase in daily PM10 or black smoke (BS) concentrations the estimated increase in the daily number of deaths for all ages was 0.6–0.7%, and the elderly had a slightly greater increase of 0.8% (Katsouyanni et al. 2001). *Black smoke* is a measure of fine black particles, which in the case of Europe are largely from traffic and more precisely diesel engines, with an aerodynamic diameter less than 4.5 μm (standards for BS have been replaced by those for PM10).

Scientific evidence increasingly indicates that the *negative health effects of PM10* may be due to the fine components such as BS, reflected in the decreased effect of PM10 when BS or other pollutants are included in the pollutant model, making exposure to BS more relevant in terms of health effects than PM10 (Bremner et al. 1999). Despite the potential by other pollutants to confound and modify the effect of the PM10-mortality association, neither S02 nor ozone could explain observed associations with particles in European cities; however, both the PM10 and BS associations were moderately confounded by N02. To illustrate the modifying effect of NO2 in PM-mortality association, in a city with a low level of long-term average N02, the predicted increase in total daily deaths, linked with an increase of 10 μg/m^3 in particulate matter, was 0.19%, whereas in a city with a high concentration of NO2 the estimated increase in mortality was 0.80% (Katsouyanni et al. 2001). In contrast, no evidence of *confounding* (confounding occurs when the observed relationship between variables resulted from another variable that is correlated with both a dependent variable, such as mortality, and an independent variable, in this case, exposure to air pollution) by NO2 was reported in the United States likely because there are few diesel cars (diesel engine exhaust emissions of NOx are much higher than from petrol/gasoline engine) there, while in many European cities, motor vehicles are a larger source of urban particles with diesel cars accounting for almost 50% of all cars.

Urban environment threatens adequate clean water supply, which is central to human well-being by affecting water quantity and quality through non-point water pollution and contamination. Leaks in water lines are common due to aging water infrastructure (Frumkin et al. 2004). The 2014 drinking water lead contamination incident in Flint, Michigan provides an example. Due to cost-cutting measures, drinking water supply was switched from the Detroit city system to the Flint River as the main water source where historically water has been of poor quality due to the presence in the area of industrial complexes, landfills, and agricultural farms. The water from the river is 19 times more corrosive compared to the water from Detroit. As a result, thousands of homes in Flint, Michigan have been exposed to contaminated water due to lead leaching from aging service pipes. Dangerous lead levels have been detected in the homes of Flint residents; the EPA advised residents to use bottled or filtered water. Lead is harmful to health, particularly for children. Lead exposure is associated with reduced fetal growth in pregnant women, and impaired cognition, behavioral disorders, and other problems in children. Exposure to high lead levels can be lethal. A recent example includes toxins discovered in groundwater near the Allen Fossil power plant in Memphis, TN in 2017. Thousands of tons of coal ash (which can contain arsenic, lead, and other toxic substances) and other coal-combustion residuals generated each year by the coal-fired 58-year-old Allen Fossil Plant are stored in huge ponds under the plant. Excessive amounts of arsenic and lead were detected about a half mile from where the Tennessee Valley Authority recently drilled wells into the Memphis Sand aquifer, the primary source of local drinking water, from which the TVA plans to pump water to cool its natural gas-fueled plant under construction. Long-term arsenic exposure leads to health problems including heart disease, diabetes, and several cancers (bladder, kidney, liver, lung, and skin cancers) among those who ingest it. The community's treated drinking water supply fortunately has not been affected as no traces of arsenic or lead have been detected.

The chapter examines the relationship between urban environment and public health focusing on health implications of urban air pollution (for commuters, residential population, school children, and pregnancy outcomes), physical inactivity, driving-related fatality and injury, and commuting-based stress. The chapter starts with sprawl-related obesity and barriers to active physical activity. Road safety is essential to reduce road traffic fatality and injury, however, several obstacles prevent the safety strategies from application within the Unites States. Transport infrastructure constitutes an important element of urban form, accordingly, transport impacts on health, including fatality and injury, are considered. To achieve a goal of getting more people engaged in active travel, controlling injury risks is vital. Pedestrian and cycling safety is examined, followed by a discussion on commuting impacts on health, especially commuting-related stress. Transportation health impacts include exposure to traffic-generated air pollution with school buses significantly contributing to childhood air pollution exposure. A case study of the link between air pollution and adverse pregnancy outcomes such as low birthweight and pre-term births is provided. An example of China is included representing the world's most populous country, which has undergone an extraordinary urbanization and a tremendous economic growth over a short time where expanding urban areas is accompanied by a substantial environmental degradation. Health impacts include exposure to air pollution, which is currently the fourth greatest risk factor in all deaths in the country causing 1–1.2 million premature deaths annually. The potential mechanism of health impacts of urbanization in China is provided with valuable lessons for developing countries with rapid urban growth.

6.2 Obesity and Inactive Lifestyles

Land use and transportation affect physical activity, in turn contributing to many health outcomes over the long term such as body weight, morbidity, and mortality. Due to increases in traffic reviewed in the preceding section, the health costs of automobile-based travel remain high (Transportation Research Board 2011). People are driving more, which contributes to physical inactivity, overweight, and obesity, and related diseases (Frumkin et al. 2004).

Most recent residential development in the United States features elements of sprawl. In sprawled communities, streets do not connect to one another; cul-de-sacs are a standard feature, block lengths are long, and destination places are farther apart from each other, while walking routes are often unavailable or unpleasant due to deficient sidewalks and crossings, thus inhibiting walking.

Due to unplanned growth, sprawling areas encroach on the prime agricultural lands, and green spaces are relatively unavailable. There are social costs of sprawl eroding *social capital* of the community (i.e., the forces that bind communities together) in the form of reduced social interactions, civic participation, and trust. There is evidence that sprawl has a direct relationship to obesity and body mass index (BMI) and an indirect relationship to physical activity (Green and Klein 2011). BMI is calculated as person's weight in kilograms divided by height in meters squared (kg/m^2). BMI is an indicator used to measure obesity among adults and children. Children are considered obese if their BMI value is greater than or equal to the 95th percentile of the sex-specific BMI growth charts.

Obesity has been growing in the US population during the recent five decades, affecting morbidity and all-cause mortality in US adults. To illustrate the impact of the obesity epidemic on mortality risk and premature deaths, an adult who is obese has at least a 20% significantly higher rate of dying of all-cause or cardiovascular disease; these rates advanced death in a range between 1.6 and 5.0 years (Borrell and Samuel 2014).

In 2011–2014, 36% of US adults and 17% of youth were obese with significant differences among racial groups: the prevalence of obesity was highest among non-Hispanic Black adults (48.1%), followed by Hispanic (42.5%), and non-Hispanic White (34.5%). The lowest rate was among non-Hispanic Asian adults (11.7%) (Ogden et al. 2015). Of concern is the increasing trend of obesity prevalence: from 1999 through 2014, more American adults and youth have become obese. The results of the 2013 national Youth Risk Behavior Survey reveal that one in seven US high school students are obese (students who were > 95th percentile BMI) (CDC 2014). The prevalence of obesity has grown between the early 1960s and 2008 for children of all age categories, but especially more so for two groups including 6–11-year-olds and 12–19-year-olds for whom the rates more than tripled from less than 5% to 19.6% and 18.1%, respectively (Ogden and Carroll 2010). The distribution of obesity prevalence by race/ethnicity and gender reveals important and significant disparities among US children and adolescents in 2007–2008, with Mexican American boys aged 12–19 years being more likely to be obese (with and obesity rate of almost 27%) compared with Non-Hispanic Whites and Non-Hispanic Blacks of the same age, while for Non-Hispanic Black girls the prevalence of obesity was significantly higher (at 29.2%) than for girls of other racial/ethnic groups (Ogden and Carroll 2010).

Researchers agree that the physical environment may facilitate active living. The reverse is also true; physical environment may support inactivity. A growing number of studies address walking as a direct source of health benefits and focus on opportunities for physical activity including walking/cycling (Sallis 2009).

Built-environment factors associated with implementations of transit-oriented development (TOD) can encourage more active and healthy lifestyles and decrease transportation-related greenhouse gas emissions (Higgins and Kanaroglou 2016). The TOD-related benefits such as lower household transportation costs or avoidance of road congestion-induced stress may attract individuals who desire to improve their quality of life (Gottholmseder et al. 2009).

6.3 Physical Activity

One of the potential solutions to the obesity occurrence is physical activity programs (CDC 2014). Accordingly, there is a substantial research on the link between transportation and public health that focuses on the determination and measurement of the impact of different factors of the built environment and urban design on physical activity and the use of active transportation such as walking and bicycling (Handy et al. 2002; Frank et al. 2004; Badland and Schofield 2005).

When walking and cycling is safe, an overall amount of walking and cycling increases, particularly among groups of vulnerable population including children, women, and elders (Buehler and Pucher 2017). Due to the low physical activity levels in most developed countries, increasing walking and cycling rates could provide regular physical activity and contribute to improved public health. Only few Americans walk or ride a bicycle for daily trips (Buehler et al. 2016). Insufficient physical activity represents a major risk factor with more than 5 million, or 9%, of premature

Fig. 6.1 Increasing regular physical activity contributes to public health: people walk along the Esplanade in NYC. (Author 2013)

deaths globally (Lee et al. 2012). When put into economic terms, the estimated cost from lack of physical activity is around $117 billion dollars or 9–11% of total US health care expenditures (Carlson et al. 2015) (Fig. 6.1).

6.3.1 Walking, Walkability, and Health Outcomes

Built environments change slowly. However, as a neighborhood becomes more walkable, what changes in health among residents and travel behavior (if any) does neighborhood walkability encourage? The definition of walkability was introduced in Chap. 5. Walkability can also be defined the extent to which the built environment facilitates walking through pedestrian comfort and safety as people are connected with different destinations "within a reasonable amount of time and effort" while providing visual interest in travel throughout the network (Southworth 2005,

p. 247). To measure walkability of neighborhoods, objective measures can be constructed. To illustrate, *Walk Score* is a walkability index that captures access to destinations and thus is useful to evaluate walking for transport. People walk for transport as they run errands on foot (e.g., get groceries, go to work, etc.). Walk Score does not measure access to other elements of the built environment that influence walking for leisure, including aesthetic quality, street traffic, or availability of walking trails. Based on the distance to various amenities (restaurants, shopping, schools, parks, and entertainment) and adjusted for *street network characteristics* (for example, areas with low intersection density and high block length get lower scores) the scores of Walk Score range from 0 to 100 (higher scores indicating better walkability) (Fig. 6.2).

New urbanism advocates walkable communities. Walkability, which is closely linked to public health, urban form, and transportation, contributes to health, well-being, and quality of life of citizens in cities. Walkability

Fig. 6.2 One solution to physical inactivity is deliberate exercise such as walking on a regular basis. (Author, Boston 2013)

provides numerous benefits including promotion of an active lifestyle among community residents and thus effective reduction of obesity and chronic diseases (Transportation Research Board and Institute of Medicine of the National Academies 2005). Walkable neighborhood design promotes the development of social capital, which in the form of social networks and community involvement has positive health consequences such as longer lives and healthier physical and mental well-being (Leyden 2003).

Studying the causal effects of environments on health outcomes can be challenging due to the problem of *self-selection*, which occurs when individuals with preferences for certain behaviors select to live in certain types of neighborhoods. Thus, an individual's neighborhood is a result of personal preferences rather than a factor of travel behavior (Krizek 2000). To overcome the problem, the impact of changing environment on health-related outcomes can be empirically assessed by examining changes occurring as part of residential relocation (Hirsch et al. 2014). Even given that longitudinal studies do not entirely overcome the self-selection effect, nevertheless, causal evidence is improved, particularly when the effect of changes in neighborhood features is examined on changes in health.

6.3.2 *Neighborhood Changes and Health-Related Outcomes*

A recent study investigated whether changes in built environment associated with residential relocation were related to concurrent changes in walking for transport or for leisure in adults (Hirsch et al. 2014). Relocation to an area with higher walkability has health benefits reflected in an increase in walking for transport and a decrease in BMI. Specifically, moving to a more walkable location (quantified in the study as a 10-point higher Walk Score) increases walking for transport (in the study, walking levels increased by 17.51 minutes per week), and lowers BMI by 0.06 (e.g., for an average woman with a height of 164.1 cm, this is equivalent to 0.36 pounds less and for an average man 178.2 cm tall, it translates to 0.42 pounds less). However, no changes in leisure walking were observed. Thus, environmental interventions can potentially influence health outcomes and cardiovascular risk effectively demonstrated by the finding that moving to a more walkable neighborhood is associated with declines in BMI.

6.3.3 *Physical Activity and Road Safety Barriers*

Barriers to walking and cycling include lacking infrastructure (bicycle trails, paths, lanes, racks, traffic signals, lighting, or crosswalks) and fatality and injury rates. Average fatality rates are lower on roads built to high design standards (such as limited-access divided highways or interstates) than those on lower-class roads (the collector and local classes). Accordingly, US safety programs focus mainly on the improvement of vehicle (e.g., vehicle design and safety features) and road safety by upgrading infrastructure and imposing safety design standards on new road construction (alignment, lane width, sight distance, and roadside clear zones); however, neglecting countermeasures targeted at changing driver behavior such as speeding and drunk driving that represent the major risk factors. However, research found no strong correlation between the infrastructure investment level and crash rates or frequencies (Noland 2003; Kopits and Cropper 2005). Inadequate technical capacities needed to apply systematic management practices is another key obstacle to transferring successful traffic safety practices implemented in other countries to the United States.

Caused by differences in political systems among developed countries, various physical characteristics of transportation systems, social and cultural factors, the Transportation Research Board (2011) identifies the following obstacles that prevent the safety strategies successfully applied in other countries from application in the United States:

1. Decentralized regulation and enforcement in the United States complicates administration: the decentralized structure of the US government results in the responsibilities for regulation and administration of traffic safety being divided among various levels of the government. There are thousands of entities with independent authority for traffic safety and for the highway system operation and maintenance. For example, the Federal Highway Administration, the lead agency within the US Department of Transportation, is responsible for designing standards for new and rehabilitated state highways built with federal aid; it is also charged with awarding states with safety capital improvement grants, while the Federal Motor Carrier Safety Administration exercises direct federal regulation of commercial truck and bus safety; it also enforces motor carrier safety regulations for commercial motor vehicle operators and drivers, regulates interstate transportation by large trucks and buses, and develops standards that states must pursue related to a commercial driver's license. State highway agencies build and operate state highways (including major intercity roads and highways, and in some states, many minor roads), while local governments' public

works departments are responsible for construction and maintenance of local streets and roads. However, highway administration in most other developed countries with better travel safety is more centralized. Regulation and enforcement is centralized where a central facility plans and administers safety initiatives. For example, in France, where the rate of improvement in traffic safety is one of the fastest in the past decade, a central facility monitors all automatic speed cameras in the country, issues citations, and collects fines, while the safety program is supervised by a cabinet-level multiagency committee that meets on a regular basis. Similarly, the Swedish traffic safety program is also centralized.

2. Rigorous safety interventions depend on public support; however, measures that are common elsewhere encounter active public opposition in the United States. For example, the public often opposes motorcycle helmet laws, seat belt laws, sobriety checkpoints, and speed and red light enforcement using automated cameras. Only 20 US states have such laws requiring motorcycle riders to wear helmets. Automated enforcement of speed limits has been a controversial measure and politically unpopular in some jurisdictions in the United States. However, both techniques contributed to fatality reductions in other high-income countries. For example, drunk driving and failure to use seat belts or motorcycle helmets incur harsh penalties in France. As a result, French rates of seat belt and motorcycle helmet use are among the highest worldwide reflecting enforcement effort. The speed control program started in France in 2002 led to dramatic reductions in fatalities and injuries on French roads between 2002 and 2005.
3. Rigorous safety interventions depend on political support; however, rigorous enforcement is only weakly supported or opposed to in legislatures and among the judiciary in the United States. Effective traffic safety laws implemented in other countries are frequently resisted by US politicians. In contrast, Sweden adopted enforcement strategies controlling alcohol-impaired driving and speed. In marked contrast to the United Sates, control of drunk driving and speed has sustained political support in Sweden. An act of the Swedish parliament established the policy of *Vision Zero*, which set the goal of zero road fatalities and injuries in 1997. The legal driver blood alcohol content limit is among the lowest globally, high-frequency random alcohol checks are implemented (380 tests per 1000 licensed drivers

in 2006), and many speed limits have been further lowered since the 1990s. Worldwide, Sweden has the safest vehicle travel reflected in the lowest fatality rate per kilometer during the past 20 years.

4. Random alcohol testing of drivers, that is, enforcement in which all drivers stopped at a roadside checkpoint are administered a breath test, is illegal in the United States as a violation of the Fourth Amendment's protection against unreasonable searches and seizures. The US police are prohibited by the Constitution from implementing frequent and regular driver sobriety testing without probable cause (that is, a reasonable suspicion that a violation has taken place) that is in common use elsewhere. Random alcohol test checkpoints are used in most European countries with the exception of the United Kingdom. In Australia, whose federal system is similar to the US structure, the states and territories are responsible for the road system and for road safety with a single state police agency having responsibility for all enforcement. For example, in the state of Victoria, whose safety program is a remarkable success story, enforcement was enhanced by a series of laws and regulations. Strategies have been implemented there aiming at controlling unsafe driver behavior. *Behavior-controlling measures* include increased random alcohol testing for drivers (currently the annual test rate is 300 per 1000 licensed drivers) and establishment of the penalty of immediate license suspension for a second drunk driving offense, introduction of the use of cameras for speed enforcement with drivers penalized points toward license suspension for speed camera violations. The probationary period for new drivers rose to three years accompanied with a zero blood alcohol level for the first three years per new license. In 2000–2004, new regulations were introduced including lower urban speed limit, higher speeding penalties, and offender monitoring including obligatory interlock devices (an interlock device prevents an alcohol-impaired person from operating a motor vehicle) on vehicles of repeat drunk drivers. As a result of the safety program in Victoria, reduction in traffic fatalities there has been greater than in Australia as a whole over the period 1988–2004; random alcohol and drug testing of drivers is routine, with the average driver tested once every few years.
5. In the United States, resources needed to carry out the safety strategies are limited, which precludes intense enforcement common in other countries such as Sweden and France where increased resources

for enforcement and automation have led to a sustained decline in the fatality rate. Resources are needed to pay for better policing or for state-level coordination, training, and publicity campaigns.

6.4 Pedestrian and Cycling Safety in Europe and the United States

Health benefits accrue from regular physical activity. A shift from driving to walking and bicycling contributes to public health. Globally, public health policies in countries set a goal of improving traffic safety for pedestrians and cyclists. Northern European countries are characterized by greater pedestrian and cycling safety because their walking and cycling is much more widespread and far better quality. To make walking and biking safe, many Northern European cities introduced rigorous safety programs and have far better infrastructure in their urban areas: much of their central areas have extensive zones that are car-free; most residential neighborhood streets are equipped with traffic-calming devices with slower speeds (a posted speed limit is 30 kilometers per hour, or 20 miles per hour, or less); both sides of almost every street have sidewalks; wide streets have pedestrian refuge islands for crossing the streets; crosswalks are clearly marked, and often raised and have special lighting; street intersections have pedestrian signals; and crosswalks have sufficient crossing times.

To make cycling safe and convenient for cyclists, Northern European cities integrated extensive systems of bikeways, both on-road and off-road, where bikeways are separate from other traffic and often include priority traffic signals and advance stop lines for cyclists at intersections. Motorists pay most attention on the visual field with other motor vehicles and away from the road edges where pedestrians and bicyclists are more likely to be found, therefore reducing car speed at intersections helps motorists to see bicyclists better (Jacobsen et al. 2015). Compared with European countries, construction of separate bike facilities in US cities only started in the 1990s. Presently, US cities lag far behind Northern European cities in terms of the extent, quality, and integration of their bikeways (Buehler and Pucher 2017). Further, mandatory traffic education is taught in schools in some European countries teaching safe walking and cycling skills. In Europe, motorist training and licensing is much stricter compared with that of the United States. Police enforcement of traffic regulations for motorists and nonmotorists (such as speeding, drunk driving, and failure to use seat belts) is also much more stringent and rigorous in Northern Europe involving high penalties, further contributing to traffic safety.

Among enforcement practices targeted at modifying driver behavior that are not in common use in the United States, there are automated speed limits enforcement (i.e., automated cameras and speed-measuring devices installed in the roadway detect and identify speeding vehicles) and alcohol-impaired driving prevention (high-frequency roadside sobriety checks) to implement laws against alcohol-impaired driving (Transportation Research Board 2011). For example, in Europe, one driver in 16 is tested yearly for blood alcohol content, while the rate of roadside alcohol testing is still higher in Australia. These countermeasures contributed to fatality reductions in the countries that use them. If implemented in the United States, sobriety testing is expected to save 1500 to 3000 lives annually, and lowering the legal blood alcohol content limit would reduce fatalities even further.

6.5 Transportation and Health: Traffic Fatalities and Injuries

With transportation infrastructure affecting urban form in a significant way, as discussed in Chap. 2, this chapter considers the links between this important element of urban form and health including road fatalities and injuries, traffic-related exposure to air pollution upon commuters and overall residential population including school children, and commuting-related stress.

According to the World Health Organization (WHO), traffic injuries and fatalities are among the world's five most important causes of unnatural death, while among persons in the group aged 15 to 29 years they were the leading cause of unnatural death in 2015, and they are estimated to become the leading cause by 2030 (World Health Organization 2015). Thus, in the United States, 28% of all deaths among young people 1 to 24 years of age were caused by motor vehicle crashes in 2006 (Heron et al. 2009).

Traffic-related fatalities and injuries represent a substantial public health threat worldwide including in the United States. From 2007 to 2008, US traffic deaths dropped by 9.3% and by 9.7% from 2008 to 2009, which are among the largest annual declines recorded. To support, in 2009, the total of traffic deaths of 33,808 was the lowest since 1950 (Transportation Research Board 2011). However, the extraordinary percentage decrease in deaths from 2007 to 2009 probably may be due to the recession that began in 2007; in the past, as the economy recovered after recessions, traffic deaths increased (Transportation Research Board 2011). Nevertheless, in aggregate, between 1977 and 2009 bike trips in the United States more than tripled, and the percentage of total trips by bike almost doubled, increasing from 0.6% to 1.0% as a consequence of overall improvement in walking and cycling safety as suggested by the large percentage reductions in per-capita pedestrian and cyclist fatality rates.

Though pedestrian and cyclist fatality rates per capita and per kilometer have declined in most countries including the United States, Canada, Australia, Japan, the United Kingdom, the Netherlands, Sweden, Denmark, Germany, France, and Spain over the period of 1990–2014, there is only slight reduction and an observed increase in serious injuries rates per kilometer between 2001 and 2009 (Buehler and Pucher 2017). The number of severe pedestrian and cyclist injuries (*severe injury* is defined as injury that requires overnight hospitalization) is much greater compared with the number of fatalities. In the United States between 2008 and 2009, the ratio of severe injuries to fatalities for pedestrians was 8 to 1, and much higher for cycling: 44 to 1. The United States has made the least progress in this regard among the 11 developed countries mentioned above reflected in the smallest reductions in fatality and injury rates.

Most high-income countries are decreasing both traffic fatalities and fatality rates (measured per kilometer of travel) faster compared with the speed of improvement in safety in the United States (Transportation Research Board 2011). To compare, while pedestrian fatality rates per capita fell by 49% in Canada, 52% in Japan, and 63–75% in seven Western European countries and in Australia, it decreased by only 35% in the United States. Similarly, while the cyclist fatality rate fell by 46% in Australia, 47% in Japan, 49% in Canada, and by 53–68% in Western Europe, in the United States it dropped by only 30% (Buehler and Pucher 2017).

Compared to Germany, fatality rates measured as pedestrian and cyclist fatality rates per 100 million kilometers walked and biked, in the United States were significantly higher for 2001–2002 and 2008–2009. During the second time period, the United States had about five times higher pedestrian fatality rates compared to that of Germany (9.7 vs 1.9) and more than three times as high a fatality rate for cyclists (4.7 vs 1.3), indicating that the United States has the most dangerous walking and cycling among the aforementioned countries (Buehler and Pucher 2017). There are also variations in injury and fatality rates by age and gender. For example, compared with the population as a whole, senior pedestrians and cyclists have much higher fatality rates, and senior pedestrians have almost twice the national average severe injury rate. Children cyclists had a much higher severe injury rate in 2008–2009 in the United States than the national average, and the rate for children increased significantly over the last decade.

While lower pedestrian and cyclist fatalities and serious injuries represent a benefit per se, levels of walking and cycling have been shown to increase particularly among vulnerable groups such as children, seniors, and women when walking and cycling conditions are safer. The situation regarding the low physical activity can be improved by increasing walking and cycling levels in most developed countries, thus contributing to better public health (Buehler and Pucher 2017). Increasing regular physical activity via higher levels of walking and bicycling is a key policy tool to address premature mortality (Rutter et al. 2013).

6.6 Transportation and Health: Commuting Impacts

Current research examines the relationship between overall life satisfaction and travel behavior (i.e., mode choice, commute length, etc.). The positive and negative perceptions of commuting depend on the mode of transportation such as car, public transport, and other transportation modes (motorcycle, bike, on foot) or a combination of different modes (Stutzer and Frey 2008). Active commuting, such as riding a bicycle or walking to work, is linked to increased physical activity and lower obesity levels. Moreover, bicyclists have been found to derive the most positive affect from riding bicycles on the way to work, followed in terms of happiness by car passengers, then car drivers, while the most negative emotions have been experienced by bus and train riders (Morris and Guerra 2015). However, the effect of the choice of mode of commuting on commuting stress is not clear with mixed research findings (Fig. 6.3).

Despite new forms of work become increasingly popular, such as of telecommuting, for many in the workforce commuting is an important daily routine (Zhou et al. 2017; Künn-Nelen 2016). Commuting offers commuters benefits by allowing travelers to get to work, or find either much better or cheaper housing, although further away from the work locations (and less attractive to people in general). Jobs that involve longer commutes have to provide their employees a higher income in order to

Fig. 6.3 Walking and biking in Manhattan. The 102-story Empire State Building is seen in the background. (Author 2013)

retain them. Anticipated income might help to overcome commuting distance more contentedly (Gottholmseder et al. 2009), feeling less stressed. Although past research also suggested that commutes represent an enjoyable experience and that individuals have ideal commuting times (Mokhtarian and Salomon 2001), different individuals evaluate the stressfulness of commuting differently based on their commuting preferences. However, people who spend a lot of time commuting and who commute longer find it a mental and physical burden and report regularly lower happiness or subjective well-being (Stutzer and Frey 2008). Reported subjective well-being measures are valid measures with people who report happiness or high subjective well-being, smiling more often, interacting with other people, and being less likely to commit suicide.

Commutes are the daily time-consuming routines in people's lives; they take time, create out-of-pocket costs, trigger stress, and interfere in the relationship between work and family. Workers in Europe commute between 29.2 minutes in Portugal and 51.2 minutes a day in Hungary. The amount of commuting time and the share of people who need to commute can be expected to increase in Europe due to most European labor markets becoming increasingly flexible (Künn-Nelen 2016). In China, a survey conducted by the Chinese Academy of Sciences found that the three largest cities in China—Beijing, Shanghai, and Guangzhou—commuters spent an average of 52, 47, and 48 minutes, respectively, on their daily journey to work travel, with some workers spending up to four hours each day on their commutes. In Beijing, this includes waits at the bus or train stations, multiple transferences, and hours-long standing in crowded buses and trains (Zhou et al. 2017). According to the recent 2011–2015 American Community Survey 5-Year Profiles, the mean travel time to work for workers who are 16 years and older in the United States, is 25.9 minutes, with South Dakota having the lowest commuting time of 16.9 minutes and commuters in New York traveling to work the longest, 32.3 minutes. Further, 8.1% of workers reported spending more than 60 minutes one way on their commutes to work every day. (ACS: www.census.gov/programs-surveys/acs/).

In the United States, commutes represent less than 20% of all trips taken, but commuting is important in terms of informing transportation planning and policy decisions as it determines peak travel demand across transportation systems including road network and transit (McKenzie 2015). However, those commuters with long journeys to and from work are compensated for the psychological stress incurred on the labor and housing market, that is, by higher salaries or by lower rents for housing, resulting in additional money for consumption. Members of their household might benefit, so the households' well-being may be improved overall. For many travelers, however, commuting appears to involve stress that does not pay off. There are both private costs (e.g., the EPA estimates that an average household spends more on driving costs than on food, that is, almost 20% of its income covers driving costs (EPA 2001)) and the social costs of commuting (congestion and environmental pollution). Private costs of commuting also include the costs of congestion often calculated as the value of time (proxied by 50% of the gross wage rate) to measure the cost of delays occurring when traveling (refer to congestion in Chap. 5's Sect. 5.1).

Commuting can constitute an unpleasant experience that can potentially manifest later on through effects on health and family life (e.g., family tensions); therefore, non-pecuniary costs may be associated with commuting. Additionally, multiple environmental stressors like noise, crowds, pollution, and thermal conditions are related to commuting, which may trigger negative emotional and physical responses. Commuting is felt to be more stressful when traffic congestion occurs or when commuters experience substantial time pressure. As a result, commuters report health-related problems including higher blood pressure, musculoskeletal disorders, reduced frustration tolerance, and greater anxiety and hostility, having a bad mood upon arrival at work in the morning and home at night, greater incidences of tardiness, absenteeism, and turnover at work, as well as negative cognitive performance (Koslowsky et al. 1995).

Given the amount of time commuters spend on their journey to work travels, commuting represents an important part in lives of workers. Thus, the quality of commute is associated with well-being (Zhou et al. 2017).

Past research has indicated several ways that commuting is related to commuters:

1. Because commuting may be stressful, commuting may affect commuters' physical health and psychological well-being.
2. As noted, poor commuting experiences are related to negative work- and family life-related consequences (e.g., commutes affect work task performance, trigger aggressive behaviors, and cause work–family conflict and negative mood).

Stress accounts for a large proportion of mental health problems (Gottholmseder et al. 2009); *commuting stress* is a unique aspect of employees' work experiences (Zhou et al. 2017). People who report themselves as highly stressed incur 46% higher health-care expenditures compared with people with lower stress levels (Gottholmseder et al. 2009). Added to the health system costs of stress are organizational costs, which include lowered labor productivity as a result of reduced work performance or greater job turnover and absenteeism with the stress-related days of absence from work accounting for more than 50% of all days of absence.

The increases in the physical spatial separation of residence and workplace locations result in longer daily travel, which is hypothesized to impact the perceived stress level (Gottholmseder et al. 2009). Various sources of commuting stress include objective factors such as commuting time, distance, speed, or commuting conditions like traffic congestion; and subjective factors, including the perceived control over the commute such as commuting modes, personal characteristics such as gender or family situation, as well as assessment of commuting time, alcohol consumption, and leisure-time activities. Although the length of the commuting trip and commuting time are among the most important factors of commuting stress, other factors listed above can all impact commuting stress. Studying the effect of commuting on individual stress perception can improve understanding of individual commuting decisions as well as appreciate the economic costs of commuting beyond the direct pecuniary costs.

In the analysis of the relation between commuting time in *passive commuters* (i.e., those commuting by car, motorcycle/moped, or public transportation) and health in the United Kingdom where commuting time has increased with the average commuting time being 43 minutes, four different types of health outcomes have been studied: *subjective health measures* (health satisfaction), *objective health measures* (health problems), *health behavior* (regular exercise), and *healthcare utilization* (the number of doctor visits) (Künn-Nelen 2016). A longer commuting time is linked to significantly lower health satisfaction and to a lower health status. Those who commute longer have more doctor visits. An interesting gender effect and mode effect has been found: women commuters and those women driving a car to work have more doctor visits. Women have a negative relationship between commuting time and regular exercise (that is, the longer the commute the less physical activity undertaken regularly by these women) and a positive relationship with sickness absence (those who commute longer have a higher probability of calling in sick).

Stutzer and Frey (2008) studied the effect of commuting on individual well-being in Germany. To obtain data on reported subjective well-being, the researchers asked participants: "How satisfied are you with your life, all things considered?" Responses ranged on a scale between 0 ("completely dissatisfied") and 10 ("completely satisfied"). Commuting time and distance were used as an approximation for the burden of commuting. The dependent variable was life satisfaction, while predictor variables included individual commuting time and distance, and variables covering socioeconomic status and demographic characteristics of the study participants. Interaction terms between commuting time and the mode of transportation have been included into the model. In their sample, on average, people commute 22 minutes one way. There were around 7% in the sample who report traveling to work taking an hour or longer. Among the findings is a large negative correlation between commuting time and individuals' well-being (Stutzer and Frey 2008). An increase in commuting time, on average, is associated with a lower reported satisfaction with life. Because different sociodemographic and socioeconomic characteristics may confound the relationship between the observed lower subjective well-being of people and longer traveling from home to work, to control for individual characteristics, a multiple regression approach is utilized. Results reveal that commuting time is significantly negatively correlated with health satisfaction and negatively impacts people's satisfaction with their spare time (Stutzer and Frey 2008).

Another study based on 698 personal interviews conducted in Vorarlberg, the most western province of Austria, analyzed the effect of commuting on individuals' stress perception. The mean travel time in the sample is about 18 minutes. The study area is one of Austria's most conspicuous commuting areas due to a substantial percent of cross-border commuting (about 8.5% of total employed population) to the neighboring countries of Switzerland, Liechtenstein, and Germany (Gottholmseder et al. 2009). The workers' perceived stress level after arriving at their place of work in the morning represented the dependent variable, which ranged from "very stressed," "stressed," "relaxed," and "very relaxed." The respondents were asked to assess their commute by categorizing commuting time as either "time well spent" or "lost time." Those who assessed their commuting as time well spent are hypothesized to perceive their commute as being less stressful since commuters can use the travel time for other activities such as working, reading, communicating, preparing for work, and so on. Due to the ordinal nature of the dependent variable, researchers used an ordered logit model to predict the effects of the predictor variables such as commuting time, commuting time assessment, commuting modes, commuting experience and predictability on the perception of stress. Predictability included commuting experience (in years) and congestion frequency based on the expectation that commuters who can predict their actual commute length may feel less stressed. The variability of the commute is related to higher levels of commuting stress, so the less predictable the commute length is, for instance, due to congestion and the like (and the more variable it is), the greater the stress. The frequency of traffic congestion was quantified as the percentage of trips on which the individuals assessed the traffic situation as "bad." Results reveal that predictability plays a role in shaping the perceived stress level after commuting in the morning due to variability where a higher variability in the expected commuting time increases stress. Findings demonstrate the effect of the physical separation between residential areas and work places in urban areas: traveling time increases the perceived stress level. An increase in commuting time of one minute (on average) reduces the probability of feeling "relaxed" or "very relaxed" by 0.1%.

6.7 Transportation and Health: Exposure to Transport-Generated Air Pollution

As a consequence of increasing use of motorized vehicles, as discussed in Chap. 5, more than a quarter (26.4%) of total greenhouse gas (GHG) emissions in 2014 was produced by the transportation sector with carbon dioxide (CO2) being the predominant GHG, making transportation the second largest producer of GHG emissions where passenger cars utilize the largest portion of fuel and produce the largest percentage of CO2 (*Passenger Travel Facts and Figures* 2016).

Transportation therefore impacts public health by exposing residents and those who travel to air pollution related to urban transportation. People's choice of different modes of surface transportation (car, bus, walking, cycling, and taxi), their proximity to the pollutant sources, and personal time-activity patterns (Kaur et al. 2007) in the urban environment impact personal exposure to air pollutants such as PM2.5, ultrafine particles, and carbon monoxide (CO), experiencing various exposure concentration levels.

Besides the mode of transport and traffic, other factors may influence exposure to transport-generated air pollution including *meteorological conditions* (such as wind speed and direction, temperature, relative humidity, and precipitation), and road and building configuration parameters (e.g., a wide and flat road versus narrow street canyons in inner-city areas) (Fig. 6.4).

In street canyons, marked departures from the simple distance decay pattern of air pollution dispersion have been observed due to the presence of tall buildings. Marked accumulation of air pollution occurs in certain places of street canyons such as on the windward side of a building so that different sides of the road may vary significantly from each other in concentrations (Hoek et al. 2008). Despite being an important variable that influences the impact of pollution sources on a local scale, wind direction has rarely been included into land-use regression models, which model spatial variation of ambient air pollution. For example, the constructed wind direction was demonstrated to be a significant predictor of NO2 concentration (Arain et al. 2007). Geometry of street canyons, which are common features of the urban landscape, hugely impacts urban pedestrian exposure to traffic pollutants in the populated urban areas with an increase in concentrations of traffic-related NOx and CO occurring inside canyons where height of the buildings is greater. Thus, careful urban planning of canyon geometry may considerably decrease inhalation of unhealthy air by urban residents (Fu et al. 2017). However, the aformentioned factors, although significant, do not fully explain exposure concentrations to the different pollutants.

Generally, the exposure studies agree that pedestrians and cyclists are exposed to lower fine particulate matter and CO concentrations compared to the passengers inside vehicles, probably due to their being slightly farther from the source of pollution as bicyclists can avoid or move through congestion while being closer to the curb where pedestrians walk, while drivers and passengers sitting within vehicles are in the direct path of traffic emissions and experience higher CO and PM exposure concentrations as the shell of a vehicle offers only weak protection (Kaur et al. 2007). For example, in the Dutch city of Arnhem, bus and car passengers are exposed to PM10 by 60% and 20% higher, respectively, than cyclists. However, all modes (cars, buses, and bicycles) have been found comparable in exposures to PM2.5 with concentrations slightly lower in electric buses than in all other modes (Zuurbier et al. 2010). The choice of routes affects exposure levels: walking and cycling along high-traffic routes exposes walkers

Fig. 6.4 Street canyon formed in Downtown Chicago. (Author 2015)

and cyclists to higher PM concentrations than on low-traffic routes (Kaur et al. 2005). To illustrate, cyclists and pedestrians who traveled on the high-traffic route in Central London were exposed to 40% higher levels of particle number counts (PNC) (Kaur et al. 2005) and 35% higher levels of soot compared with these activities on the low-traffic route, while the differences in PM10 and PM2.5 exposure among cyclists who traveled along high- versus low-traffic roads were negligible (Zuurbier et al. 2010).

6.8 Transportation and Health: Commuters' Exposure to Particulate Matter Air Pollution

Urban travelers experience high concentrations of air pollutants as residents spend substantial time per day traveling. For example, according to the WHO (2005b), in Europe, people spend 1–1.5 hrs./day traveling. Thus, travelers' exposure studies are important to better understand, manage, and reduce the health risks posed by fine particulate matter and carbon monoxide in the urban environment (Zuurbier et al. 2010). Additionally, as a substantial amount of traveling takes place during morning and evening rush hours, daily commuting adds significantly to total exposure to transport-related air pollution with the resultant short-term cardiovascular and respiratory health effects.

Commuters' exposure to air pollution may differ depending not only on the choice of modes of transport, but also on the choice of routes and fuel types used. Zuurbier et al. (2010) studied commuters' exposure to particulate matter air pollution for different modes of transport, fuel type, and route. To evaluate impact by transport mode and fuel, data on diesel and electric buses, and gasoline- and diesel-fueled cars have been used. In addition, to assess route's impact upon pollution exposure, two bicycle routes with different traffic intensities (low- and high-traffic intensity) were evaluated in the medium-sized city of Arnhem in the Netherlands. Traffic ranged between car-free conditions to 30,000 vehicles/day with average traffic intensity of approximately 5000 vehicles/day. Air pollution was measured by taking medians for particle number counts (PNCs), concentrations of PM2.5 (particulate matter ≤ 2.5 μm in aerodynamic diameter), PM10, and soot (elemental carbon, black carbon).

Black carbon is a primary pollutant and a component of ambient particulate matter that is produced from the combustion of carbonaceous material (i.e., material containing carbon or its compounds, such as fossil fuel, biomass burning) and has a number of adverse public health and climate impacts (Briggs and Long 2016). "Black carbon" is sometimes used interchangeably with "soot," for example, by the Intergovernmental Panel on Climate Change (IPCC 2007, pp. 135, 184, 188, 814) and others. Fossil fuel combustion sources contribute more than 70% and more than 78% of black carbon emissions in Europe and the United States, respectively. Its health effects include lung cancer and cardiovascular and all-cause mortality (Grahame et al. 2014) suggesting that black carbon is a more appropriate indicator of air pollution health effects than outdoor PM2.5 given it has stronger associations from epidemiological studies (Janssen et al. 2012). Airborne elemental carbon has been commonly used as a surrogate measure of ambient air black carbon; studies report good correlations between the two measurements when measuring emissions from the same source (Briggs and Long 2016). Soot, as defined by Buseck et al. (2012, p. 24824) represents "the solid, carbon-rich material that condenses from the vapour phase during combustion."

Researchers compared exposure levels during commuting for different pollutants and urban locations by mode of transportation and compared exposure to pollution emitted by all modes of transport with urban background levels. All modes of transport produce significantly higher concentrations of all measured pollutants than urban background levels. The highest PNC exposure is in diesel buses and for cyclists along the high-traffic intensity route, while the lowest PNC exposure is in electric buses. Urban location and building configuration affect more PNC pollution concentration: for example, the city center where high buildings are standing on both sides of a road, a street canyon effect is generated contributing to higher concentrations of PNC compared to suburban locations. During cycling, certain modes of surface transportation can contribute significantly to cyclists' exposure to PNCs: as a cyclist is passed by mopeds and trucks, as well as travelling on busy traffic road, short-term peaks in PNC pollution concentration have been observed. PM10 exposure is highest from diesel buses and lowest along the high- and low-traffic bicycle routes. There was less clear pattern of higher PM2.5 concentrations in the city center. The soot exposure is highest in gasoline-fueled cars, diesel cars, and diesel buses, while lowest exposure to soot is along the low-traffic bicycle route (Zuurbier et al. 2010).

To further assess the potential health consequences, assessment of the inhaled air pollution doses was conducted. To compute the amount of inhaled air pollution, researchers measured the heart rates of 34 volunteers during commuting by bicycle, car, and bus and estimated minute ventilation (defined as ventilation rate, or volume of air, per minute). Because of their increased physical activity, the minute ventilation of cyclists was twice those of car and bus passengers; the inhaled air pollution doses were highest for cyclists. Compared with car and bus passengers, in cyclists the difference in the amount of inhaled dose was especially high for PNC and PM2.5, while that for PM10 and elemental carbon (calculated using the soot levels) was not as great (Zuurbier et al. 2010). Bus passengers inhaled higher PM10 doses than car passengers, whereas passengers of both diesel buses and cars inhaled comparable elemental carbon (EC) doses, while passengers of electric buses inhaled the lowest PM10 doses. Besides transportation mode, fuel type, and choice of route, the doses of inhaled air pollutants are influenced by traveling time, which in turn depends on the route and on the time of day. While usually cars travel faster, cyclists and buses may travel faster than cars during peak hour due to the use of separate bus lanes, which may result in different exposure levels.

6.8.1 *Transportation and Health: School Buses and Childhood Air Pollution Exposure*

In urban communities, school buses add significantly to childhood air pollution exposures (Adar et al. 2008). Studies note that children who regularly commute by school bus are especially exposed to high concentrations of air pollutants aboard school buses and near roadways. Specifically, riding diesel school buses adds to childhood fine particulate matter exposures: children are exposed to air pollution both from surrounding roadway traffic and the bus they are riding on. In their study based on the Seattle, Washington metropolitan area, onboard fine particulate matter (PM2.5) air pollution (conceptualized as a sum of pollution from surrounding vehicles and ambient background pollution, and self-pollution from the diesel bus itself) was shown to be higher than roadway levels of pollutants: the average in-cabin concentration was found to be four times higher than the ambient (outdoor) background level (ambient concentrations were measured at remote central monitoring stations as hourly averages) and two times higher than roadway concentrations (i.e., the levels conceptualized as a sum of pollution from surrounding vehicles and ambient background pollution, without self-pollution from the diesel bus) (Adar et al. 2008) (Fig. 6.5).

Fig. 6.5 School buses add to childhood air pollution exposures. (Author 2017)

Bus characteristics (bus age, diesel oxidative catalysts) and roadway concentrations are likely to impact in-cabin concentrations. In addition, school buses running low-sulfur diesel fuel generate pollution from their diesel exhaust. Diesel emissions consist of a mixture of volatile gases (e.g., nitrogen oxides, or NOx) and particulate matter (e.g., soot, black carbon). Diesel emissions are harmful to human health, particularly children. Diesel exhaust generates small-sized particles that when inhaled can penetrate deeper into the lungs of children who, compared with adults, have faster respiration rates and smaller airway passages and inhale ultra-fine particulates. Further, diesel engines are the second-largest producers of particulate pollution making diesel exhaust one of the worst contributors to air pollution. Zuurbier et al. (2010) found that retrofitting diesel buses with diesel particulate filters does not result in a significant reduction of pollutant concentrations in terms of soot: small and not significant differences in median soot concentrations have been found on the diesel buses with and without filter, with soot concentrations in diesel buses comparable to those in electric buses.

Prolonged exposure to the in-cabin pollution may worsen asthma and trigger allergies in asthma-prone children who routinely commute by diesel bus. Additionally, self-pollution while in transit occurs when the bus doors often open to let passengers on and off the vehicle, during which air pollution from outside is let in. In case of diesel buses, infiltration of bus exhaust from the bus itself is likely when a portion of the bus's own tailpipe emissions adds to the in-bus exposure by reaching the front door under the proper wind conditions (Zuurbier et al. 2010). Adar et al. (2008) conclude that self-pollution may contribute a substantial percent (up to 30%) of air pollutants, at least US school buses.

6.9 Transportation and Health: Residential Traffic Exposure

Within the United States, routes of major roadways have been often directed through areas with less political and economic power where lower-income communities resided. Epidemiological studies indicate that both short-term and long-term exposure to elevated levels of traffic (e.g., increased traffic density and/or proximity to major roadways) have been associated with several adverse health outcomes. Specifically, exposure may negatively affect the pulmonary system and increase risks of cardiovascular disease through systemic inflammation and increased blood pressure. Studies identify traffic density as a significant predictor of several types of air pollutants including nitrogen oxides and dioxide (NOx, NO2), particulates < 2.5 μm in aerodynamic diameter (PM2.5), the soot content of PM2.5, and volatile organic chemicals.

Controlling for income/poverty ratio and educational level, residential exposure to traffic pollution has been modeled to predict cardiovascular disease in adults. Using health, demographics, and blood data of the adult US Puerto Rican population residing primarily in the Greater Boston area, Rioux et al. (2010) examined an association between traffic exposure indicators including residential proximity to roads with > 20,000 vehicles/day with counts reflecting traffic volumes in both directions and traffic density, and cardiovascular disease. Diabetes and obesity have been used as modifying variables. Body mass index (BMI) was calculated as weight in kilograms divided by height in meters squared. To better measure residential traffic exposure, participants' addresses were geocoded.

Residential traffic exposure was assessed using several different traffic indices for local impacts near busy roads, as well as traffic density measured as vehicle miles traveled per square mile. Only roadways with at least 20,000 vehicles/day were considered in the study.

Several traffic exposure indices included roadway proximity:

(a) residential proximity $\leq$ 100 m of a roadway; (b) residential proximity $\leq$ 200 m of a roadway ; (c) number of roadways (0- used as a reference level, 1, or $\geq$ 2) within 200 m of a residence; and (d) indicator dummy variables of the distance gradient: 100-m, 200-m buffers as the areas within 100 m or 200 m of either side of a roadway, and the distance at greater than 200 m of a roadway (used as a reference level).

Additionally, several traffic density exposure levels were determined, measured as vehicle miles traveled per square mile (VMT/mi^2), at the level of traffic analysis zones (TAZ) (the lowest density level used as a reference level), and traffic density level of the residential TAZ assigned to study participants. Four raster-based exposure levels were defined conducting a raster-based spatial density analysis with density values within 10 × 10 m cells calculated over a 1000-mile radius from the center of each TAZ, ranging from the lowest to the highest traffic density (the lowest density level used as a reference).

Researchers concluded that exposure to traffic at roadway volumes as low as 20,000–40,000 vehicles/day may be associated with an increase in the risk of cardiovascular disease via negative effects on blood pressure and inflammation. Individuals with elevated body mass index, that is, BMI $\geq$ 30, may be more prone to the effects of traffic exposure.

6.10 Social Stress and Air Pollution

Chronic stress may be related to the health effects of air pollution. In real-world settings, stress and pollution exposures are often tightly confounded. Epidemiologic research indicates that chronic exposures to social stressors such as poverty, violence, fear of violence, and perceived stress may influence individual vulnerability to air pollution exposures (Hood 2010). For example, exposure to social stressors (e.g., low socioeconomic status) and exposure to environmental pollution are frequently correlated, with lower-income communities often located proximate to highways and industry. People with lower socioeconomic status (SES) appear to be more strongly affected by air pollution compared with people of higher SES status. Therefore, it is important to separate the effects of social and environmental exposures (i.e., traffic-related pollution) because of their potential spatial correlation (Clougherty et al. 2010). Chronic psychological stress is associated with changes in several functions including immune, inflammatory, and endocrine function. Immune-related inflammatory responses influence airway diseases and respiratory inflammation, while early-life stress may lead to childhood asthma.

However, epidemiologic methods alone cannot fully separate the effects of stress and pollution exposures. A 2010 study examined the role of chronic stress in modification of the effect of fine particle air pollution on respiratory function in animals. Clougherty et al. (2010) explored the influence of chronic stress on pollution susceptibility. In the study, rats were randomly assigned to four exposure groups including two groups of stressed animals with one group exposed to concentrated ambient particulates (CAPs) and another group exposed to filtered air with no particulate matter, and two non-stressed groups with a similar design, thus stress and pollution exposures were controlled for to better separate their effects on respiratory function. While both stressed and non-stressed groups, which were exposed to particulate matter, exhibited a significant response to the exposure, all stressed animals showed significantly greater increases in respiratory frequency compared with non-stressed animals. Animals that had been under chronic stress revealed a rapid, shallow breathing pattern (lower expiratory flows and volumes) compared with non-stressed animals (Clougherty et al. 2010). The authors conclude that when stress and pollution exposures occur simultaneously, in combination they may affect inflammation in a different way compared with an effect that either exposure can have alone, thus demonstrating that chronic stress may increase vulnerability to effects of air pollution on respiratory illnesses.

6.11 Measures to Reduce Air Pollution Exposure

Measures can be implemented to reduce personal exposure to transportation-related air pollution and improve health in urban areas. Cyclists are impacted by air pollutants emitted by surrounding traffic. Although it may not be possible to avoid all contact with motorized traffic, cyclists can choose traveling on a low-traffic route. Health effects of cycling overwhelm the health risks of increased pollution loads making cycling beneficial to health. Cycling is also beneficial to clean outdoor air by not emitting any air pollutants, thus,cycling should be encouraged. City planners should create bicycle lanes with as little as possible exposure to motorized traffic (Zuurbier et al. 2010).

To reduce childhood exposures to pollution it is important to understand the major pollution sources. With air pollution exposure being significantly lower in electric-powered buses than in diesel buses, using clean buses may improve both urban air quality and health of bus passengers. Policies that increase the use of clean buses, such as electric buses, should be encouraged (Zuurbier et al. 2010). A large fraction of school-aged children is exposed to traffic pollution while in transit, especially as children commute to school by school bus. The use of clean diesel technologies is indicated in the study by Adar et al. (2008) who found that the presence of diesel oxidation catalysts on new school buses is associated with a reduction of fine particles onboard school buses. The reductions, however, have been shown mostly for buses with their windows closed as self-pollution is highest when the bus windows are closed. Retrofitting buses with diesel oxidative catalysts (DOCs) designed to reduce tailpipe emissions may reduce some components of PM2.5.

Exposure studies provide impetus to the development of strategies to control and reduce urban air pollution. In Europe, BS has been historically measured as the routine indicator of particles, however, standards have been replaced with PM10 measurements representing all particles with an aerodynamic diameter less than 10 μm, comprising a mixture of primary and secondary particles coming from different origins with varying physical and chemical characteristics and various levels of toxicity. The suggestions that traffic particles are more toxic compared with particles originating from other sources justify continuation of monitoring of BS (BS are fine black particles with an aerodynamic diameter less than 4.5 μm), or monitoring of carbon. Evidence indicates that BS is more specific for traffic-related particles (i.e., it originates mostly from traffic) and as noted earlier, exposure to BS is more relevant to health effects than PM10. Associations between air pollutants including CO, particles, NO2, and SO2 and daily mortality were detected in London in the 1990s. In London, the fine fraction comes from primary emissions from mobile sources or secondary sources, the significant associations were found between particles and respiratory deaths, and between NO2 and O3 and cardiovascular deaths in 1992–1994 (Bremner et al. 1999). The monitoring of BS should be encouraged for potential health risks assessment and not abandoned (Katsouyanni et al. 2001).

Since air pollution represents an urgent public health problem on a global level (Greenstone et al. 2015), it is important that the health effects associated with air pollutants including particulate matter be studied rigorously to raise awareness in people and to protect them from the adverse impacts of air pollutants (Deepa et al. 2017). Investigations on the health impacts of pollution may assist the authorities to make informed decisions regarding appropriate actions and policies on air pollutants reduction.

6.12 Air Pollution and Births

Unequal distribution of pollution-emitting facilities across urban areas caused environmental justice studies to comment on geographic inequalities in adverse health outcomes due to uneven environmental exposures. Health studies examine pregnancy outcomes such as low birthweight and pre-term births due to exposure to environmental hazards. *Low birthweight* (LBW) is defined as infants with a weight less than 2500 g (5.5 pounds) at birth. *Preterm births* (short gestation) are births at less than 36 weeks of gestation; prematurity is the main cause of death during the first four weeks of life (Goldenberg et al. 2008). LBW is used as an important public health indicator (UNICEF and WHO 2004), resulting in poor health outcomes including neonatal mortality and morbidity (Barker 1992). LBW risk factors include the mother's long-term malnutrition, ill health, maternal lifestyles, socioeconomic conditions, and hard physical work during pregnancy.

Yet another important causation and health risk to pregnancy outcomes is the physical environment (WHO 2004), particularly exposure to contaminated environment and toxic substances, which increases incidence of LBW and prematurity (Valero de Bernabe et al. 2004). Infants from heavily exposed areas have on average a lower weight than those born in non-contaminated areas even when controlling for other confounding factors (Bobak et al. 2001). Exposure to carbon monoxide may lead to preterm births (Goldenberg et al. 2008).

Studies analyze the links between unequal exposure to environmental hazards and negative health outcomes using a "hot spot" approach based on proximity to noxious land uses (Maantay 2007). In New York City, hospitalization rates were higher for people within identified hotspots compared to those living beyond these areas. Asthma hot spots have been identified in New York City (Corburn et al. 2006). Other studies utilized pollution concentration- and facility density-based approachs to identify pollution-impacted areas to examine asthma hospitalization rates in children at the zip code level controlling for socioeconomic factors (Grineski 2007).

Antipova and Larsen (2017), in a recent case-control study, examined the impact of environmental exposure on negative birth outcomes in Shelby County, TN. Specifically, the authors analyzed differences in birth outcomes between areas of high and low exposure to environmental pollution. The study uses estimates of air pollution exposure from stationary emission sources and models both LBW and preterm births. The scale of analysis was at the individual level.

The study used a comprehensive National Emissions Inventory (NEI) database from all air emissions sources published by the EPA to map air polluting facilities using the Geographic Information System (GIS). There were 69 stationary facilities reported for Shelby County in 2011. A dataset of individual-level births in the county for 2011 was purchased from the Tennessee Department of Health in 2016 (Fig. 6.6).

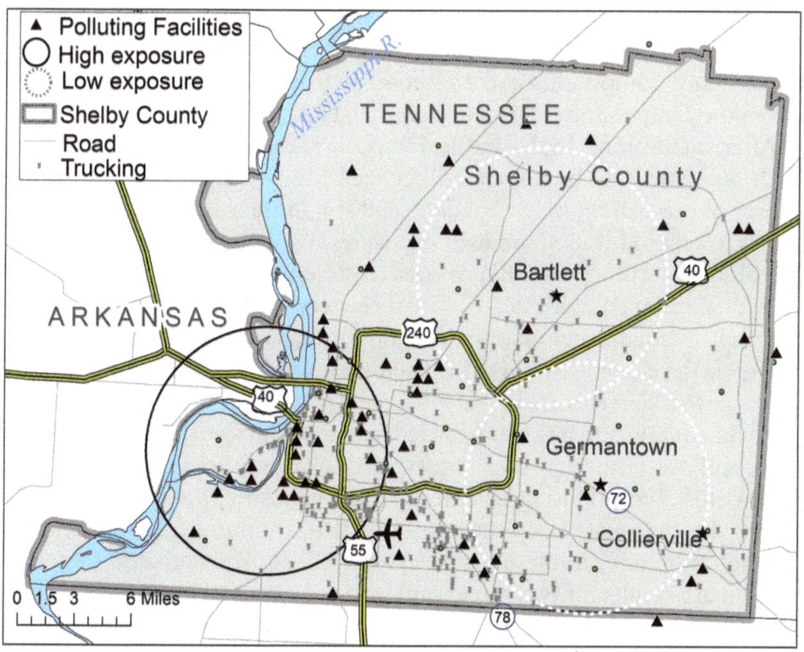

Fig. 6.6 High- and low-exposure areas identified across Shelby County, TN. (Author 2017)

Since person-level birth data provided maternal residential zip code information but lacked individual addresses due to confidentiality concerns, it was not possible to directly quantify the impact of residential proximity to pollution-emitting facilities. To measure proximity to polluters, researchers use broader geographical areas such as zip codes (DeSoto 2009). Thus, by knowing maternal zip code of residence, one can identify areas of high and low exposure to pollution by aggregating facilities over these geographical areas and identifying high and low clusters of exposure. The former are the areas with an excess of polluting facilities and therefore represent higher exposure, while the latter have lower risk of exposure than expected. The degree of exposure to polluters was measured by identifying clusters of pollution-emitting facilities with higher or lower than expected counts. Births within the detected clusters were designated as "High exposure" and "Low exposure," respectively. In agreement, a prior study identified a residential region located in Southwest Shelby County (spatially overlapping the "High exposure" area) suffering significant environmental stress due to stationary and mobile pollution sources whose emissions pose negative health risks to local residents. Average levels of many compounds exceed those in many industrial communities across the United States (Jia and Foran 2013).

After establishing high- and low-exposure areas, the rates of negative birth outcomes can be compared between the two areas to examine the impact of exposure. Ozdenerol et al. (2005), in their study of health outcomes in relation to contaminant exposures, used cluster analysis to identify high occurrence of LBW births clustered in space and time using individual point data. To detect spatial clusters of polluting facilities' locations and evaluate their statistical significance, we used the SaTScan 9.4 software developed by Martin Kulldorff together with Information Management Services Inc.

A chi-square test has been carried out to compare the probability of LBW and pre-term births for two types of exposure to air pollution, one within clustered polluting facilities ("High exposure") and another outside of the clusters ("Low exposure"). It also computes confidence limits for the odds ratio. Table 6.1 displays the chi-square statistics. Only the preliminary results for two races combined are reported here; the full analysis will be published in a separate publication (Fig. 6.7).

To summarize, the impact of high environmental exposure on LBW and premature births was assessed in Shelby County, TN. Residential proximity to polluters was proxied by locating within the identified clus-

Table 6.1 Results of the chi-square test for LBW and pre-term births

			Response (Pregnancy outcome)		
Exposure	*LBW*	*Normal birth*	*X2*	*OR*	*CI*
Pollution exposure	253	1615	22.2 (***<0.0001)	1.47	1.25; 1.72
No pollution exposure	521	4883			
Exposure	*Pre-term*	*Normal birth*	*X2*	*OR*	*CI*
Pollution exposure	253	1590	6.79 (**=0.009)	1.23	1.05; 1.44
No pollution exposure	601	4656			99

Notes: *OR* odds ratio, *CI* confidence interval

**level of significance = 0.01; ***level of significance= 0.001

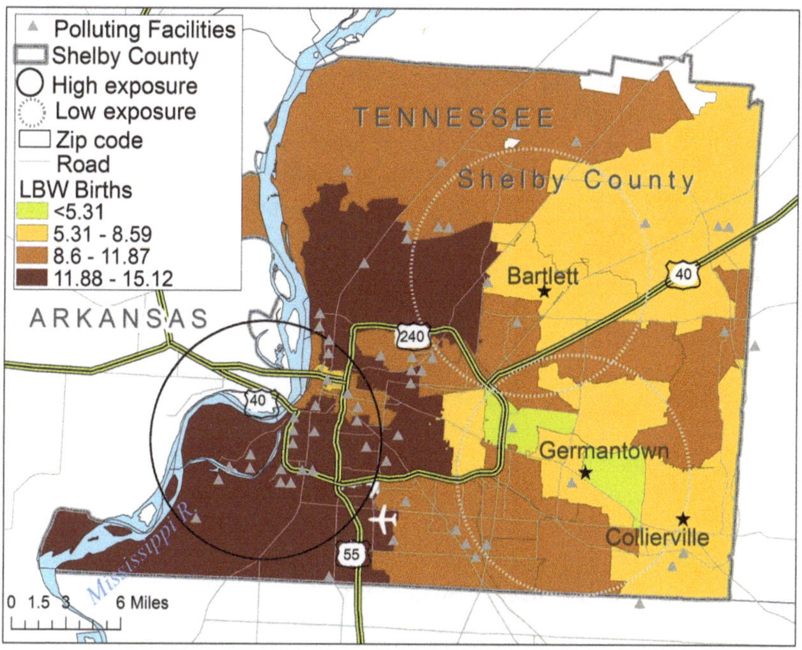

Fig. 6.7 Low Birthweight (LBW) rates superimposed on high- and low-exposure areas. (Author 2017)

ters of pollution-emitting facilities (designated "High exposure"). The study supports the hypothesis that areas of high environmental exposure experience negative birth outcomes reflected in higher rates of both LBW and pre-term births within the "High exposure"-designated areas compared with those of "Low-exposure" areas.

6.13 Health Benefits of Air Quality Improvements

Harmful air emissions can be reduced and ambient air quality can be improved considerably by implementing a range of measures. Despite compelling evidence existing about the harmful effects of PM2.5 on longevity, even in the most polluted cities average life expectancy could be increased by approximately 20 months if average annual level of PM2.5 is reduced to the WHO-recommended value of 10 μg/m^3 (WHO 2013). Additionally, to facilitate a gradual shift to lower concentrations in highly polluted cities, the WHO established guideline values targeting each air pollutant. Meeting these WHO-recommended values may minimize health effects including significant reduction of risks for air pollution-caused acute and chronic health effects.

Sustained, long-term intervention is linked with lower concentrations of air pollution with attendant health improvements for the population expected almost immediately after a decrease in air pollution levels (WHO 2013). A follow-up analysis of the Harvard Six Cities study including Watertown, MA, Steubenville, OH, Topeka, KS, Kingston and Harriman, TN, and tracts in St. Louis, MO, finds evidence for a strong association between chronic PM2.5 exposure and early mortality and indicates increases in mortality ranging from 3% to 26% per 10 μg/m^3 increase in PM2.5 (Lepeule et al. 2012).

Reductions in ambient air pollution may indicate decreases in personal exposures. While chronic exposure to fine particulate matter (PM2.5) is linked with negative human health impacts including morbidity and mortality, health benefits have been documented shortly after decreases in ambient air pollution related, for example, to a decrease in industrial activities including the closure of a steel mill in Utah Valley, UT (Pope et al. 1992), the banning of coal burning in Dublin, Ireland (Clancy et al. 2002), and a reduction in sulfur oxides through the imposition of fuel restrictions in Hong Kong, China (Hedley et al. 2002); all have been accompanied with reductions in mortality after the regions experienced substantial reductions in particulate matter concentrations. Pope et al.

(2007) explore the mortality effects of air pollution during the copper smelter strike in 1967–1968. They analyze associations between mortality and ambient particulate matter air pollution specifically focusing on *sulfate particles*, a type of fine particulate matter with an aerodynamic diameter ≤ 2.5 μm cut point. In the 1960s, approximately 90% of all sulfate emissions in the Southwest United States was attributable to copper smelters.

Pope and colleagues (2007) report in their retrospective study the 2.5 $\mu g/m^3$ decrease in sulfate particle concentrations observed during an eight-month copper smelters' strike from July 1967 through the beginning of April 1968 accompanied by a 2.5% decrease in the number of deaths in the North American Southwest including New Mexico, Arizona, Utah, and Nevada with Phoenix and Tucson, AZ, Las Vegas, NV, Salt Lake City, UT, and Albuquerque, NM being most affected by the strike. During the period of the strike, the region experienced improved visibility and about a 60% decrease in the level of suspended sulfate particles. The researchers used the monthly mortality counts controlling for the US non-event-related mortality trends by including all six nationwide cause-of-death variables (such as influenza/pneumonia, cardiovascular disease, and other respiratory system illnesses). Due to the ubiquitous regional exposure and the high proclivity of sulfate particles to penetrate indoors, the decrease in ambient air pollution results in large reductions in personal exposures (Pope et al. 2007).

6.14 China

6.14.1 Urbanization

Recent large-scale urbanization occurring in the newly industrialized Asian economies has been caused by economic growth and political changes. Rapid economic growth is accompanied by fundamental changes in urban, demographic, ecological, and social systems interacting in complex and unpredictable ways (Phua et al. 2015). When economic development makes an emphasis on gross domestic product (GDP) growth while environmental protection is disregarded, as was the case in China in the past few decades, such development is accompanied with a consumption of a great amount of energy, generating large urban air pollution diffusing to the surrounding areas (Han et al. 2014).

The fast and expanding urbanization contributes to various urban ills across many contemporary cities. The primary problems include traffic congestion and environmental degradation. These problems arise due to concentrated anthropogenic activities, such as use of transport vehicles, operation of industrial plants, and urban construction. Urban and rural residents with rising affluence may adopt unhealthy Western-style, richer diets high in sodium with low intake of fruit, vegetables, and whole grains, and less active, sedentary lifestyles with attendant increases in non-communicable illnesses such as cardiovascular disease, cancer, and diabetes. A growing number of vehicles and associated outdoor air pollution and traffic-related accidents represent other important risk factors that will affect the future health of the urban population in many cities across the world (Gong et al. 2012).

Global health-related knowledge and experience would be relevant to the challenges that many countries undergoing urbanization processes are facing (Holdaway et al. 2015). For example, despite impressive economic growth and development China experienced over the recent three decades, the extraordinary pace of urbanization is associated with negative health consequences: its human development has been ranked below that of the most developed economies. Environmental pollution including air, water, and soil pollution negatively impacts health of the exposed population. Urbanization causes the rapid rise of non-communicable diseases including cardiovascular and chronic respiratory diseases, cancers, and diabetes. Moreover, China's growing non-communicable diseases, due to their high occurrence (they account for more than 80% of annual deaths in the country) and very high mortality rates (mortality rates are higher than those in the United States, the United Kingdom, Japan, and France), and an enormous potential to trigger severe, yet avoidable, economic loss, have been named as key risk factors to the global economy (Marquez et al. 2011).

In developing countries such as China, extraordinary urbanization led to the tremendous social, economic, and environmental changes, leading, in turn, to substantial changes in the country's disease profile, especially in exposure to various health risk factors and their extent. Massive rural-to-urban migration drives urbanization and changes the spatial distribution of China's population while exposing the labor force to higher accident rates and road injuries (Yang et al. 2013) and occupational diseases, with vulnerable population groups disproportionately suffering from infectious and communicable diseases and other health problems (Phua et al. 2015). To illustrate, the chief mortality causes in China in 2010 were stroke, ischemic heart disease, and chronic obstructive pulmonary disease (Yang et al. 2013). To address health problems accompanying rapid urbanization research that could contribute to our understanding of urban population exposures is needed.

In China, urbanization is accompanied by environmental pollution. While a reduction of some types of pollutants has occurred, other pollutants including particular matter (PM) and ground ozone (O3) have been increasing. O3 forms due to the reactions of NOx and volatile organic compounds (VOCs) in the presence of sun. VOCs is a diverse group of compounds emitted by various sources (automobiles, solvents, and oil and gas production). The nitrogen oxides (NOx) ((NOx ≡ NO + NO2) are gases such as nitric oxide (NO) and nitrogen dioxide (NO2) that play a big role in air pollution, significantly contributing to the formation of smog, acid rain, and tropospheric ozone. The majority of NOx are from "mobile sources," that is, mainly from automotive vehicles. Local emission sources for PM2.5 include on-road sources (i.e., automotive vehicles), coal combustion, and economic activity including industrial and construction activities (UNEP 2016). When a country has a tremendous global role in pollution emissions (China emits the highest amounts of CO2), SO2, and NOx in the world) and attendant huge decades-long health damage and economic losses (for example, yearly losses in economy amount to between 1% and 7% of China's GDP), understanding its "air pollution policies is of worldwide relevance" (Jin et al. 2016). To lessen the impact of these and other problems that cities are facing, understanding health consequences of urban growth is critically important.

6.14.2 *Demographic Profile*

Demographic transition in the Asian countries is reflected in steep fertility declines and profound child mortality decreases as well as simultaneous increases in life expectancy at birth. Despite huge improvements in health in China including much longer lifespans (e.g., life expectancy in males changed from 60 years to 73 years between 1970 and 2010, and female life expectancy increased from 63 years to 79 years during the same time interval), the achievement may be undermined by rising rates of lung cancer, obesity, and diabetes with the latter predicted to become the most prevalent disease, and cases of lung cancer estimated to increase fivefold during the next 50 years (Marquez et al. 2011). Additionally, longer lifespans result in population aging with severe economic and social implications to meet the needs of elderly people in the context of decreasing labor force. The share of elderly population who is 65 years and older in China's megacities including Beijing, Guangzhou, Shanghai, Shenzhen, Wuhan, among other big cities, is already greater than 10% of their urban populations compared with the national average of 8.5% (McKinsey & Company 2009).

The changes in the age structure and the tremendous increases of the older population are estimated to increase the prevalence of non-communicable diseases by at least 40% by 2030. However, premature mortality due to non-communicable diseases within this age category can be avoided by postponing disability due to the risk factors, and healthy aging can be achieved (Marquez et al. 2011). Yet another concern is the sharp increase in incidences of HIV/AIDS with deaths rising from 200 to 36,200 between 1990 and 2010 despite a general decline of communicable, maternal, neonatal, and nutritional disorders by about 60% (Yang et al. 2013).

6.14.3 Active Travel and Health

Worldwide, motor vehicle use increased most dramatically in China, India, and other developing countries, where traffic-related fatalities and injuries today is a major public health threat (Transportation Research Board 2011). Developing countries undergoing fast urbanization acknowledge the importance of walkability in the new urban development and thus advocate using a combination of walking and public transport and avoiding using more private cars, especially in the context of China's current population and limited land use (Tong et al. 2016). In China, to counteract urban sprawl and high personal transportation dependency, many cities are attempting to transition from overspread to contracted urban structures with municipalities defining a rigid growth boundary around the city and strongly advocating intermodal transportation and compact neighborhoods (Tong et al. 2016).

6.14.4 Chinese Large-Scale Migration and Hukou

As the planned economy was introduced in 1953, a dual system of rural and urban households was set up in the early 1950s codified in the Chinese system of household registration, or the *hukou* system (*hukou* is a system of official residence registry in China established in 1958), which differentiates households into urban and rural. The chief function of the *hukou* system included allocation of resources to selected population groups including rationing-based distribution (food and clothes available only at one's place of permanent residence) between the 1950s and 1970s, which effectively deterred migration until rationing was abolished with the transition to a market economy in the 1990s (Gransow 2012). The *hukou* system regulated access to public goods, subsidies, and other privileges. For example, the urban population accounted for only about 15% of the Chinese population in the early 1950s. The *hukou* system entitled city dwellers to many forms of support enabling them to live in the city with access to urban educational facilities, health care and pension provision, subsidized housing, and employment, while rural households were entitled to a plot of land but lacked the financial means, privileges, and access to public goods and services enjoyed by urban households. Another principal function of the *hukou* system was regulation of internal migration (in China it is predominantly represented by rural-to-urban migration). Although rural residents have been allowed to work in cities since the early 1980s, they were excluded from a variety of desirable urban jobs and could mostly get employed in the private or informal sectors only if they could find accommodation (Yao and Kim 2015). Rural migrants were also ineligible for urban welfare including housing, health care, and children's education.

Prior to the 1990s, the system aimed at stopping migration from rural to urban areas. Since the 1990s, rural migrant workers have been given freedom of migration and employment opportunities and so could find jobs in urban areas, but due to their lack of urban *hukou*, their mobility and rural migrants' individual factors such as education, lack of language proficiency, and lack of knowledge of welfare programs, they were not covered by the social insurance systems, which are mainly created for urban citizens (Yao and Kim 2015). Since the 1990s, in order to prevent the emergence of slums in Chinese big cities and megacities and prevent the "Latin Americanization" of Chinese cities, the aim changed to channeling the flows of migrants toward smaller towns and restricting their influx into the large metropolitan areas (Gransow 2012). However, small and medium-sized towns and cities do not attract migrants as much as the big cities as they do not have the same necessary assets and resources.

Internal migration (internal migrants are defined as those who remained for long periods of time outside their permanent place of residence) in China is accompanied by a rapid process of urbanization. Prior to 2003, undocumented internal migrants who had come to a city and had no income, no home, and no papers were subject to detention and forcible return to their home villages. Since 2003, the system has been abolished. Despite the difficulties associated with the process of switching agricultural *hukou* to non-agricultural *hukou*, pushed by multiple reasons and driven by economic opportunities offered by the cities, nearly a fifth of total population has migrated to urban areas. The primary destination regions of internal migration for rural workers have been large metropolitan areas along the east coast in China including those in the Yangtze Delta (the largest metropolitan area in China), the Pearl River Delta, Shanghai, and the Beijing/Tianjin region where a great many new jobs have been created due to foreign direct investment and export-oriented Chinese economy as well as a thriving construction sector and the revived service and catering industry; these factors explain the rise in internal migration since the 1980s (Gransow 2012). Inland regions including central and western provinces also attract internal migrants, although smaller flows. For example, in 2009, large and medium-sized cities accounted for 63% of all employed rural migrant workers. This population, with many migrants often undocumented in the context of China, is known as *floating* or *transient population*. Enhanced by the massive rural-to-urban migration across the country and natural city population growth, China's urban population is increasing quickly; in 2011, it reached 51% of the total population, for the first time overtaking the rural proportion. As of 2016, urban population accounts for 56% of China's population, and this percentage will increase over time, while China's government expects an additional 300 million people to move to cities during the following three decades (Zheng and Kahn 2017).

According to the United Nations Development Programme (UNDP) estimates, as much as 60% of the total population will be living in urban areas by 2030 with a big share of the increased population being migrants (UNDP 2009). The speed of urbanization in China is truly extraordinary: the share of the urban population has more than tripled during the past three decades, and the number of cities with at least 500,000 people grew between 1978 and 2007 from 40 to 140. McKinsey's report (2009) states that there will be 221 cities with one million-plus inhabitants in China by 2025 versus 35 such cities in Europe today, with implications for social services as well as pressures on arable land, water, energy consumption, and the environment. In fact, the rate of urbanization in China is accelerating with urban boundaries expanding while farmland is converted to urban land. Much of the rural-to-urban conversion is taking place both along the coast in the east and in inland provinces including Sichuan. To illustrate the rapid rate of urbanization, the ratio of the growth of urban area in provinces between 2000 and 2010 exceeds that for the earlier period of between 1990 and 2000 (Gong et al. 2012).

A substantial shift in the attitude toward rural-to-urban migration has occurred since the late 1970s from the strict migration regulation to the much relaxed economy-based migration policy. The new migration policy encouraged and promoted migration to increase China's industrialization and global competitiveness as well as to reduce the existing city-countryside disparity and create more equal wealth distribution. Because migration to urban areas is deemed important to stimulate economic growth (almost three quarters of Chinese internal migration is labor migration), the cities are adjusting to the influx of migrant populations by expanding housing (constructed both by the government and by private enterprises) and basic infrastructure in urban areas that the China's governmental development strategy had pledged commitment to. Thus, although the formation of large-scale slums, unlike other developing countries, has been largely avoided, crowded substandard housing for rural-to-urban migrants who do not hold urban *hukou* remains a major issue (Gong et al. 2012).

6.14.5 Urbanization and Health: Potential Mechanism in China

Urbanization affects health via a complex and multipath mechanism. On the one hand, urbanization offers substantial health benefits including greater access to health care and basic infrastructure. On the other hand, there are substantial urbanization-related health risks within the physical environment (e.g., high particulate matter levels such as PM2.5), individual behaviors and practices such as consumption of tobacco and alcohol and inactive lifestyle, the physical layout of the road system, and urban layout, which reduce quality of life and contribute to excess mortality and morbidity (Yang et al. 2013). Policies aimed at changing road design and vehicle safety equipment as well as modifying urban layout may reduce road injuries and fatalities and improve physical activity. Specifically, infrastructure and transportation intervention measures can be implemented whereby accessibility to health services and physical activities may be improved by better road, transport, and housing planning such as provision of cycling and walking opportunities and construction of safer and more livable communities. Further, poor environment causes as much as 25% of all preventable illnesses such as cancer at the global level, so in order to transform unhealthy environmental conditions stringent environment standards should be enforced. Also, developers could be encouraged to incorporate physical exercise facilities in real estate projects. Anti-air pollution legislation should be adopted and enforced, while media campaigns could be run promoting healthy lifestyles (Marquez et al. 2011).

The potential pathways of urbanization negatively influencing health include the following:

- Urban residents are facing the risks of injury and illness because of environmental exposure to chemical, biological, and physical hazards. For example, high prevalence of smoking among men aged 15–69 years at 54% in China is among the highest in the world, and a high rate of second-hand smoke exposure at 72.4% greatly increases the burden of diseases such as cancer due to behavioral risks and chemical exposure: smoking increases several types of cancer including stomach, colorectal, and lung cancer, which are listed in the top causes of premature mortality in China (Yang et al. 2013). Reducing these risk factors even by a small amount could produce both substantial health benefits for the population, and promote sustainable development of economy and society.
- The residents who change their occupational activities, socioeconomic status, and social structures may suffer from non-communicable chronic diseases including neuropsychiatric disorders and cardiovascular diseases. To illustrate, rates of chronic disability are high with mental and behavioral disorders representing prominent causes of chronic disability in 2010 in China. Given a shortage of human resources, the growing rates of chronic disability will determine the types of health workers China will require in the next decades (Yang et al. 2013).
- Urban health-care delivery is struggling as it tries to accommodate the rural-to-urban migrants, many of whom come from previously isolated rural communities whose increasingly felt urban presence in places of concentrated commerce and recreation may lead to transmission of traditional rural infectious illnesses to cities, which in turn may lead to the dispersion of communicable diseases across the country (Gong et al. 2012). Examples are the 2002–2003 outbreak of severe acute respiratory syndrome (SARS) and the 2013 H7N9 epidemic with human cases. Another example is the HIV epidemic. China's first recognized HIV outbreak occurred in 1989 in Ruili in Yunnan province. Since then, the HIV epidemic substantially expanded into other provinces circulating now in the rest of China as

well, spreading from rural areas to cities and from coastal communities to those deeper inland China, and shifting from ethnic minorities such as Yunnan's Jingpo and Dai people living along the border of Myanmar to the majority Han population (Xiao et al. 2007).

- Migrant workers come to their hometowns occasionally, for instance, to participate at Spring Festival, so infectious illnesses acquired elsewhere may be spread to the local population (Xiao et al. 2007).
- Yet another major problem is soaring health care expenditure imposing substantial burdens on low-income households. For instance, Ramesh and Wu (2009) provide some data from the 2003 national health services survey by China' Ministry of Health regarding financial burden placed on poor households: as many as 60% of respondents from the poorest income quintile residing in urban areas did not seek medical care when ill.

6.14.5.1 China: Access to Urban Health Care and Risk Factors

Although still low, access to urban health-care services is slowly improving across urban areas with a substantial share of the urban population not covered by medical insurance and other social benefits. According to the main urban health-care insurance called "The Basic Health Insurance Scheme," which was established in 1998, registered urban residents are provided with basic health services, but informal sector workers and migrant workers are not. In 2006, only 28% of the urban residents, or about 160 million people, had access to the basic health services, which is the main urban health-care insurance scheme (Ramesh and Wu 2009; Hu et al. 2008). In addition to low health care coverage, access to health care is declining because health expenditures are rapidly rising. Health service fees that households paid in 2008 were more than 18 times what they were in 1990 while the average cost for hospital admission is almost equal to per capita annual income, placing a significant burden on patients and their families and prompting Chinese media to run a new slogan: "It's too difficult to see a doctor, and too expensive to seek health care!" (Hu et al. 2008, p. 68). Since 2009, the country has been undergoing a series of transformational health reforms that attempt to provide all citizens with access to basic health-care services, establish a primary care system (Suessmuth and Wang 2010), make health care more affordable by decreasing high out-of-pocket payments, and improve health equity in utilization of health care among low-income people (Zhou et al. 2015). Reinforced by its commitment to achieve universal health coverage by 2020, government health spending is now sharply rising (Ramesh and Wu 2009).

With accelerating urbanization in China, population exposure to major risk factors is increasing. Major risk factors for disease include environmental pollution and physical inactivity (Gong et al. 2012). In 2002, 23% of people in China were overweight and more than 7% were obese. Especially alarming is the growing overweight and obesity rates among children and adolescents aged 7 to 18 years in urban areas for whom the rate of overweight and obesity is sometimes higher than that in developed nations. In the northern coastal cities, due to increased physical inactivity and fat-rich fast food diets, the prevalence of overweight and obesity reached almost 33% for boys and 18% for girls (Ji and Cheng 2008). Other substantial risk factors include occupational and traffic hazards, individual behaviors such as adoption of richer diets that are low in fruit and whole grains and high in sodium, and tobacco and alcohol consumption. These risk factors increase mortality and morbidity in the rural (which has been urbanizing) and urban populations.

6.14.6 China: Exposure to Air Pollution

Air quality is seen as the key challenge and it has become the focal point of concern by both the general public and governments in many burgeoning cities (UNEP 2016). There is an existing significant link between pollution emissions and health damage: worldwide, 3.7 million deaths were attributable to ambient air pollution in 2012, where low- and middle-income countries containing 82% of the world population account for 88% of these deaths, and high-income countries of Europe, the United States, the Western Pacific, and Eastern Mediterranean contribute another 12% of air pollution-caused mortality (WHO 2014). Therefore, it is crucial to analyze air pollution control policies as some policies are more efficient than others. For example, Jin et al. (2016), in their analysis of the different policy changes in the context of air pollution problems in China, show that the early pre-2005 policies were ineffective at reducing emissions. It has been suggested that ineffectiveness was due to both weak implementation and enforcement of environmental regulations during this time period. Jin et al. (2016) explain that local leaders at the municipal level, including mayors and vice-mayors, and local environmental protection bureaus (EPBs), are responsible for local environment issues, however, under normal conditions do not pay sufficient attention due to lacking capacity (such as insufficient fiscal budget and monitoring and enforcement personnel) but pay attention to environmental issues when these become "severe and urgent." For example, after the severe PM2.5 crisis during the winters of 2012–2013, the air quality standards were updated to the international level and strengthened, with the current standards for coal-fired power plants even more rigorous compared to those in the United States and Japan (Jin et al. 2016). Applying the policies after 2005, the national goal of reducing emissions of total SO2 associated with coal burning and industry processes by 10% was achieved, but not for PM2.5 and O3 whose concentration grew worse (Jin et al. 2016).

In many developed and developing countries, the problems of air pollution due to vehicle emissions and associated adverse health outcomes have emerged and increased in urban areas. Although in developed countries ambient air quality has been improved considerably as various measures have been implemented, current concentrations of air pollution still represent a considerable threat to human health and the surrounding environment (WHO 2013). Van Donkelaar et al. (2010) found that Central and Eastern Asia exceed the WHO Air Quality PM2.5 Interim Target-1 (35 $\mu g/m^3$ annual average) exposing 38% and 50% of the population, respectively, to the air pollutant. According to the WHO guidelines, PM2.5 concentrations at the level of 35 $\mu g/m^3$ and higher may increase risk of mortality by approximately 15% (WHO 2005a). Worldwide, 80% of the population live in regions where ambient concentrations exceed the Air Quality Guideline (AQG) of 10 $\mu g/m^3$ and who thus have an increased likelihood of air pollution-related health impacts. The annual US standard is 12 $\mu g/m^3$ for PM2.5.

Settled areas with the highest concentrations of population and economic activities such as the *megacities* (a megacity is defined as a city where the population is 10 million or more) are facing the highest air pollution-related challenges. For instance, premature respiratory system-related deaths in China's cities have been linked to air pollution, while the increase in cancer was partly attributed to various types of pollution making the control of ambient and household air pollution (e.g., exposure to second-hand smoke in inadequately ventilated homes due to tobacco consumption, mold, formaldehyde, or coal smoke caused by using coal for heating, cooking, and lighting) one of the public policy priorities for China (Yang et al. 2013).

In the context of uneven economic growth, there is substantial regional diversity in emission distribution in China, with eastern and central areas experiencing a disproportionate share of air pollution severely affecting health status of urban residents of these regions (Jin et al. 2016). However, air pollution spread out across other regions including Northern China and significant levels are not limited to major cities (Rohde and Muller 2015). Aggravation occurs when weather conditions are not conducive for pollutant diffusion, with air pollution at the regional level becoming an important issue since high concentration of PM2.5 and O3 impact public health in a more significant way compared to the effect caused by *primary pollutants* (primary pollutants such as SO2, CO, VOCs, NO, lead, and mercury are emitted by their sources directly, and *secondary pollutants* such as O3, NO2, and acid rain form when primary pollutants react in the atmosphere)). PM2.5, especially, is a very hazardous air pollutant that causes lung-related diseases: when inhaled, small-sized particles have the potential of reaching the deep regions of the respiratory tracts and causing damage to the lungs by inducing durable inflammation. Lungs are most directly affected by air pollution from particulate matter: there is a statistically significant association between each 10 $\mu g/m^3$ increase in the level of PM2.5 and a 15–27% increase in lung-cancer mortality in persons who never smoked (Lepeule et al. 2012).

Particulate matter is the key cause of air pollution. It consists of a mix of solid and liquid particles suspended in the air. The air quality is determined by the amount of particulate matter produced by vehicular and industrial pollution (Deepa et al. 2017), and with rapid increase in number of vehicles and industrialization, air quality is getting worse gradually across the metropolitan cities. Ubiquitous exposure disrupts the daily life of people residing there. The number of people who suffer from cardiovascular and lung-related diseases due to poor air quality is increasing. People may be exposed to PM-related air pollution short term (hours, days) and long term (months, years), and while no evidence of a safe level of exposure at which no adverse health effects occur exists, exposure to PM2.5 decreases the life expectancy of the population (WHO 2013). The assessment of population exposure to particulate matter in Eastern Europe, Caucasus, and Central Asia indicates that health effects include worsening of asthma, respiratory symptoms, and an increase in hospital admissions and deaths from lung cancer and those caused by cardiovascular and respiratory system diseases. Exposure over the long term to PM2.5 increases the long-term risk of death due to heart and lung diseases by 6–13% per 10 μg/m of PM2.5 (WHO 2013). In China, air pollution is currently the fourth greatest risk factor in all deaths (Fang et al. 2015). The most severely affected are people with pre-existing lung or heart disease as well as elderly people and children. In the context of China, it has been estimated that on average, between 1990 and 2015, due to exposure to ambient air pollution, particularly PM2.5 and O3, as many as 1–1.2 million premature deaths occurs in China annually (Lim et al. 2013).

6.14.6.1 China: Spatial Variation in Air Pollution

Han et al. (2014) examined air quality impacts of urbanization by studying the spatial pattern of PM2.5 concentrations in Chinese cities. Specifically, PM2.5 concentrations at the level of Chinese prefectures have been compared between urban areas and surrounding regions. Urban areas due to a high concentration of human activity may produce a greater amount of air pollutants including PM2.5 concentrations compared to surrounding rural areas. At the same time, different urban regions differ in terms of PM2.5 concentrations due to the imbalance in China's urbanization development (Fang and Liu 2009) and natural effects such as Asian dust storms (Han et al. 2013).

6.14.6.2 Urbanization and Air Quality

Han et al. (2014) investigated the impact of urbanization on urban PM2.5 concentrations at the Chinese prefectures. The prefectures distributed along the east coast and southeast of Sichuan province, including the cities located in the Beijing-Sichuan and Shanghai-Guangxi belts, had annual PM2.5 concentrations greater than the AQG of 10 $\mu g/m^3$ established by the WHO (WHO 2005a, b), and the most severe particulates pollution was found in the southeast of the Sichuan province while the surrounding non-urban areas had lower PM2.5 levels. At the same time, prefectures located in Western China had better air quality than that in surrounding non-urban areas.

Three indicators of urbanization, including urban population, city size (i.e., built-up area), and urban secondary industry fraction (heavy industry, real estate), substantially impact urban air quality. Urban population may be used to reflect the magnitude of human activities with large population inducing heavy air pollution (Marshall, et al. 2007). In China's cities, the more people the higher the PM2.5 concentrations (Han et al. 2014). For example, cities with 0.5–1 million people have significantly lower concentrations of PM2.5 than that in cities with 1–3 million people, and the smallest cities with less than 0.5 million people had significantly lower PM2.5 levels than that in largest cities with more than 10 million people. In a similar way, economic structure also impacts urban air quality: the higher the industry concentration the higher the PM2.5 concentrations (Han et al. 2014).

6.14.7 Impact of Urban Form on Air Quality

Urban form is defined as the land-use configuration of the urban landscape across the space (She et al. 2017). Urban form describes the spatial structure of land use and transport system within an urban area (Ou et al. 2013). Urban form dramatically impacts urban functioning and has wide implications for the carbon emissions of a city (e.g., how efficiently resources are consumed and the amount of pollution cities create). Urban form impacts air pollution by influencing travel behavior through vehicle kilometers traveled, which in turn impacts automotive (tailpipe) emissions because if urban area becomes less dense and land uses more segregated, residents increasingly rely on single-occupant vehicles and drive further to reach various destinations. Accordingly, the per capita emission of tailpipe pollutants including CO2, NOx, PM, and CO increase (Bereitschaft and Debagge 2013).

Urban form can impact air quality by affecting local meteorology, for example, by forming the ***urban heat island effect*** and influencing the ***energy efficiency of buildings***. To illustrate the former, man-made constructed impervious surfaces (roads, buildings) mainly absorb rather than reflect the sun's radiation, while the natural cooling of shading and evapotranspiration is reduced due to the displacement of natural vegetation and increasing uncanopied area, and many urban activities (motor vehicle travel, space heating) create waste heat, together resulting in 1 °C–3 °C air temperature increase in a city versus the surrounding rural area, that is, urban heat island effect (Ewing and Rong 2008). To illustrate the latter, residents of sprawling communities tend to reside in single-family detached houses (which require more energy), which are also more likely to be big houses (also requiring more energy); both larger house size and house design (detached versus attached) result in higher residential energy use (Ewing and Rong 2008). Urban sprawling development characterized by lower-density, decentralized land-use patterns seems to generate more surface heat than denser urban forms (Stone and Rodgers 2001). Further, due to enhanced heat island formation, sprawling cities may experience a higher quantity of ozone exceedances than more compact cities (Stone 2008), thus potentially influencing public health. Additionally, arrangements of streets and building density and height impact pollutant dispersion.

Urban form is commonly measured with residential and building density, street accessibility, land-use mix, and centrality. Additionally, urban form can be measured using composite measures including urban continuity and shape complexity; both are significantly associated with air pollutant emissions. Urban continuity represents the extent to which the urban landscape is fragmented and can be interpreted as a measure of "leapfrog" development; less fragmented urban areas with more contiguous and less "leapfrog" development have higher urban continuity. Urban shape complexity measures urban area-to-perimeter ratio and indicates the "jaggedness" of the urban boundary; suburban and exurban areas often have highly irregular boundaries (Bereitschaft and Debagge 2013). In the assessment of the effect of urban form on air quality, residential density was found to be a significant determinant of concentrations of ozone and PM2.5, the nonpoint source emission of NOx, and the on-road (i.e., mobile) emission of CO2, while urban continuity and shape complexity were significantly related to O3 precursor and PM2.5 emissions (Bereitschaft and Debbage 2013). Urban design can influence NO2 concentrations: cities that have highly contiguous built-up areas have lower NO2 levels, while leapfrog development is associated with higher NO2 concentrations (Bechle et al. 2011). Metro Atlanta and the Riverside–San Bernardino region of western Los Angeles (the "Inland Empire"), ranked as one of the largest, most sprawling (i.e., with at least 500,000 people) US metropolitan areas, had some of the highest average yearly amounts of O3 and PM2.5 between 1998 and 2002 (Bereitschaft and Debbage 2013).

Population density is commonly employed by studies as a measure of urban form, however, when used alone it may turn out insufficient to comprehensively characterize urban form. On the other hand, Bereitschaft and Debbage (2013) find little evidence that the composite sprawl index based on four different measures of urban form can predict air pollution better than sprawl indexes based entirely on population density; an increase in residential density and street accessibility was linked to a significant decrease in the outdoor O3 and PM2.5 levels. Further, residential density was the only urban form metric significantly associated with CO2 emissions from mobile vehicles sources at the 5% significance level.

Tremendous urban growth in China in a short time resulted in a severe environmental challenge, particularly air pollution, in many urban areas of China. Human activity and the surrounding environment directly affect the concentrations of a major urban air pollutant, PM2.5. High levels of PM, especially PM2.5, reduce visibility (Hyslop 2009) and impact public health (Pope and Dockery 2012). The concern is enhanced due to an increase in annual quantity of PM in Chinese cities (Wang and Hao 2012). In their recent study focused on the Yangtze River Delta, the largest metropolitan region in China, located where the Yangtze River enters the East China Sea, She et al. (2017) empirically investigated the link between urban form and air quality and challenges faced by the densely populated area. The region represents one of the most highly developed, densely populated regions. Accounting for just 2.1% of the national land area, it contains 11% of the population. The region is economically vibrant, contributing as much as 21.7% of the Chinese national economy. It suffers from rapid urbanization in China: 25 cities in total are found in the region. The region specializes in finance, export processing, heavy manufacturing with major companies, and is known for a high amount of air pollutant emissions seriously threatening its sustainable development.

Landscape metrics are often used to characterize urban land-use patterns. Developed in the late 1980s, landscape metrics are useful for the analysis of urban environments, for instance, to quantify the shape and pattern of vegetation in urban areas. She et al. (2017) explored the link between urban form and air quality. Therefore, six landscape metrics based on built-up areas have been employed to analyze the spatial distribution of urban elements and to characterize the urban forms of 25 cities in the Yangtze River Delta. Urban form was measured using satellite-derived land cover data and consisted of six landscape metrics reflecting the size, shape, regularity, fragmentation, and traffic organization of urban patches (She et al. 2017). Specifically, landscape metrics of urban forms consisted of *Total Urban Area* measure, which represented the expansion of urban land use with low values indicative of high city fragmentation; *Number of Urban Patches*, which showed the density of landscape with high values indicating a scattered/fragmented pattern; *Largest Patch Index* characterized the extent of the fragmentation of settlement patches measured as the percentage of urban landscape area comprised by the largest patch; and *Mean Perimeter-Area* ratio described the shape of landscape signifying the extent of irregularity of the perimeter of settlement patches, with higher values reflective of lower overall urban land-scape regularity. This metric is useful to characterize urban landscape configuration regarding the complexity of patch shape. Next, *Mean Euclidean Nearest Neighbor Distance* was used to describe the extent of sparseness of settlement patches conceptualized as spatial connection between the urban patches and measured taking the average distance between any two nearest neighboring urban patches. Finally, *Traffic Coupling Factor* characterized road network and reflected the interaction in urban traffic organization and spatial structure, measured as a ratio of total buffer areas of urban road area and total urban built-up area, where higher values indicate more harmonious urban form development (She et al. 2017).

There are six categories to characterize the *air quality*: good (0–50), moderate (51–100), unhealthy for sensitive groups (101–150), unhealthy (151–200), very unhealthy (201–200) and hazardous (>300). The value 100 is considered acceptable for public health.

She et al. (2017) measured air pollution using six criteria pollutants (SO2, NO2, coarse particulate matter (PM10), fine particulate matter (PM2.5), CO, O3) and air quality index (AQI) using 2015 ambient air quality data from 129 monitoring stations. Air quality index is based on the real-time observation of the six criteria air pollutants listed previously. Higher value of AQI means lower air quality. Results indicate that in 2015, three pollutants, PM2.5, PM10, and O3, were primary pollutants in the Yangtze River Delta as they had the maximum AQI value, and therefore were the largest contributors to the poor air quality. The annual average air quality was "moderate" from a human health perspective with air quality index at 79, with winter having the highest AQI and summer the lowest AQI. In general, the southern areas in the Yangtze River Delta (Zhejiang province) have better air quality compared with the northern parts of the region (Jiangsu province) (She et al. 2017).

Ou et al. (2013) quantified the relationships between urban forms and carbon emissions for the four fastest-growing, most important cities in China: Beijing, Shanghai, Tianjin, and Guangzhou, located in the eastern coastal areas of China. Being the most developed metropolis areas, their rapid economic development causes a range of environmental problems, including air pollution and growing carbon emissions with the resultant greenhouse effect. To sustain economic growth, all four cities are consuming an enormous amount of natural resources: the total area of urban land use in the four cities increased 192% between 1990 and 2010, in the process transforming natural ecosystems, farmland, and water into urban areas. To assess carbon emissions, *carbon dioxide equivalent emissions* have been used, which is defined as the carbon content of greenhouse gases that would have the same global warming potential. Since carbon emissions mostly come from the fossil energy sources, carbon emissions were calculated only from fossil energy consumption. Ou et al. (2013) found that the growth of urban areas speed up carbon emissions; similarly, an increase in urban form fragmentation or irregularity is likely to produce more carbon emissions. In contrast, "good" urban form that is compact might reduce carbon emissions. Carbon emissions might be further reduced with increases in the coupling degree between urban spatial structure and traffic organization. Mononuclear structure may accelerate carbon emissions

(Ou et al. 2013). Researchers conclude that urban forms in China, instead of disperse, single-nuclei development, should develop in a compact, multiple-nuclei pattern if carbon emissions are to be reduced.

Urban form provides a significant effect on urban air quality mainly through the size and shape of urban patches. Urban air quality is impacted most significantly by mean perimeter-area ratio, mean Euclidean nearest neighbor distance, total urban area, and number of urban patches. The larger the urban area and the bigger the number of urban patches, the greater are both PM10 and PM2.5; the amount of the two important pollutants is negatively related to mean perimeter-area ratio, mean Euclidean nearest neighbor distance. Compared with monocentric development, the polycentric urban form appears to be more beneficial to public health as it is associated with better air quality because the polycentric urban structure might reduce commuting by decreasing distances to work and thus result in less automobile pollution, positively affecting urban air quality. Therefore, urban design strategies giving support to polycentric development may effectively minimize the air pollution within metropolitan areas. The configuration of land use can be used to describe the urban air quality with important implications to enhance or preserve the forested areas to mitigate the air pollution. At the 25 km-scale of buffer distance, air quality is highly correlated with forest coverage, with higher rates of forest coverage contributing to better air quality (She et al. 2017).

6.14.8 Beijing

Beijing is the political, cultural, and educational capital of China. Geographically, the capital city is located in the northern part of the North China Plain. With an area of 16,410 km^2, it includes 14 urban and suburban districts and two rural counties, with a population of 19.62 million in 2010 (Ou et al. 2013). Its GDP was 1,411.36 billion RMB in the same year. The city attracts rural migrants; the city's floating population rate is greater than 25% (Yao and Kim 2015). In terms of social insurance, compared with other big cities such as Shenzhen, Suzhou, and Chengdu, rural migrant workers working in Beijing have the lowest coverage of pension insurance at just 5.5% due to high occupational mobility. To illustrate, rural migrant workers work on a particular construction project in the construction industry. Upon the project's completeness, they either return to the countryside or move to another city. Another explanation is economic structure with a small-sized manufacturing sector, so the construction (which due to a high risk of workplace injuries and accidents overwhelmingly employs males, or about one third of male migrants) and service industries employ most rural migrant workers. Within the services sector, male and female rural migrants work in trading and small businesses, restaurant services, domestic services, and other services (Yao and Kim 2015).

Air pollution is especially concentrated in a northeast corridor that extends from near Shanghai to north of Beijing. The urban–non-urban differences in PM2.5 concentrations in Beijing, in Shanxi province, are greater than 10 μg/m^3, the AQG recommended by the WHO (WHO 2005a, b). Han et al. (2014) find that the urban–non-urban differences in PM2.5 in the Beijing-–Sichuan belt was larger than that in the Shanghai–Guangxi belt. Beijing experiences severe fog and haze; the average PM2.5 concentration in the winter of 2013–2014 was 108 μg/m^3 (Zheng and Kahn 2017).

Beijing, the capital city of the world's most populous country, has suffered from acute air pollution. At the turn of the century, the situation became much worse due to rapid urban expansion, a thriving construction industry, growing industrial zones, an increase in registered vehicles on the road (e.g., in Beijing, over 15 years, the number of registered vehicles increased by 300%), and steep growth of a population that was using more coal (between 1998 and 2013, the resident population increased by 70%). To offset pollution and improve air quality, stringent measures were introduced in 1998, which included a ban on leaded gasoline, introduction of catalytic converters in cars, promotion of clean energy, introduction of electric vehicles, control of vehicle emissions, and upgrading of industrial structure.

Thus, air pollution control became the top priority, especially because the city hosted major events such as the Summer Olympic Games in 2008 and the Asia-Pacific Economic Cooperation (APEC) Summit in 2014. To control the pollution, China's government implemented many short-run regulations; as a result, Beijing's PM2.5 levels reached its lowest values when the Olympic Games were held. The US Embassy in Beijing has been publishing PM2.5 data close to its downtown location since 2008 (Zheng and Kahn 2017). Since then, the PM2.5 level has averaged about 100 μg/m^3, getting worse during winter because of coal-based heating. Its air quality is aggravated by topography as mountains surround the city on three sides essentially trapping air pollution over the city, and stagnant meteorological conditions occurring several times annually leading to heavy air pollution episodes such as in winter 2015 with high concentrations of air pollutants posing harm to public health (UNEP 2016). Beijing suffers from nearby industrial emissions migrating into its territory. Air pollution can disperse across regions in Beijing, which is adjacent to the Hebei Province, which produces a big share of the world's steel; as much as 18% of the PM2.5 air pollution is attributed to emissions from this nearby dirty area (Zheng and Kahn 2017).

During the last 10 years, Beijing (alongside with Shanghai), despite population growth and significant per-capita income, has experienced improvements in urban air: Beijing's PM10 concentrations have decreased by 39 per (Zheng and Kahn 2017). The Clean Air Action Plan 2013–2017 proposes to limit total coal consumption in Beijing and aims to reduce concentrations of fine particulate matter by 25% by 2017 (UNEP 2016). Cleaner energy was introduced in the selected areas of Beijing: for instance, old one-story buildings in the traditional residential areas in the city center were renovated whereby coal-burning heating systems were converted to electric heating systems. As a result of these efforts, during the period 2003–2013 there was a decline in emissions of 630 tons of PM2.5, 870 tons of PM10, 2,070 tons of SO2, and 790 tons of NOx.

In light of many urban challenges including traffic congestion, growth of vehicles, and worsening air quality, another way to optimize urban planning is to develop and promote public transport with an emphasis on rail-based and ground transportation systems and slow active modes of transportation such as walking and bicycling. The percentage of public transport grew between 2000 and 2013 from 26% to 46%. Beijing has adopted a policy of car license distribution to withstand worsening traffic congestion and urban pollution according to which potential car buyers have to take part in a random lottery. Traffic control measures have been introduced such as the "odd-and-even license policy" based on the last digit of a passenger car's license plate. According to this long-term policy, in effect since 2008, during a week passenger cars cannot drive one weekday.

Chinese cities have launched health policies to respond to the major health problems that migrants are challenged with. To control growth of the migrant population, major cities initiated the requirement of health certification for migrant workers. Beijing adopted the approach in the 1990s, so migrant workers had to have health certification indicating that they were in good health in order for them to legally work in Beijing. In 2006, the policy changed with community health-service centers offering basic health-care services, which include free diagnosis and treatment for tuberculosis and HIV/AIDS, to migrant workers while children of the floating population receive access to health-care services, as do registered urban citizens (Gong et al. 2012).

Bibliography

Adar, S.D., M. Davey, J.R. Sullivan, M. Compher, A. Szpiro, and L.-J.S. Liu. 2008. Predicting Airborne Particle Levels Aboard Washington State School Buses. *Atmospheric Environment* 42 (33): 7590–7599.

American Community Survey (ACS). 2011–2015. *American Community Survey 5-Year Profiles*. At: https://www.census.gov/programs-surveys/acs/.

Antipova, A., and D. Larsen. 2017. *Analysis of the Relationship Between Environmental Exposure and Maternal Health in Shelby County, TN*. Presented at 62nd Annual Meeting of the Association of American Geographers (AAG), April 5–9, 2017, Boston, MA.

Arain, M.A., R. Blair, N. Finkelstein, J.R. Brook, T. Sahsuvaroglu, B. Beckerman, L. Zhang, and M. Jerrett. 2007. The Use of Wind Fields in a Land Use Regression Model to Predict Air Pollution Concentrations for Health Exposure Studies. *Atmospheric Environment* 41: 3453–3464.

Badland, H., and G. Schofield. 2005. Transport, Urban Design, and Physical Activity: An Evidence-Based Update. *Transportation Research Part D* 10 (3): 177–196.

Barker, D.J.P., ed. 1992. *Fetal and Infant Origins of Disease*. London: BMJ Books.

Baron, P. 1995. Transportation in Germany: A Historical Overview. *Transportation Research Part A Policy and Practice* 29 (1): 9–20.

Bechle, M.J., D.B. Millet, and J.D. Marshall. 2011. Effects of Income and Urban Form on Urban NO2: Global Evidence from Satellites. *Environmental Science and Technology* 45: 4914–4919.

Bereitschaft, B., and K. Debbage. 2013. Urban Form, Air Pollution, and CO2 Emissions in Large U.S. Metropolitan Areas. *The Professional Geographer* 65 (4): 612–635. https://doi.org/10.1080/00330124.2013.799991.

Bernabe, V., T. Soriano, R. Albaladejo, M. Juarranz, et al. 2004. Risk Factors for Low Birth Weight: A Review. *European Journal of Obstetrics & Gynecology and Reproductive Biology* 116: 3–15.

Bobak, M., M. Richards, and M. Wadsworth. 2001. Air Pollution and Birth Weight in Britain in 1946. *Epidemiology* 12 (3): 358–359.

Borrell, L.N., and L. Samuel. 2014. Body Mass Index Categories and Mortality Risk in US Adults: The Effect of Overweight and Obesity on Advancing Death. *American Journal of Public Health* 104 (3): 512–519.

Brauer, M., et al. 2015. Ambient Air Pollution Exposure Estimation for the Global Burden of Disease 2013. *Environmental Science & Technology* 50 (1): 79–88.

Bremner, S.A., H.R. Anderson, R.W. Atkinson, A.J. McMichael, J.M. Bland, D.P. Strachan, and J. Bower. 1999. Short-Term Associations Between Outdoor Air Pollution and Mortality in London 1992–94. *Occupational and Environmental Medicine* 56: 237–244.

Briggs, N.L., and C.M. Long. 2016. Critical Review of Black Carbon and Elemental Carbon Source Apportionment in Europe and the United States. *Atmospheric Environment* 144: 409–427.

Buehler, R., and J. Pucher. 2017. Trends in Walking and Cycling Safety: Recent Evidence from High-Income Countries, Market with a Focus on the United States and Germany. *American Journal of Public Health* (0), e1–e7. At: http://ajph.aphapublications.org/doi/full/10.2105/AJPH.2016.303546. Accessed 28 Aug 2017.

Buehler, R., T. Götschi, and M. Winters. 2016. Moving Toward Active Transportation: How Policies Can Encourage Walking and Bicycling. Active Living Research, San Diego, CA. At: https://activelivingresearch.org.

Buehler, R., J. Pucher, D. Merom, and A. Bauman. 2011. Active Travel in Germany and the U.S. Contributions of Daily Walking and Cycling to Physical Activity. *American Journal of Preventive Medicine* 41 (3): 241–250.

Buseck, P.R., K. Adachi, A. Gelencse, E. Tompa, and M. Posfai. 2012. Are Black Carbon and Soot the Same? *Atmospheric Chemistry and Physics Discussions* 12: 24821–24846.

Carlson, S.A., J.E. Fulton, M. Pratt, Z. Yang, and E.K. Adams. 2015. Inadequate Physical Activity and Health Care Expenditures in the United States. *Progress in Cardiovascular Diseases* 57 (4): 315–323.

Centers for Disease Control and Prevention (CDC). 2014. *The Obesity Epidemic and United States Students.* Atlanta, GA: Centers for Disease Control and Prevention, Division of Adolescent and School Health, 2011. At: https://www.cdc.gov/healthyyouth/data/yrbs/pdf/us_obesity_combo.pdf. Accessed 8 July 2017.

Clancy, L., P. Goodman, H. Sinclair, and D.W. Dockery. 2002. Effect of Air-Pollution Control on Death Rates in Dublin, Ireland: An Intervention Study. *Lancet* 360: 1210–1214.

Clougherty, J.E., C.A. Rossi, J. Lawrence, M.S. Long, E.A. Diaz, R.H. Lim, B. McEwen, P. Koutrakis, and J.J. Godleski. 2010. Chronic Social Stress and Susceptibility to Concentrated Ambient Fine Particles in Rats. *Environmental Health Perspectives* 118: 769–775. At: https://ehp.niehs.nih.gov/0901631/. Accessed 2 Aug 2017.

Corburn, J., J. Osleeb, and M. Porter. 2006. Urban Asthma and the Neighbourhood Environment in New York City. *Health Place* 12 (2): 167–179.

Deepa, M., M. Rajalakshmi, and R. Nedunchezhian. 2017. Impact of Air Pollution on Respiratory Diseases: Correlation and Classification by Multivariate Data Analysis. *Data-Enabled Discovery and Applications* (Accepted): DOI https://

doi.org/10.1007/s41688-017-0004-z. At: https://link.springer.com/article/10.1007/s41688-017-0004-z/fulltext.html. Accessed 7 Aug 2017.

DeFries, R.S., T. Rudel, M. Uriarte, et al. 2010. Deforestation Driven by Urban Population Growth and Agricultural Trade in the Twenty-First Century. *Nature Geoscience* 3: 178–181.

DeSoto, M.C. 2009. Ockham's Razor and Autism: The Case for Developmental Neurotoxins Contributing to a Disease of Neurodevelopment. *NeuroToxicology* 30: 331–337.

Dockery, D.W., C.A. Pope, X. Xu, J.D. Spengler, J.H. Ware, M.E. Fay, B.G. Ferris Jr., and F.E. Speizer. 1993. An Association Between Air pollution and Mortality in Six U.S. Cities. *The New England Journal of Medicine* 329 (24): 1753–1759.

EPA (Environmental Protection Agency). 2001. *Commuter Choice Leadership Initiative: Facts and Figures*, EPA 420-F-01-023. Washington, DC: EPA.

Ewing, R., and F. Rong. 2008. The Impact of Urban Form on U.S. Residential Energy Use. *Housing Policy Debate* 19 (1): 1–30.

Fang, C., and X. Liu. 2009. Temporal and Spatial Differences and Imbalance of China's Urbanization Development During 1950–2006. *Journal of Geographical Sciences* 19: 719–732.

Fang, C., H. Liu, G. Li, D. Sun, and Z. Miao. 2015. Estimating the Impact of Urbanization on Air Quality in China Using Spatial Regression Models. *Sustainability* 7 (12): 15570–15592.

Frank, L.D., M.A. Andresen, and T.L. Schmid. 2004. Obesity Relationships with Community Design, Physical Activity, and Time Spent in Cars. *American Journal of Preventive Medicine* 27 (2): 87–96.

Frumkin, H., L. Frank, and R. Jackson. 2004. *Urban Sprawl and Public Health*. Washington, DC: Island Press.

Fu, X., J. Liu, G.A. Ban-Weiss, J. Zhang, X. Huang, B. Ouyang, O. Popoola, and S. Tao. 2017. Effects of Canyon Geometry on the Distribution of Traffic-Related Air Pollution in a Large Urban Area: Implications of a Multi-canyon Air Pollution Dispersion Model. *Atmospheric Environment* 166: 111–121.

Goldenberg, R.L., J.F. Culhane, J.D. Iams, and R. Romero. 2008. Epidemiology and Causes of Preterm Birth. *Lancet* 371 (9606): 75–84.

Gong, P., S. Liang, E.J. Carlton, et al. 2012. Urbanisation and Health in China. *Lancet* 379: 843–852.

Gottholmseder, G., K. Nowonty, G.J. Pruckner, and E. Theurl. 2009. Stress Perception and Commuting. *Health Economics* 18 (5): 559–576.

Grahame, T.J., R. Klemm, and R.B. Schlesinger. 2014. Public Health and Components of Particulate Matter: The Changing Assessment of Black Carbon. *Journal of the Air & Waste Management Association* 64 (6): 620–660. https://doi.org/10.1080/10962247.2014.912692.

Gransow, B. 2012, December. *Internal Migration in China – Opportunity or Trap? Focus Migration*, Policy Brief No. 19. Osnabrück: Institute for Migration Research and Intercultural Studies (IMIS) of the University of Osnabrück.

Green, C.G., and E.G. Klein. 2011. Promoting Active Transportation as a Partnership Between Urban Planning and Public Health: The Columbus Healthy Places Program. *Public Health Reports* 126 (Suppl 1): 41–49.

Greenstone, M., J. Nilekani, R. Pande, N. Ryan, A. Sudarshan, and A. Sugathan. 2015. Lower Pollution, Longer Lives: Life Expectancy Gains If India Reduced Particulate Matter Pollution. *Economic and Political Weekly* 50 (1): 40–46.

Grineski, S. 2007. Characterizing Children's Asthma Hospitalizations on the Texas-Mexico Border. *Journal of Asthma* 44 (9): 783–787.

Han, L., A. Tsunekawa, M. Tsubo, and W. Zhou. 2013. An Enhanced Dust Index for Asian Dust Detection with MODIS Images. *International Journal of Remote Sensing* 34: 6484–6495.

Han, L., W. Zhou, W. Li, and L. Li. 2014. Impact of Urbanization Level on Urban Air Quality: A Case of Fine Particles (PM2.5) in Chinese Cities. *Environmental Pollution* 194: 163–170.

Handy, S.L., M.G. Boarnet, R. Ewing, and R.E. Killingsworth. 2002. How the Built Environment Affects Physical Activity: Views from Urban Planning. *American Journal of Preventive Medicine* 23 (2): 64–73.

Hedley, A.J., C.M. Wong, T.Q. Thach, S. Ma, T.H. Lam, and H.R. Anderson. 2002. Cardiorespiratory and All-Cause Mortality After Restrictions on Sulphur Content of Fuel in Hong Kong: An Intervention Study. *Lancet* 360: 1646–1652.

Heron, M., D.L. Hoyert, S.L. Murphy, J. Xu, K.D. Kochanek, and B. Tejada-Vera. 2009. Deaths: Final Data for 2006. *National Vital Statistics Reports* 57 (14), April 17.

Higgins, C.D., and P.S. Kanaroglou. 2016. A Latent Class Method for Classifying and Evaluating the Performance of Station Area Transit-Oriented Development in the Toronto Region. *Journal of Transport Geography* 52: 61–72.

Hirsch, J.A., A.V. Diez Roux, K.A. Moore, K.R. Evenson, and D.A. Rodriguez. 2014. Change in Walking and Body Mass Index Following Residential Relocation: The Multi-Ethnic Study of Atherosclerosis. *American Journal of Public Health* 104 (3): e49–e56.

Hoek, G., R. Beelen, K. de Hoogh, D. Vienneau, J. Gulliver, P. Fischer, and D. Briggs. 2008. A Review of Land-Use Regression Models to Assess Spatial Variation of Outdoor Air Pollution. *Atmospheric Environment* 42: 7561–7578.

Holdaway, J., P. Levitt, J. Fang, and N. Rajaram. 2015. Mobility and Health Sector Development in China and India. *Social Science & Medicine* 130: 268–276.

Hood, E. 2010. Stress and the City: Measuring Effects of Chronic Stress and Air Pollution. *Environmental Health Perspectives* 118: a258–a258. At: https://ehp.niehs.nih.gov/118-a258b/. Accessed 2 Aug 2017.

Hu, S., S. Tang, Y. Liu, Y. Zhao, M.L. Escobar, and D. de Ferranti. 2008. Reform of How Health Care Is Paid for in China: Challenges and Opportunities. *Lancet* 372 (9652): 1846–1853. At: http://www.who.int/management/district/6.Financing.pdf. Accessed 12 July 2017.

Hyslop, N. 2009. Impaired Visibility: The Air Pollution People See. *Atmospheric Environment* 43: 182–195.

Intergovernmental Panel on Climate Change (IPCC): Climate Change 2007: The Physical Science Basis, Contribution of Working Group I to the Fourth Assessment Report of the Intergovernmental Panel on Climate Change. Ed. S. Solomon, D. Qin, M. Manning, Z. Chen, M. Marquis, K.B. Averyt, M. Tignor, and H.L. Miller. Cambridge/New York: Cambridge University Press, 996 pp. At: https://www.ipcc.ch/pdf/assessment-report/ar4/wg1/ar4_wg1_full_report.pdf. Accessed 5 Sept 2017.

Jacobsen, P.L., D.R. Ragland, and C. Komanoff. 2015. Safety in Numbers for Walkers and Bicyclists: Exploring the Mechanisms. *Injury Prevention* 21 (4): 217–220.

Janssen, N.A.H., M.E. Gerlofs-Nijland, T. Lanki, R.O. Salonen, F. Cassee, G. Hoek, P. Fischer, B. Brunekreef, and M. Krzyzanowski. 2012. *Health Effects of Black Carbon*. Copenhagen: World Health Organization (WHO), 96 p.

Ji, C.Y., and T.O. Cheng. 2008. Prevalence and Geographic Distribution of Childhood Obesity in China in 2005. *International Journal of Cardiology* 131 (1): 1–8.

Jia, C.R., and J.A. Foran. 2013. Air Toxics Concentrations, Source Identification, and Health Risks: An Air Pollution Hot Spot in Southwest Memphis, TN. *Atmospheric Environment* 81: 112–116.

Jin, Yana, Henrik Andersson, and Shiqiu Zhang. 2016. Air Pollution Control Policies in China: A Retrospective and Prospects. *International Journal of Environmental Research and Public Health* 13 (12): 1219–1241.

Katsouyanni, K., G. Touloumi, E. Samoli, A. Gryparis, A. Le Tertre, Y. Monopolis, G. Rossi, D. Zmirou, F. Ballester, A. Boumghar, H.R. Anderson, B. Wojtyniak, A. Paldy, R. Braunstein, J. Pekkanen, C. Schindler, and J. Schwartz. 2001. Confounding and Effect Modification in the Short-Term Effects of Ambient Particles on Total Mortality: Results from 29 European Cities Within the APHEA2 Project. *Epidemiology* 12: 521–531.

Kaur, S., M.J. Nieuwenhuijsen, and R.N. Colvile. 2005. Pedestrian Exposure to Air Pollution Along a Major Road in Central London, UK. *Atmospheric Environment* 39 (38): 7307–7320.

———. 2007. Fine Particulate Matter and Carbon Monoxide Exposure Concentrations in Urban Street Transport Microenvironments. *Atmospheric Environment* 41 (23): 4781–4810.

Kopits, E., and M. Cropper. 2005. *Why Have Traffic Fatalities Declined in Industrialized Countries? Implications for Pedestrians and Vehicle Occupants*, Policy Research Working Paper 738. Washington, DC: World Bank, August.

Koslowsky, M., A.N. Kluger, and M. Reich. 1995. *Commuting Stress: Causes, Effects, and Methods of Coping*. New York: Plenum Press.

Krizek, K.J. 2000. Pretest-Posttest Strategy for Researching Neighborhood-Scale Urban Form and Travel Behavior. *Transportation Research Record: Journal of the Transportation Research Board* 1722: 48–55.

Künn-Nelen, A. 2016. Does Commuting Affect Health? *Health Economics* 25 (8): 984–1004. https://doi.org/10.1002/hec.3199.

Lee, I.M., E.J. Shiroma, F. Lobelo, P. Puska, S.N. Blair, and P.T. Katzmarzyk. 2012. Effect of Physical Inactivity on Major Non-Communicable Diseases Worldwide: An Analysis of Burden of Disease and Life Expectancy. *Lancet* 380 (9838): 219–229.

Lepeule, J., F. Laden, D. Dockery, and J. Schwartz. 2012. Chronic Exposure to Fine Particles and Mortality: An Extended Follow-up of the Harvard Six Cities Study from 1974 to 2009. *Environmental Health Perspectives* 120 (7): 965–970.

Leyden, K.M. 2003. Social Capital and the Built Environment: The Importance of Walkable Neighborhoods. *American Journal of Public Health* 93 (9): 1546–1551.

Lim, S.S., T. Vos, A.D. Flaxman, G. Danaei, K. Shibuya, H. Adair-Rohani, M.A. AlMazroa, M. Amann, H.R. Anderson, and K.G. Andrews. 2013. A Comparative Risk Assessment of Burden of Disease and Injury Attributable to 67 Risk Factors and Risk Factor Clusters in 21 Regions, 1990–2010: A Systematic Analysis for the Global Burden of Disease Study 2010. *Lancet* 380: 2224–2260.

Maantay, J. 2007. Asthma and Air Pollution in the Bronx: Methodological and Data Considerations in Using GIS for Environmental Justice and Health Research. *Health & Place* 13: 32–56.

Marquez, P., J. Langenbrunner, and S. Wang. 2011. *Toward a Healthy and Harmonious Life in China: Stemming the Rising Tide of Non-communicable Diseases*. The World Bank, pp. 1e48 (63426). At: https://openknowledge.worldbank.org/bitstream/handle/10986/27357/634260WP00Box30officia l0use0only090.pdf?sequence=1&isAllowed=y. Accessed 16 July 2017.

Marshall, J., H. Sander, and S. Manson. 2007. Air Pollution and Urban Form: Evidence From Satellite Data. In: *5th WSEAS International Conference on Environment, Ecosystem and Development, Tenerife, Spain, 14–16 December 2007*, pp. 463–465.

McKenzie, B. 2015. *Who Drives to Work? Commuting by Automobile in the United States: 2013*. American Community Survey Reports. Report Number: ACS-32. August 13, 2015. At: https://www.census.gov/content/dam/Census/library/publications/2015/acs/acs-32.pdf. Accessed 27 Aug 2017.

McKinsey & Company. 2009. *Preparing for China's Urban Billion*. At http://www.mckinsey.com/global-themes/urbanization/preparing-for-chinas-urban-billion. Accessed 16 July 2017.

Mokhtarian, P.L., and I. Salomon. 2001. How Derived Is the Demand for Travel? Some Conceptual and Measurement Considerations. *Transportation Research Part A: Policy and Practice* 35 (8): 695–719.

Morris, E.A., and E. Guerra. 2015. Mood and Mode: Does How We Travel Affect How We Feel? *Transportation* 42: 25–43.

Noland, R.B. 2003. Traffic Fatalities and Injuries: The Effect of Changes in Infrastructure and Other Trends. *Accident Analysis and Prevention* 35 (4): 599–611.

Ogden, C., and M. Carroll. 2010. *Prevalence of Obesity Among Children and Adolescents: United States, Trends 1963–1965 Through 2007–2008*. Centers for Disease Control and Prevention, National Center for Health Statistics. At: http://www.cdc.gov/nchs/data/hestat/obesity_child_07_08/obesity_child_07_08.pdf. Accessed 8 July 2017.

Ogden, C.L., M.D. Carroll, C.D. Fryar, and K.M. Flegal. 2015. *Prevalence of Obesity Among Adults and Youth: United States, 2011–2014*, NCHS Data Brief No. 219. Hyattsville: National Center for Health Statistics. 2015. At: https://www.cdc.gov/nchs/data/databriefs/db219.pdf. Accessed 15 Aug 2017.

Ou, J., X. Liu, X. Li, and Y. Chen. 2013. Quantifying the Relationship Between Urban Forms and Carbon Emissions Using Panel Data Analysis. *Landscape Ecology* 28: 1889–1907.

Ozdenerol, E., B.L. Williams, S.Y. Kang, and M.S. Magsumbol. 2005. Comparison of Spatial Scan Statistic and Spatial Filtering in Estimating Low Birth Weight Clusters. *International Journal of Health Geographics* 4: 19.

Passenger Travel Facts and Figures. 2016. U.S. Department of Transportation, Bureau of Transportation Statistics. At: https://www.rita.dot.gov/bts/sites/rita.dot.gov.bts/files/PTFF%202016_full.pdf. Accessed 15 Aug 2017.

Phua, K.H., K. Sheikh, S.-L. Tang, and V. Lin. 2015. Editorial – Health Systems of Asia: Equity, Governance and Social Impact. *Social Science & Medicine* 145: 141–144.

Pope, C., III, and D. Dockery. 2012. Health Effects of Fine Particle Air Pollution: Lines That Connect. *Journal of the Air & Waste Management Association* 56: 709–742.

Pope, C.A., III, J. Schwartz, and M.R. Ransom. 1992. Daily Mortality and PM10 Pollution in the Utah Valley. *Archives of Environmental Health* 47: 211–217.

Pope, C.A., III, D.L. Rodermund, and M.M. Gee. 2007. Mortality Effects of a Copper Smelter Strike and Reduced Ambient Sulfate Particulate Matter Air Pollution. *Environmental Health Perspectives* 115 (5): 679–683. At: https://www.ncbi.nlm.nih.gov/pmc/articles/PMC1867960/. Accessed 29 July 2017.

Pope, C.A., M. Ezzati, and D.W. Dockery. 2009. Fine-Particulate Air Pollution and Life Expectancy in the United States. *The New England Journal of Medicine* 360 (4): 376–386. https://doi.org/10.1056/NEJMsa0805646.

Ramesh, M., and X. Wu. 2009. Health Policy Reform in China: Lessons from Asia. *Social Science & Medicine* 68 (12): 2256–2262.

Rioux, C.L., K.L. Tucker, M. Mwamburi, D.M. Gute, and D. Brugge. 2010. Residential Traffic Exposure, Pulse Pressure and C-reactive Protein: Consistency and Contrast Among Exposure Characterization Methods. *Environ Health Perspectives* 118 (6): 803–811.

Rohde, R.A., and R.A. Muller. 2015. Air Pollution in China: Mapping of Concentrations and Sources. *PLOS ONE* 10 (8): e0135749. At: https://doi.org/10.1371/journal.pone.0135749.

Rutter, H., N. Cavill, F. Racioppi, et al. 2013. Economic Impact of Reduced Mortality Due to Increased Cycling. *American Journal of Preventive Medicine* 44: 89–92.

Saelens, B.E., J.F. Sallis, and L.D. Frank. 2003. Environmental Correlates of Walking and Cycling: Findings from the Transportation, Urban Design, and Planning Literature. *Annals of Behavioral Medicine* 25 (2): 80–91.

Sallis, J.F. 2009. Measuring Physical Activity Environments: A Brief History. *American Journal of Preventive Medicine* 36 (4 Suppl): S86–S92. https://doi.org/10.1016/j.amepre.2009.01.002.

Sallis, J.F., L.D. Frank, B.E. Saelens, and M.K. Kraft. 2004. Active Transportation and Physical Activity: Opportunities for Collaboration on Transportation and Public Health Research. *Transportation Research A* 38 (4): 249–268.

She, Q., X. Peng, Q. Xu, L. Long, N. Wei, M. Liu, W. Jia, T. Zhou, J. Han, and W. Xiang. 2017. Air Quality and Its Response to Satellite-Derived Urban Form in the Yangtze River Delta, China. *Ecological Indicators* 75: 297–306.

Southworth, M. 2005. Designing the Walkable City. *Journal of Urban Planning and Development* 131 (4): 246–257.

Stone, B. 2008. Urban Sprawl and Air Quality in Large US Cities. *Journal of Environmental Management* 86: 688–698.

Stone, B., and M. Rodgers. 2001. Urban Form and Thermal Efficiency: How the Design of Cities Influences the Urban Heat Island Effect. *Journal of the American Planning Association* 67 (2): 186–198.

Stutzer, A., and B.S. Frey. 2008. Stress That Doesn't Pay: The Commuting Paradox. *The Scandinavian Journal of Economics* 110 (2): 339–366.

Suessmuth-Dyckerhoff, C., and J. Wang. 2010. China's Health Care Reforms, 2010. *Health International* 10: 54–66. McKinsey & Company: McKinsey's Health Systems and Services.

Tong, X., Y. Wang, and E.H.W. Chan. 2016. International Research Trends and Methods for Walkability and Their Enlightenment in China. *Procedia Environmental Sciences* 36: 130–137.

Transportation Research Board. 2011. *Achieving Traffic Safety Goals in the United States: Lessons from Other Nations*, TRB Special Report 300. At: http://online-pubs.trb.org/onlinepubs/sr/sr300.pdf. Accessed 28 Aug 2017.

Transportation Research Board and Institute of Medicine of the National Academies. 2005. *Does the Built Environment Influence Physical Activity? Examining the Evidence*, Special Report 282. Washington, DC: Transportation Research Board.

U.S. Environmental Protection Agency (EPA). 2013. Federal Register, Part II. Rules and Regulations. *National Ambient Air Quality Standards for Particulate Matter* 78 (10), January 15. At: https://www.gpo.gov/fdsys/pkg/FR-2013-01-15/pdf/2012-30946.pdf. Accessed 5 Sept 2017.

UNDP. 2009. *Overcoming Barriers: Human Mobility and Development*, Human Development Report 2009. New York: UNDP.

United Nations Children's Fund and World Health Organization, (UNICEF and WHO). 2004. *Low Birthweight: Country, Regional and Global Estimates.* New York: UNICEF.

United Nations Environment Programme (UNEP). 2016. *A Review of Air Pollution Control in Beijing: 1998–2013.* Nairobi: United Nations Environment Programme (UNEP). At: http://www.bjepb.gov.cn/bjepb/323265/340674/4389403/2016052511264687868.pdf. Accessed 8 July 2017.

van Donkelaar, A., R. Martin, M. Brauer, R. Kahn, R. Levy, C. Verduzco, and P. Villeneuve. 2010. Global Estimates of Ambient Fine Particulate Matter Concentrations from Satellite-Based Aerosol Optical Depth: Development and Application. *Environmental Health Perspectives* 118 (6): 847–855. At: https://ehp.niehs.nih.gov/0901623/. Accessed 29 July 2017.

Wang, S., and J. Hao. 2012. Air Quality Management in China: Issues, Challenges, and Options. *Journal of Environmental Sciences* 24: 2–13.

World Health Organization (WHO). 2004. *Technical Consultation, 'Towards the Development of a Strategy for Promoting Optimal Fetal Growth'*, Report of a Meeting (Draft). Geneva: World Health Organization.

———. 2005a. *WHO Air Quality Guidelines for Particulate Matter, Ozone, Nitrogen Dioxide and Sulfur Dioxide (Global Update 2005): Summary of Risk Assessment*, 9–11. Geneva: World Health Organization.

———. 2005b. *Health Effects of Transport-Related Air Pollution.* Copenhagen: World Health Organization Regional Office for Europe.

———. 2013. *Health Effects of Particulate Matter—Policy Implications for Countries in Eastern Europe, Caucasus and Central Asia*. Joint WHO/ Convention Task Force on Health Aspects of Air Pollution.

———. 2014. *Burden of Disease from Ambient Air Pollution for 2012*. Available Online: http://www.who.int/phe/health_topics/outdoorair/databases/ AAP_BoD_results_March2014.pdf?ua=1. Accessed 19 July 2017.

———. 2015. *Global Status Report on Road Safety 2015*. Geneva: World Health Organization.

Xiao, Y., S. Kristensen, J. Sun, L. Lu, and S.H. Vermund. 2007. Expansion of HIV/AIDS in China: Lessons from Yunnan Province. *Social Science & Medicine* 64 (3): 665–675.

Yang, G., Y. Wang, Y. Zeng, G.F. Gao, X. Liang, M. Zhou, X. Wan, S. Yu, Y. Jiang, M. Naghavi, T. Vos, H. Wang, A.D. Lopez, and C.J.L. Murray. 2013. Rapid Health Transition in China, 1990–2010: Findings from the Global Burden of Disease Study 2010. *Lancet* 381: 1987–2015.

Yao, J., and B.-C. Kim. 2015. Social Insurance Participation of Rural Migrant Workers Based on Gender Dimension: Evidence from Four Chinese Cities. *Asian Social Work and Policy Review* 9: 57–69.

Zhang, Q., and R. Crooks. 2012. *Toward an Environmentally Sustainable Future: Country Environmental Analysis of the People's Republic of China*. Mandaluyong City: Asian Development Bank.

Zheng, S., and M.E. Kahn. 2017. A New Era of Pollution Progress in Urban China? *The Journal of Economic Perspectives* 31 (1): 71–92.

Zhou, Z., Y. Su, B. Campbell, Z. Zhou, J. Gao, Q. Yu, et al., 2015. The Financial Impact of the 'Zero-Markup Policy for Essential Drugs' on Patients in County Hospitals in Western Rural China. *PLoS ONE* 10 (3): e0121630. https://doi.org/10.1371/journal.pone.0121630. At: http://journals.plos.org/plosone/article?id=10.1371/journal.pone.0121630. Accessed 8 July 2017.

Zhou, L., M. Wang, C.-H. Chang, S. Liu, Y. Zhan, and J. Shi. 2017. Commuting Stress Process and Self-Regulation at Work: Moderating Roles of Daily Task Significance, Family Interference with Work, and Commuting Means Efficacy. *Personnel Psychology*: 1–32. https://doi.org/10.1111/peps.12219.

Zuurbier, M.,G. Hoek, M. Oldenwening, V. Lenters, K. Meliefste, P. van den Hazel, and B. Brunekreef. 2010. Commuters' Exposure to Particulate Matter Air Pollution Is Affected by Mode of Transport, Fuel Type, and Route. *Environmental Health Perspectives* 118: 783–789. At: https://ehp.niehs.nih.gov/0901622/#t2. Accessed 1 Aug 2017.

CHAPTER 7

Conclusions

7.1 Introduction

Whether contemporary cities are livable communities where residents enjoy high quality of life is shaped by their living conditions, healthy environment, equal opportunities of obtaining education and employment, access to multiple transportation, and a non-discriminatory housing market. Should the federal government be involved in ensuring the equality? Chapter 1 discusses the federal government's role in dealing with housing discrimination. The federal government passed the Fair Housing Act in 1968. Central to the enforcement of the law is the US Department of Housing and Urban Development's (HUD) responsibility for investigating and resolving claims of housing discrimination. One indicator that housing discrimination is reduced is the percentage of the population who have sufficient knowledge of fair housing legislation. Therefore, it is crucial in ensuring equal opportunity in housing that the general public is aware and knowledgeable of federal fair housing laws (Abravanel and Cunningham 2002).

In 2000–2001, HUD sponsored a national survey of 1001 persons designed to represent all adults in the nation. This first national survey evaluated the degree of public awareness and approval of fair housing law and persons' perceptions of whether they had ever experienced housing discrimination. The second HUD-sponsored survey in 2005 incorporated questions about the federal government's role and fair housing responsibilities. Survey participants were asked whether or not they thought the federal government should have responsibility for enforcing

A. Antipova, *Urban Environment, Travel Behavior, Health, and Resident Satisfaction*, https://doi.org/10.1007/978-3-319-74198-7_7

the Fair Housing Act. While Blacks and Hispanics show more support, only 27% of the adult population are of the opinion that the federal government should not be responsible to investigate fair housing complaints and take legal action on behalf of victims.

What Is the Attitude Towards the Federal Government's Involvement in Social Policies? There is a relationship between public approval of the federal government's responsibility for fair housing law enforcement and attitude on the broader involvement of the federal government's in certain social policies. Of the general public, 60% believe the federal government should do more when addressing issues related to education, housing, and the like. These people also provide more support compared with others to the federal government having responsibility for enforcing the Fair Housing Act.

Public support of the law and fair housing practices increased over time with the public disapproving of discriminatory treatments such as a landlord's opposition to constructing a wheelchair ramp at the door of an apartment building and to renting to a person with mental illness, as well as charging a higher down payment because of ethnicity, among other instances (Abravanel 2006). Knowledge of the Fair Housing law is associated with positive attitude for the government having responsibility for discriminatory claims investigation and taking legal action: the higher the degree of awareness, the higher the amount of support. The 2005 survey reported that 17% of the adult population believed to have experienced discriminatory home purchase or rental experiences (therefore, possibly illegal) at some point in their lives.

What Housing Experiences Are Discriminatory? Several new findings relating to perceived discrimination have emerged from the 2005 survey that should be highlighted. First, 30% of the respondents who reported that they perceived discrimination stated they experienced it only once, while the rest (two-thirds) believed that discriminatory actions against them occurred on multiple occasions. This high frequency deserved additional investigation. Second, about 50% of all respondents who claimed having experienced discrimination when trying to buy or rent housing may have no bases (that is, categories of housing discrimination based on race, ethnicity, religion, disability, familial status, etc.) prohibited by the federal Fair Housing Act or by state or local legislation that extend beyond the Act, while having other factors used as basis for discrimination such as being a student, age, being on welfare, lacking sufficient employment or income, etc., not covered by legislation. Third, a gap was found between people's willingness to act when confronted with perceived discrimination and the actual rates of real action. This last finding may have important implications for the form of the primary strategy for promoting the filing of claims: merely providing information about the filing process may prove less effective compared to publicizing rulings in housing discrimination cases, which may lead to public recognition that taking action is likely to produce results (Abravanel 2006).

What Are the Types of Discrimination Offenders on the Housing Market? Mostly, the alleged discrimination victims (70%) were trying to rent rather than buy a home (26%). Accordingly, regarding the type of offender who allegedly discriminated, an owner or representative of an apartment building was identified most frequently by the majority of the respondents as being responsible for housing discrimination, followed by homeowners, real estate agents, and lenders/loan officers.

Are There Other Discrimination Reasons Besides Those Prohibited Under the Act? The 2005 survey incorporated additional questions probing supplemental information: survey participants were asked to supplement their answers with descriptions of the nature and a believed basis for the discrimination along with the prohibited bases of the discrimination under the Act (that is, race, color, religion, sex, national origin, familial status, and disability). Other reasons of perceived discrimination not covered by the Fair Housing Act have been provided such as having a criminal history, having a pet, being a student , not having enough income or employment, lacking a good work or credit history, age, welfare status, and appearance (tattoos), so that multiple reasons could have applied in the instances of discrimination. However, not all allegations of housing discrimination are valid. Though the character of the survey (a mostly close-ended telephone survey) prevents from making a conclusion as to whether or not discrimination allegations have merit, these inquiries could be used as an additional refinement to identify discrimination claims that are likely to *be prohibited by* the federal Fair Housing Act *bases* and separate those from claims that do not fall within the terms of the law (Abravanel 2006).

Finally, Do People Support Fair Housing Law? Apart from the general public's awareness and knowledge of fair housing law, do people support its basic principles? The public's opinion regarding the discriminatory conduct of landlords, sellers, real estate agents, and mortgage lenders is that they should not engage in such behavior. However, the sizes of groups vary depending on the type of unlawful conduct. For example, when asked about the public support of basic tenets of fair housing law, a smaller percentage of the population (58%) is opposed to the unlawful conduct of advertising a religious preference for a rental, while many more people disapprove of higher fees charged due to ethnicity (84%) (Abravanel and Cunningham 2002). Acceptance of discriminatory restrictions because of a buyer's religion, nationality, or race has decreased. There was a steady steep decline in the attitude of White Americans who were willing to permit restrictive home sales practices on the basis of a buyer's race: the proportion of White Americans changed from more than 60% in the early 1970s–1980s to 29% in 1996. Interestingly, the 2005 survey revealed a statistically significant gender effect: women appear to be more likely to lend high degree of support for fair housing law than men are (Abravanel 2006).

Measures to Control Congestion Chapter 2 addresses urban built environment, or urban forms. Urban sprawl, a typical development in the United States, results in congestion among other adverse effects. Land-use regulations are commonly applied throughout the world to control population density as well as to effectively combat urban externalities including traffic congestion. In the United States, most metropolitan areas have zoning ordinances in place. Land-use regulations control building size or height through floor area ratio (FAR) regulation, and city size can be controlled by imposing the urban growth boundary (UGB). However, as a policy against traffic congestion it is insufficient to use land-use regulations alone. Cordon pricing has already been implemented in several cities including London, Milan, Oslo, Singapore, and Stockholm for its potential to mitigate traffic congestion. Unlike congestion pricing where a price should vary at each point of driving time, according to the scheme, a vehicle pays a fixed toll when it crosses a specified cordon line in the inbound and/or outbound direction. However, despite the simple toll-collecting method of cordon pricing, governments are tasked with rationally determining the toll rate and the location of the cordon line. Several cities including Milan, Oslo, Singapore, and Stockholm impose cordon pricing and land-use regulations simultaneously (Kono and Kawaguchi 2017).

Chapter 2 describes "good urban forms." Strategic urban development, including neighborhood design, smart growth planning, New Urbanism, and transit-oriented developments (TODs), has the potential to modify travel behaviors by making the living environment high density and walkable, containing mixed commercial, retail, and residential land uses near transit centers (Boschmann and Brady 2013). TODs are land-use developments that are high-density and pedestrian-friendly, with mixed land uses, and promote a greater sense of livability around the stations (Ratner and Goetz 2013).

7.2 How Changing Demographics and Travel Behavior Are Related

One aspect of urban form (TODs) might be an important modifying factor in changing travel habits in older adults. In light of the rapidly changing demographics in the Western developed countries, with the increasing share of the older population, the changing travel behaviors and needs of this population group in cities need to be investigated. Boschmann and Brady (2013) contributed to the existing research on the links between urban form and travel behavior by examining the effect of TOD residences on the travel behaviors of older adults in a metropolitan area. Compared with the previous cohorts of this age category, older adults today are healthier, more educated, wealthier, and better housed, so they want to experience active livelihoods after retirement.

Livability of communities is reflected in all community members enjoying long healthy lives. Older people who soon will represent a substantial part of the US urban (and rural, for that matter) society, desire to keep their quality of life including the capacity to age independently, maintain autonomy, and manage self-care. They also desire to continue living in the same neighborhoods where they lived for many years. Since incorporating changes to individual behaviors can impact the environmental damages of transportation in the long term, ways to reduce the need for travel in cities via changing travel behavior, including the total daily trips taken, trip distance, and alternative modal use, need to be sought. In this light, active aging is important. Active aging occurs when physical activity is integrated into daily routines such as walking, yard work, and so on, thus contributing to more non-work travel within cities. Although still limited, walking is becoming more important. However, compared with younger adults, older people are less likely to use public transit, which mostly meets the needs of younger commuters while it lacks flexibility, safety, reliability, comfort, and convenient service hours and routes for older travelers.

Empirical research indicates that residing in or near a TOD does significantly impact travel behaviors including those of older adults: TOD residents exhibit less automobile dependence and drive less, and make shorter but more trips, offering some empirical evidence for urban land-use development and transportation planning that TODs do support more sustainable travel behaviors among travelers of this age group (Boschmann and Brady 2013). They also found that compared with transit, walking is a more significant non-car alternative mode for older adults living in a TOD. Despite these developments being called "transit-oriented" because of their closeness to public transportation, their other design features including focus on pedestrians, centrality, high density, commercial and retail amenities, quality housing may attract residents to a TOD rather than transit access itself (Chatman 2013). This may be valid since despite rail transit offering good access to certain amenities such as cultural and sporting entertainment venues in central downtown or regional shopping malls, access may be limited to other destinations that meet daily needs including groceries, pharmacies, banks, or the major medical centers, making transit access a less important attraction factor. Chatman (2013) examined whether households choosing new housing near rail stations have different patterns of auto ownership and use.

7.3 What Are Housing/Working Perceptions and Motivations of Transit Access?

One US city, Denver, Colorado, aggressively pursues TOD. To understand who lives or works in TODs and why, the Denver Regional Council of Governments (DRCOG) conducted the Who is TOD Survey in 2009, 2010, and 2016 among residents, businesses, and employees who live or work close to existing rail-transit stations (DRCOG 2014). The resident survey extracted information on demographics, travel behaviors, and perceptions of residents regarding mode choice and home location decisions at varying distances from rail-transit stations. Business owners and managers whose businesses are located within one mile of existing rail-transit stations participated in a business survey. Researchers were interested to understand what motivated businesses to select their current location as well as their attitudes and perceptions about parking and transit access, and what strategies and incentives employers use to manage transportation demand. The third component, an employee survey, solicited information from employees of businesses that participated and allowed their employees to take part in the survey. Employees shared travel behavior information regarding commute distance, influence of transit access, and mix of uses on employment decisions, mode choice, and location and cost of parking. The results of the Who is TOD Survey reveal that living and working benefits of transit-oriented, walkable, mixed-use communities attract large numbers of people (Who is TOD in Metro Denver? 2010).

In order to understand better how access to transit impacts housing preferences, researchers surveyed households across the Denver metropolitan region with a sampling emphasis on those within one mile of a rail-transit station and compared the characteristics, travel behaviors, and perceptions of households located close to transit stations to those further away (Who is TOD in Metro Denver? 2010). Next, based on assessment of household and intersection density by census block groups, households within one mile were classified into three areas, namely, downtown, urban, and suburban. In terms of demographics, people living in transit-oriented communities tend to have smaller-sized households without children, therefore attracting younger adults and older empty nesters (Boschmann and Brady 2013). In agreement with this general observation, residents of these communities in downtown Denver are younger overall, least likely to have children, and have smaller households. Compared to residents in a suburban transit area or in the greater metro Denver area, they tend to have a lower income. Home ownership rates are lower among downtown residents than among other metro Denver residents. For all trip purposes, their travel behavior differs from that of urban and suburban residents in terms of using alternative transportation modes: downtown residents walk, bike, and use transit to buy groceries (48% versus 30% and 13%, respectively), do shopping (56% versus 45% and 25%), visit restaurants or bars (63% versus 44% and 20%), utilize open space and parks (90% versus 80% and 79%), and access entertainment (65% versus 46% and 16%).

The Who is TOD Survey revealed demand for communities that offer a mix of uses and transit within walking distance among three housing market segments. The first market segment, *Access Focused*, comprising 21% of metro Denver residents, is concerned with access: this population desires easy access to amenities including transit, shopping, restaurants, and parks, while being less concerned about their living space. This segment tends to be younger with the average age of 41 years, are generally single or with two adults per household, are least likely to have children, and are most likely to live within one mile of a rail-transit station, downtown, or urban area. Compared to the other two market segments, the first market segment population is most likely to commute by alternate mode (61% of commute days), and their work commute is the shortest with 6.3 miles on average. The second market segment, *Home Focused*, represents 25% of metro Denver residents, is more interested in the size of their house and yard, and were looking for a quiet neighborhood with low crime rates. The demographic composition of this market segment is different; they tend to be older with an average age of 48 years, with households consisting of a single person or two adults, and are unlikely to have children. In contrast to the first segment, those focused mostly on their home are unlikely to live within one mile of a rail-transit station and they have a twice longer work commute (12.3 miles on average) and commute by alternate mode on only 30% of commute days. The final market segment, *Family Focused*, accounts for 54% of metro Denver residents. The housing preferences of this market segment include safe neighborhoods with good schools and enough space in the house for everyone. The average age of individuals in this segment is 48 years, with larger households consisting of two adults, most likely including children. Members of this market segment are least likely to live within one mile of a rail-transit station (only 8%), have the longest commutes to work (13.3 miles average), and commute by alternate mode 32% of commute days (Who is TOD in Metro Denver? 2010).

7.4 What Is the Role of Urban Built Environment in Pollution Exposure?

The urban built environment discussed in Chap. 2 is linked to pollution exposure, which is dealt with in Chaps. 5 and 6. In high-density urban areas, residential areas are important human living environments where different landscape elements can contribute substantially towards reductions of airborne particle concentrations in residential environments. A recent study by Fan et al. (2017) examined the effect of landscape structure parameters such as vegetation and building cover on the concentrations of different-sized airborne particles in 18 residential areas in Beijing, China, with the results indicating that the distribution of particle concentrations has great spatial variation in urban environments and where the net concentration differences were all significantly different among the studied areas. Particularly, percentage of vegetation cover and building cover can greatly influence airborne particle levels in residential environments. Greater vegetation coverage and lower building construction were found to be linked to a decrease in airborne particle concentrations. Thus, landscape structure can be used for predicting the potential particle reduction at local scale (Fan et al. 2017).

Similarly, the global and regional emission of *biogenic volatile organic compounds* (BVOCs) is greatly impacted by land use and land-cover change. So, applying different land-based strategies such as afforestation and avoided deforestation, or bioenergy, helps to mitigate anthropogenic climate change including the emission of greenhouse gases such as atmospheric CO2 or CH4 and also other carbon trace gases such as BVOCs (Szogs et al. 2017). BVOC emissions can be significantly decreased by expanding land area under crops or grassland. The choice of crops is crucial, using woody bioenergy crops or afforestation may lead to increasing the emissions of BVOCs.

Building configuration, for example, a street canyon, strongly affects human exposure to traffic pollutants in large urban areas. With street canyons being universal in the populated urban areas, canyon geometry influences the distribution of traffic-related air pollution, which can adversely affect human health in street canyons. For example, vehicle exhausts produced from traffic emissions first scatter inside street canyons, after which air pollutants leave the street canyon and move further over the urban area. The effects of street canyon geometry on the distribution of NOx and CO from traffic flows were studied over the center of Beijing. Building heights and street widths influence pedestrian exposure to pollution inside and outside canyons: greater building heights resulted in heavier pollution inside canyons leading to greater inhalation of unhealthy air by urban residents, while the opposite was observed outside canyons where pollution was lower at the pedestrian level (Fu et al. 2017). Increasing street widths may both reduce emissions and improve ventilation and thus decrease pollutant concentrations. Thus, taking into consideration traffic demand and local-scale weather patterns as well as careful planning, street layout and canyon geometry may ameliorate adverse health impacts by significantly reducing inhalation of unhealthy air by urban residents (Fu et al. 2017).

Another recent study examined whether policies implemented at a busy street canyon in the capital city of Sweden, Stockholm to reduce vehicle exhaust emissions were effective. Between 2006 and 2013, the vehicle fleet underwent key changes: light-duty vehicles have been dieselized and vehicle engines have been technologically improved. Concentrations of NOx, black carbon, and other particles were measured in 2006 and 2013. Though NOx concentrations remained stable, these changes lead to lower ambient concentrations including those of black carbon and particle emission factors. For example, a big decrease in particle concentrations was observed between these two years, including an average of 60% for black carbon, because of decreases in both gasoline and diesel emissions (Krecl et al. 2017).

7.5 What Are the Trends in Urban Air Pollution and Its Health Effects?

Chapters 5 and 6 address urban effects on travel behavior and health, respectively. Recent rapid increase in vehicles and industrialization leads to air quality worsening because transportation- and industry-related pollution increase concentrations of air pollutants among which particulate matter (PM2.5, PM10) determines the air quality and negatively affects human health (Deepa et al. 2017). *Ambient particulate matter* is a type of air pollution that consists of a mixture of tiny particles with different physical and chemical characteristics suspended in air. Despite the health effects of the particles are not yet fully understood, the amounting evidence indicates that particles with a smaller aerodynamic diameter (that is, 10 or even 2.5/μm or smaller) are associated with adverse health effects.

Because different locations differ in terms of the composition of particles, their health effects differ with uncertainties remaining on the nature of the exposure and the mechanisms of the impact. In Europe, black smoke (BS) measurements, or concentrations of black particles with aerodynamic diameter <4.5 μm, have a long history and have historically been used to represent particles. In Europe, standards for BS recently have been replaced by PM10 standards. PM10 measurements in the United States have been used as an indicator of all particles with an aerodynamic diameter less than 10 μm, a mix of primary and secondary particles coming from different sources with different characteristics and toxicity levels. However, there is lack of historical fine particulate matter with diameter <2.5 μm (PM2.5) data in many countries including the United States where monitoring of PM2.5 began in 1999 while many other countries started its monitoring even later (Li et al. 2017).

Fine particulate matter (PM2.5) is of particular interest because its measurements have been directly linked to estimates of health risks. The lack of PM2.5 data prevents epidemiological studies from investigating long-term exposure of PM2.5 and health outcomes such as cancer. PM10 measurements need to be first converted to PM2.5 for such estimates (WHO 2016). However, annual mean PM2.5 measurements are not available in every country, and many low- and middle-income countries are still lacking these important measurements. As of its 2016 update, the second version of the WHO's Urban Ambient Air Pollution database contains PM2.5 measurements for 339 cities in low- and middle-income countries versus 1241 cities and towns with PM2.5 measures in high-income countries, while there is a better situation regarding annual mean PM10 measurements. These are available for as many as 586 cities in low- and middle-income countries, and 1639 cities and towns with PM10 measurements could be accessed in high-income countries (WHO 2016). Therefore, one solution to this problem is that developing countries, even those with urgent need for economic growth, need to expand the accuracy and coverage of pollution monitoring, which will enable greater public awareness of pollution. To illustrate the need for building more monitoring stations with functional and well-calibrated monitors in more locations: Kolkata, the Indian city with the most monitoring stations, has only 20 monitoring stations compared to Beijing, which has 35 (Greenstone et al. 2015).

7.5.1 What Are the Global Health Impacts of Air Pollution?

Air pollution became a *global public health problem* (Greenstone et al. 2015). Exposure studies provide impetus to the development of strategies to control and reduce urban air pollution. In India, the population is exposed to dangerously high concentrations of air pollution, and the severity of impact by air pollutants is steadily and consistently growing across all metropolitan cities impacting the lives of millions of urban residents including children, especially in the national capital of Delhi, one of the worst affected Indian and world cities, where there is an increase in the number of people who suffer from cardiovascular and lung-related diseases associated with poor air quality caused by a growing migrant population looking for job opportunities and a rapid increase in vehicles in the national capital (Deepa et al. 2017). In India, large areas in the north of the country, which include both rural and urban areas, do not comply with the Indian standards on air quality by a wide margin. This pattern illustrates how fine particles can travel over long distances far from where they are initially emitted, placing people who live far from major urban industrial and transport sources at high health risks (Greenstone et al. 2015).

Export of other air pollutants over long distances is well documented, with an example of ozone (O3) whose export and long-range influence is well studied. O3 is an important ground-level pollutant; its concentrations and emissions of O3 precursors have increased significantly over recent decades in East Asia. East Asian ozone is transported both eastward across the Pacific Ocean influencing during springtime North America and southward to tropical Southeast Asia. Its export to tropical Southeast Asia during the Northeast Monsoon season (monsoon season there lasts November to February) is especially strong during "cold surges," with the largest changes in the level of key pollutants such as ozone and carbon monoxide occurring over much of the Indochinese Peninsula and surrounding seas when the concentrations of these pollutants are elevated by more than 40% (~15 ppb) and more than 60% (~80 ppb), respectively, on average, during cold surges (Ashfold et al. 2017). Thus, East Asian pollution influences air quality in both North America and Southeast Asia.

The WHO's Urban Ambient Air Pollution database, updated in 2016 (WHO 2016), contains air quality data represented by annual means for PM10 and/or PM2.5 data mostly for urban areas covering almost 3000 human settlements in 103 countries between 2008 and 2015. The WHO established Air Quality Guidelines with annual mean values of 20 $\mu g/m^3$ (for PM10) and 10 $\mu g/m^3$ (for PM2.5). Some world cities in the Eastern Mediterranean region, including Riyadh (in Saudi Arabia) and Ma'ameer (in Bahrain), are the worst ranked in terms of concentrations of fine particulate matter of 10 microns or less (PM10) with cities exceeding the values of 350 $\mu g/m^3$ and 250 $\mu g/m^3$, respectively, in the period 2011–2015 (WHO 2016). Delhi has PM10 levels at about 230 $\mu g/m^3$ during the same time, however, among mega-cities of more than 14 million habitants with available PM10 levels for the last available year in the period 2011–2015, it is ranked the worst (WHO 2016).

In China, a fast urbanizing country where 4000 deaths/day or 17% of all deaths can be attributed to PM2.5 air pollution, the most intense pollution was found in the east including the northeast corridor stretching from near Shanghai to north of Beijing, however, substantial concentrations are not limited to major cities and are widespread across northern and central China (Rohde and Muller 2015). Research demonstrates that both PM10 and BS are predictors of daily deaths across European cities, where BS was associated with a 0.6% increase in the total daily number of deaths per 10 $\mu g/m$ increase in the one-day BS levels and per 10 $\mu g/m^3$ increase in the daily PM concentrations; the percentage increase in total daily number of deaths associated with this increase is 0.7% (Katsouyanni et al. 2001). In the context of the United States, the analysis of data from the 90 largest US cities indicated overall estimated increase per 10 μg/m^3 increase in PM level was 0.5% (Katsouyanni et al. 2001). Other air pollutants, including SO2, ozone, and N02 concentrations may confound the health effects of PM. A study based on 29 cities found that NO2, an indicator of pollution originating from vehicle exhausts, is the most important confounder: The higher the N02 levels the larger the effect observed on total daily deaths among the elderly across the European cities, indicating that particles related mostly to traffic are more toxic than particles originating from other sources (Katsouyanni et al. 2001). Among other confounding variables there is temperature with increases PM-mortality association perhaps because of the greater average population exposure with more people spending more time outdoors and keeping their windows open due to warmer weather.

The prospective cohort study examined the relationship between long-term exposure to air pollution and mortality in terms of the range of exposure to air pollutants in six US cities including Steubenville, St. Louis, Harriman, Watertown, Topeka, and Portage, selected as representative of the range of particulate air pollution in the country (Dockery et al. 1993). Except for the ozone concentrations, Steubenville had the highest ambient concentrations for all measures of air pollution making it the most polluted city in the sample while Portage and Topeka had the lowest levels (and the highest mean ozone concentrations) (Dockery et al. 1993). To illustrate, average annual concentration of fine particles in Steubenville was 18.8 μg/m^3 higher compared to Portage, and the difference in the level of annual average inhalable particles was 28.3 μg/m^3 between the two cities. The results showed that the most polluted city of Steubenville, Ohio, had the highest mortality followed by St. Louis, while the lowest mortality was in Portage and Topeka. Steubenville had 26% higher mortality rates due to all causes while mortality rates due to cardiopulmonary causes were 37% higher compared with the least polluted city of Portage, Wisconsin. Increased mortality was also linked to less than a high school education and increased body mass index (BMI), particularly in women. Mortality was more linked to inhalable, fine and sulfate particles (e.g., lung cancer was associated with particulate air pollution). The association between long-term exposure to particulate air pollution and elevated mortality was observed even when controlling for sex, age, smoking, and educational status, and occupational exposure to dust, gases, and fumes.

However, the researchers note that observed excess mortality may be due to the cumulative effect of a lifetime particulate air pollution exposure (Dockery et al. 1993).

The impact of air pollutants (CO, particles, NO2, SO2) was studied on 1990s mortality in London, where respiratory deaths were more strongly affected in terms of having the greatest and most significant relative risks, while particles and SO2 seemed to be the most important pollutants associated with deaths due to respiratory diseases (Bremner et al. 1999). Cardiovascular deaths were impacted to a lesser degree with NO2 and O3 the most important pollutants for the cardiovascular deaths model. The influence of PM10 was sharply reduced after BS or other pollutants were included in the pollutant model, making the effect of BS seemingly more robust than that of PM10 in the model (Bremner et al. 1999).

7.5.2 *How Does School Travel Relate to Pollution Exposure?*

An important aspect of pollution is exposure to harmful effects of air pollutants on school students during their school bus rides. In the United States, almost 55 million elementary and secondary students travel to and from school each day (National Center for Education Statistics 2009). In 2009, American students traveled to school on average 4.4 miles. Compared with high school students, elementary students had shorter average trip lengths (3.6 miles versus 5.5 miles), and on average, trips to school lasted 17.3 minutes. Among all travel modes reported by the National Household Travel Survey (NHTS), which documents America's travel patterns since 1969, regarding children's school travel, 39.4% of school-aged children (grades K-8, or 5–14 years of age) commuted by school bus in 2009 (McDonald et al. 2011); more in middle school (aged 12–14 years) than in elementary school (aged 5–11 years): 42.8% versus 37.9%, respectively.

Because previous studies reported that children living on or going to schools closer to busy streets had increased risks of asthma and respiratory symptoms, and knowing that diesel school buses are an important source of children's daily particulate exposures (Behrentz et al. 2005), the Seattle metropolitan area-based study explored fine particulate matter (PM2.5) aboard school buses. The mean travel time of all trips to and from school was 22 minutes with approximately four stops per trip. However, children who ride diesel school buses are exposed to particulate air pollution from the bus running low-sulfur diesel fuel and other roadway traffic with traffic congestion and self-pollution into the cabin (a mean self-pollution concentration of 7 μg/m^3 has been observed) contributing to onboard concentrations (Adar et al. 2008). As children routinely ride school buses they are exposed to onboard particulate air pollution; the PM2.5 concentrations are four times the ambient background levels at central monitoring stations and approximately two times the levels on surrounding roadways. Newer buses that have emission-reducing technologies such as diesel oxidation catalysts (DOCs) were shown to substantially decrease in-cabin concentrations; more so with the bus windows closed (Adar et al. 2008).

7.5.3 Commuting Has Important Effects on Health, and Work Experience

Chapter 6 discusses the links between travel behavior and health. A recent study examined the impact of commuting experience on work experience, specifically, the link between workers' morning commutes and their subsequent self-regulation at work (Zhou et al. 2017). When self-regulation at work is effective, workers control and direct their resources to work-related tasks; in contrast, when the process is impaired, it could lead to various negative consequences including increased deviant behavior and decreased work engagement. Commuting strain refers to the adverse psychological responses (e.g., anger, frustration) to the commuting stressors (i.e., challenging events, conditions, and poor commute experiences in the commute environment including threats such as reckless behaviors of other drivers, traffic control, congestion, crowded bus, and the like) experienced by the individual commuters on their daily commutes. In other words, commuting stressors trigger the stress process and induce strain. Certain commuting stressors (traffic jam, bus running behind schedule) delay the time of arrival at the destination (e.g., their workplaces) with commuters feeling worried that due to these stressors they can potentially arrive late at work. Because commuters perceive threats of tardiness and pay loss or other disciplinary actions as a consequence of their late arrival to work, workers may feel distressed on their commute. Researchers wanted to test the hypothesis that commuting stressors experienced during the commute produce commuting strain (that is, negative psychological responses) that is likely to weaken individuals' self-regulation at work (Zhou et al. 2017).

There is evidence that commuting influences physical health and psychological well-being and therefore may affect work performance. Commuting strain is positively related to workplace aggression. Commuting research provides evidence that commuting can negatively affect workers' performance. During the commute, a commuter may experience challenging factors that can delay their arrival time at work (e.g., traffic jam) and threaten their desired safety and comfort during the commute (e.g., overcrowded bus, other driver's unsafe behavior on the road) creating commuting strain (defined as the adverse psychological responses (e.g., anger, frustration) to these negative commuting experiences (Zhou et al. 2017). The potential pathway of the link between commuting and work is that commuting experience impacts subsequent work experience through commuting strain. As a worker arrives at work in the morning, they need to carry out different work-related tasks, however, with commuting strain being unlikely to disperse as soon as work starts, the worker regulates these negative emotions by suppressing thoughts on negative commuting experience with the risk of losing attention on the important job-related task.

As commuting is often associated with greater amounts of fatigue, commuting is predicted to negatively affect health measured by objective and subjective levels of stress. Regarding various travel modes, compared to commuters using public transportation, health of car drivers is more strongly affected by commuting time. A longer commuting time is linked to lower health satisfaction, lower health status, and a higher BMI among those who drive a car on their way to work (Künn-Nelen 2016). Commuters by car visit the doctor more often. No significant relation between commuting time and any of these health measures has been found for commuters using public transportation. An interesting finding related to gender is that women experience higher negative effects of longer commuting times and report lower health satisfaction, a lower probability of regular exercise, and a higher BMI; calling in sick more frequently; and having more doctor visits than male commuters (Künn-Nelen 2016). Since commuting is important in determining peak travel demand during rush hours, with urbanization leading to longer distances between homes and work places, the economic costs of commuting must also take into consideration commuting-induced stress and stress-related health problems when analyzing commuting costs.

7.6 What Are the Links Between Health and Active Transportation?

Unhealthy trends in both physical activity and obesity have been identified: presently, the United States is facing a physical inactivity epidemic (Green and Klein 2011). In the United States, more than one third of adults and 17% of youth were obese in 2011–2014 (Ogden et al. 2015). Compared with men overall, women have higher obesity prevalence, while non-Hispanic Black and Hispanic adults are more likely to be obese compared to other racial and Hispanic origin groups. Yet only a small fraction of the adults meets the physical activity guidelines (Green and Klein 2011). Research findings suggest that encouragement of active transportation may help obesity prevention efforts.

Active travel makes communities healthy and more sustainable by reducing environmental pollution. Cycling in childhood impacts cycling later in life: if kids start cycling in childhood, they are more likely to cycle later in their life, so it may influence the likelihood of their cycling throughout their life (Clayton and Musselwhite 2013). Overall, cycling as a family is indicated as a positive experience. The multitude of sensory feelings including sights, smells, sounds, fluidity, and speed of body movement obtained from cycling rides adds substantially to the experience of cycling as a family activity. When there is a reason for the cycle (e.g., going to a certain destination), the family members seem to enjoy it more since the family cycle is purposeful and is validated. Good infrastructure results in an increased amount of cycling, although improving infrastructure does not always lead to increased active modes of travel including cycling, therefore intervention of a different kind is needed to overcome existing social norms. For example, public artwork installed along cycle routes may improve young people's perceptions and cycling experiences as "fun cycling." Other investments in cycling infrastructure may include providing safe and high-quality cycle lanes and routes.

Cycling barriers for all ages are perceptions and feelings of being unsafe and vulnerable when one is exposed to heavy traffic. People feel unsafe to cycle when infrastructure is poor and is lacking in segregated, dedicated, or prioritized cycling roads. Parents are concerned over road safety and less worried about safety from crime. Parents often err on the personal security side and trade off ensuring kids' safety or fostering kids' independent mobility. Parents' attitudes or behavior towards cycling may impact whether kids will engage in cycling or not. Kids' cycling attitude/behavior is linked to parental confidence in cycling. Factors that may help increase cycling include higher density, segregation from traffic, and proximity (Clayton and Musselwhite 2013).

7.7 There Are Safety Aspects of Active Travel

Walking and cycling occurs more within safer environments. However, compared with most Western European countries, traffic fatality rates in the United States is much higher. For example, in 2014, the per-capita fatality rate was 2–3 times higher, whereas in the Netherlands and Denmark, the fatality rates for their populations as a whole are very low (in 2010, pedestrian fatality rate per 100 million kilometers was 1.2 in the Netherlands and 2.5 in Denmark compared to 9.7 in the United States), while their levels of walking and cycling are very high (Buehler and Pucher 2017).

Areas with high numbers of injuries particularly need interventions to decrease motor vehicle collisions with pedestrians and bicyclists. To illustrate, planners in San Francisco concluded that 60% of pedestrian collisions occurs on just 6% of the streets, and the highest priority was made to addressing those streets (Jacobsen et al. 2015). Design of roads shall be such that motorists are "reminded" that people are possibly walking and bicycling, incorporating simple reminders such as more visible crosswalks, bicycle lanes marked on major streets, curbs at intersections extended beyond parked cars bringing a pedestrian closer to the motorists' zone of attention. Vulnerable road users including pedestrians and cyclists need be separated from motorized traffic and the speeds of the motorized traffic need to be decreased to enable motorists to have more time to see people walking and bicycling. Safer streets attract more people to walk and bicycle (Jacobsen et al. 2015). Graves et al. (2014) describe that five cities in the United States and Canada that fulfilled bike-share programs succeeded at increasing the amount of bicycling while simultaneously reducing the yearly number of bicycle injuries by an average of 28%. Five cities used as a control cities that did not implement bike-sharing increased bicyclist injuries by an average 2%. Additionally, traffic laws punish motorists for traffic violations more severely in Europe explaining the much lower risk of walking and bicycling in Germany and the Netherlands as their laws protect people walking and bicycling, while in the United States the responsibility for collision avoidance is shifted to the pedestrian (Jacobsen et al. 2015).

Across the large US cities, Portland clearly stands out as the most successful regarding the implementation of the combination of all cycling policies. Portland's success is reflected in the numbers, as cycling levels increased sixfold since 1990, whereas in New York it only doubled. The main focus of Portland's policy package of infrastructure, programs, and policies to promote cycling has been to continuously increase and improve the bikeway network comprising bike paths, lanes, and bike boulevards via residential neighborhoods. The city has steadily increased bikeway safety, convenience, and connectivity, redesigning many intersections on a yearly basis by installing bike boxes (advance stop lines) at the intersections as well as priority signage and advance green lights for cyclists (Pucher et al. 2011).

7.7.1 What Can Be Done to Prevent Obesity?

Due to the alarming data regarding the national obesity epidemic, one underused resource existing in metropolitan areas is the adoption of novel programs emphasizing community design changes to address sedentary lifestyle and promote physical activity by making positive changes in the built environment (Green and Klein 2011).

Having a walking and biking infrastructure changes the living environment where we live and facilitates active living in our communities. Green and Klein (2011) make the following recommendations to metropolitan areas considering an investment in community-based obesity prevention in order to be successful:

1) To convince reluctant individuals and agencies, establish good relationships with partner agencies. It is crucial for success of such a program that the value of the program is communicated to the agency and community.
2) As any concerns or issues emerge, it is important to stay in regular contact with individuals at each agency through e-mail or regular meetings to address and share problems or successes. This proactive approach is needed for program maintenance.
3) Since the adoption of a novel program may be resisted by individuals and agencies, health agencies should be ready to protect the public health role in development and built-environment decisions.
4) With an urban planner being the key link between public health and planning, he/she should have a strong background in both fields.

On an optimistic note, although active transportation including walking and biking represents a small share of passenger travel, 11.5% of trips and 1.0% of miles traveled, the number of Americans who are walking and bicycling is growing: 10.4% of trips were done by walking, and biking accounted for 1.0% of trips in 2009 (*Passenger Travel Facts and Figures* 2016). Nationally, walk and bicycle commutes make up 3.4%, accounting for 2.8% and 0.6% of workers, respectively. The number of commuters who bike to work has almost doubled since 2000. A greater share of workers who walk or bike to work is in urban areas. High walking and biking shares occur in several small and medium-sized cities, especially where university or college presence is substantial.

Among the large cities (with population equal or greater than 200,000), top walk-commuting cities in 2010–2014 were Boston, MA, Washington, DC, Pittsburgh, PA, San Francisco, CA, and New York, NY, while cities in the same category with the highest-ranked bike commuting were Portland, OR, Madison, WI, Washington, DC, Minneapolis, MN, and San Francisco,

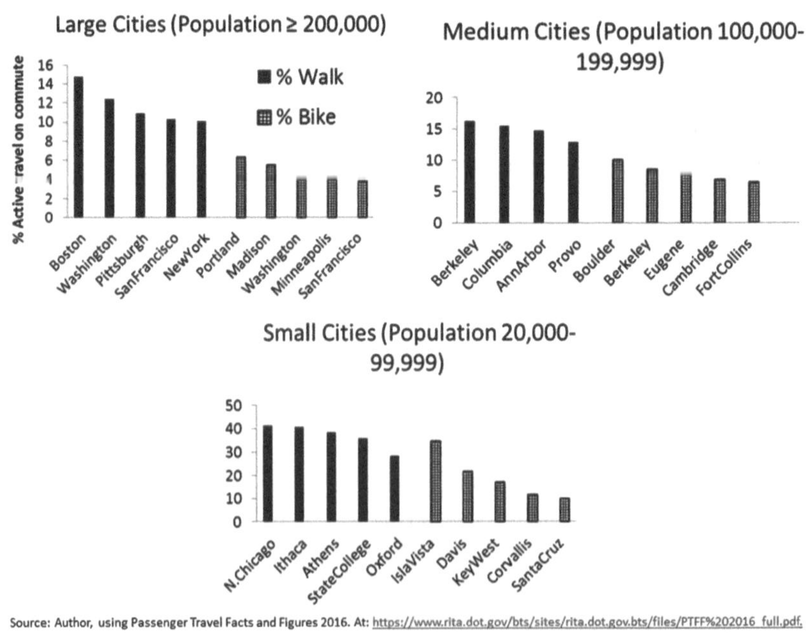

Source: Author, using Passenger Travel Facts and Figures 2016. At: https://www.rita.dot.gov/bts/sites/rita.dot.gov.bts/files/PTFF%202016_full.pdf.

Fig. 7.1 Top 5 walk and bike commuting cities by city size: 2010–2014

CA. Among medium cities (with population of 100,000–199,999), top active transportation cities in 2010–2014 were Cambridge, MA, Berkeley, CA, Columbia, SC, Ann Arbor, MI, and Provo, UT, while medium-sized cities with the highest-ranking bike commutes were Boulder, CO, Berkeley, CA, Eugene, OR, Cambridge, MA, and Fort Collins, CO. Small cities (with population 20,000–99,999) with the largest share of walk commutes in 2010–2014 were North Chicago, IL, Ithaca, NY, Athens, OH, State College, PA, Oxford, OH, and small cities with high share of bike commutes were Isla Vista, CA, Davis, Ca, Key West, Ca, Corvallis, OR, and Santa Cruz, CA (see Fig. 7.1).

Bibliography

Abravanel, M.D. 2006. *Do We Know More Now? Trends in Public Knowledge, Support and Use of Fair Housing Law*. Prepared for the U.S. Department of Housing and Urban Development. HUD USER. Washington, DC: The Urban Institute. At: https://www.huduser.gov/Publications/pdf/FairHousingSurveyReport.pdf. Accessed 7 July 2017.

Abravanel, M.D., and M.K. Cunningham. 2002. *How Much Do We Know? Public Awareness of the Nation's Fair Housing Laws*. Washington, DC: The Urban Institute. At: https://www.huduser.gov/portal//Publications/pdf/hmwk.pdf. Accessed 26 Mar 2018.

Adar, S.D., M. Davey, J.R. Sullivan, M. Compher, A. Szpiro, and L.-J.S. Liu. 2008. Predicting Airborne Particle Levels Aboard Washington State School Buses. *Atmospheric Environment* 42 (33): 7590–7599.

Ashfold, M.J., M.T. Latif, A.A. Samah, M.I. Mead, and N.R.P. Harris. 2017. Influence of Northeast Monsoon Cold Surges on Air Quality in Southeast Asia. *Atmospheric Environment* 166: 498–509.

Behrentz, E., L.D. Sabin, A.M. Winer, D.R. Fitz, D.V. Pankratz, S.D. Colome, et al. 2005. Relative Importance of School Bus-Related Microenvironments to Children's Pollutant Exposure. *Journal of the Air and Waste Management Association* 55 (10): 1418–1430.

Boschmann, E.E., and S.A. Brady. 2013. Travel Behaviors, Sustainable Mobility, and Transit-Oriented Developments: A Travel Counts Analysis of Older Adults in the Denver, Colorado Metropolitan Area. *Journal of Transport Geography* 33: 1–11.

Bremner, S.A., H.R. Anderson, R.W. Atkinson, A.J. McMichael, J.M. Bland, D.P. Strachan, and J. Bower. 1999. Short-Term Associations Between Outdoor Air Pollution and Mortality in London 1992–94. *Occupational and Environmental Medicine* 56: 237–244.

Buehler, R., and J.P. Pucher. 2017. Trends in Walking and Cycling Safety: Recent Evidence from High-Income Countries, with a Focus on the United States and Germany. *American Journal of Public Health* 107 (2): 281–287.

Chatman, D.G. 2013. Does TOD Need the T? On the Importance of Factors Other Than Rail Access. *Journal of the American Planning Association* 79 (1): 17–31.

Clayton, W., and C. Musselwhite. 2013. Exploring Changes to Cycle Infrastructure to Improve the Experience of Cycling for Families. *Journal of Transport Geography* 33: 54–61.

Deepa, M., M. Rajalakshmi, and R. Nedunchezhian. 2017. Impact of Air Pollution on Respiratory Diseases: Correlation and Classification by Multivariate Data Analysis. Data-Enabled Discovery and Applications. Accepted https://doi.org/10.1007/s41688-017-0004-z. At: https://link.springer.com/article/10.1007/s41688-017-0004-z/fulltext.html. Accessed 7 Aug 2017.

Denver Regional Council of Governments (DRCOG). 2014. At: https://drcog.org/about-drcog/about-drcog. Accessed 22 Aug 2017.

Dockery, D.W., C.A. Pope III, X. Xu, J.D. Spengler, J.H. Ware, M.E. Fay, B.G. Ferris Jr., and F.E. Speizer. 1993. An Association Between Air Pollution and Mortality in Six U.S. Cities. *The New England Journal of Medicine* 329 (24): 1753–1759.

Fan, S., Xi Li, J. Han, Y. Cao, and L. Dong. 2017. Field Assessment of the Impacts of Landscape Structure on Different-Sized Airborne Particles in Residential Areas of Beijing, China. *Atmospheric Environment* 166: 192–203.

Fu, X., J. Liu, G.A. Ban-Weiss, J. Zhang, X. Huang, B. Ouyang, O. Popoola, and S. Tao. 2017. Effects of Canyon Geometry on the Distribution of Traffic-Related Air Pollution in a Large Urban Area: Implications of a Multi-Canyon Air Pollution Dispersion Model. *Atmospheric Environment* 166: 111–121.

Graves, J.M., B. Pless, L. Moore, et al. 2014. Public Bicycle-Share Programs and Head Injuries. *American Journal of Public Health* 104: e106–e111.

Green, C.G., and E.G. Klein. 2011. Promoting Active Transportation as a Partnership Between Urban Planning and Public Health: The Columbus Healthy Places Program. *Public Health Reports* 126 (Suppl 1): 41–49.

Greenstone, M., J. Nilekani, R. Pande, N. Ryan, A. Sudarshan, and A. Sugathan. 2015. Lower Pollution, Longer Lives: Life Expectancy Gains if India Reduced Particulate Matter Pollution. *Economic and Political Weekly* 50 (1): 40–46.

Jacobsen, P.L., D.R. Ragland, and C. Komanoff. 2015. Safety in Numbers for Walkers and Bicyclists: Exploring the Mechanisms. *Injury Prevention* 21 (4): 217–220.

Katsouyanni, K., G. Touloumi, E. Samoli, A. Gryparis, A. Le Tertre, et al. 2001. Confounding and Effect Modification in the Short-Term Effects of Ambient Particles on Total Mortality: Results from 29 European Cities Within the APHEA2 Project. *Epidemiology* 12: 521–531.

Kono, T., and H. Kawaguchi. 2017. Cordon Pricing and Land-Use Regulation. *Scandinavian Journal of Economics* 119 (2): 405–434.

Krecl, P., C. Johansson, A.C. Targino, J. Ström, and L. Burman. 2017. Trends in Black Carbon and Size-Resolved Particle Number Concentrations and Vehicle Emission Factors Under Real-World Conditions. *Atmospheric Environment* 166: 155–168.

Künn-Nelen, A. 2016. Does Commuting Affect Health? *Health Economics* 25 (8): 984–1004. https://doi.org/10.1002/hec.3199.

Li, L., A.H. Wu, I. Cheng, J.-C. Chen, and J. Wu. 2017. Spatiotemporal Estimation of Historical PM2.5 Concentrations Using PM10, Meteorological Variables, and Spatial Effect. *Atmospheric Environment* 166: 182–191.

McDonald, N.C., A.L. Brown, L.M. Marchetti, and M.S. Pedroso. 2011. U.S. School Travel, 2009. An Assessment of Trends. *American Journal of Preventive Medicine* 41 (2): 146–151.

National Center for Education Statistics. 2009. Table 2: Enrollment in Educational Institutions. In *Digest of Education Statistics*. Washington, DC: U.S. Department of Education.

Ogden, C.L., M.D. Carroll, C.D. Fryar, and K.M. Flegal. 2015. *Prevalence of Obesity Among Adults and Youth: United States, 2011–2014*, NCHS Data Brief, no 219. Hyattsville: National Center for Health Statistics. At: https://www.cdc.gov/nchs/data/databriefs/db219.pdf. Accessed 15 Aug 2017.

Passenger Travel Facts and Figures. 2016. *US Department of Transportation*. Bureau of Transportation Statistics. At: https://www.rita.dot.gov/bts/sites/rita.dot.gov.bts/files/PTFF%202016_full.pdf. Accessed 2 June 2017.

Pucher, J., R. Buehler, and M. Seinen. 2011. Bicycling Renaissance in North America? An Update and Re-appraisal of Cycling Trends and Policies. *Transportation Research Part A: Policy and Practice* 45 (6): 451–475.

Ratner, K.A., and A.R. Goetz. 2013. The Reshaping of Land Use and Urban form in Denver Through Transit Oriented Development. *Cities* 30 (1): 31–46.

Rohde, R.A., and R.A. Muller. 2015. Air Pollution in China: Mapping of Concentrations and Sources. *PLOS ONE* 10 (8): e0135749. At: https://doi.org/10.1371/journal.pone.0135749.

Szogs, S., A. Arneth, P. Anthoni, J.C. Doelman, F. Humpenöder, A. Popp, T.A.M. Pugh, and E. Stehfest. 2017. Impact of LULCC on the Emission of BVOCs During the 21st Century. *Atmospheric Environment* 166: 73–87.

Who Is TOD in Metro Denver? Benchmarking the Evolution of TOD in Metro Denver. A Summary of Key Findings from the Business, Employee and Resident Surveys. Denver Regional Council of Governments (DRCOG). June 2010. At: https://drcog.org/sites/drcog/files/resources/Who%20Is%20TOD%-20Summary%20Report.pdf. Accessed 23 Aug 2017.

World Health Organization (WHO). 2016. *WHO's Urban Ambient Air Pollution Database—Update 2016*. At: http://www.who.int/phe/health_topics/outdoorair/databases/AAP_database_summary_results_2016_v02.pdf?ua=1&ua=1. Accessed 7 Aug 2017.

Zhou, L., M. Wang, C.-H. Chang, S. Liu, Y. Zhan, and J. Shi. 2017. Commuting Stress Process and Self-Regulation at Work: Moderating Roles of Daily Task Significance, Family Interference with Work, and Commuting Means Efficacy. *Personnel Psychology*: 1–32. https://doi.org/10.1111/peps.12219.

Index

A. Antipova, *Urban Environment, Travel Behavior, Health, and Resident Satisfaction*, https://doi.org/10.1007/978-3-319-74198-7

FSC
www.fsc.org
MIX
Papier | Fördert gute Waldnutzung
FSC® C083411

Kolibri 360 GmbH
Ferdinand-Jühlke-Straße 7
99095 Erfurt, Deutschland
produktsicherheit@kolibri360.de